Water Use, Management, and Planning in the United States

Water Use, Management, and Planning in the United States

Stephen A. Thompson
Department of Geography
Millersville University
Millersville, Pennsylvania

Academic Press
San Diego London Boston New York Sydney Tokyo Toronto

Cover photographs: Hoover Dam–Boulder Canyon Project-NV-AZ—An Aerial view of Hoover Dam looking upstream at the face of Hoover Dam. Courtesy of Bureau of Reclamation, March 31, 1996, by Andrew Pernick.
Waves © Digital Stock

This book is printed on acid-free paper. ♾

Academic Press
a division of Harcourt Brace & Company
525 B Street, Suite 1900, San Diego, California 92101-4495, USA
http://www.apnet.com

Academic Press
24-28 Oval Road, London NW1 7DX, UK
http://www.hbuk.co.uk/ap/

Library of Congress Catalog Card Number: 98-87241

International Standard Book Number: 0-12-689340-3

PRINTED IN THE UNITED STATES OF AMERICA
98 99 00 01 02 03 BB 9 8 7 6 5 4 3 2 1

This book is dedicated to Ben and Lynne.

CONTENTS

PREFACE xiii

1

THE PHYSICAL SYSTEM

Early Ideas about Hydrology 1
Water Measurement 3
The Hydrologic Cycle 5
Water Storage 5
Water Cycling Processes 6
Precipitation 12
Infiltration and Soil Moisture 16
Runoff 17
Groundwater 20
Water Balance 22
Climate Change 24
Computer Simulation of Climate Change 25
Evidence for Climate Change 27

2

A History of Water Development in the United States

1800–1900: A Time of Resource Exploitation 32
1900–1921: The Progressive Period 39
1921–1933: Post-World War I 42
1933–1943: The New Deal 45
Other New Deal Activities 48
1943–1960: Congressional Control of Water Resource Activities 50
New Water Resource Issues 52
Other Water Resource Activities 55
1960–1980: The Environmental Period 59
Water Quality 62
The National Water Commission 64
Traditional Water Agency Activities 65
1980 to the Present: A Time of Changing Focus 66

3

Water Quantity Law: The Legal System

How Laws Are Made 72
The Purpose of Law 73
State Law: Surface Water 74
The Riparian Doctrine 75
The Prior Appropriation Doctrine 80
Regulated Riparianism 91
State Law: Groundwater 92
Absolute Ownership—The English Doctrine 93
Reasonable Use—The American Doctrine 94
Correlative Rights 94
Prior Appropriation 94
Federal Law 96
Federal Powers Related to Water Resources 96
Federal Reserved Rights 98
Nonreservation Indian Reserved Rights 103
Other Federal Reserved Rights 104
Other Federal Interests That Affect Water Resources 104
Transboundary Water Resources 105
Judicial Decisions 105
Interstate Compacts 106
Legislation 112

4

WATER USE IN THE UNITED STATES

Water Availability: An International View 120
A Comment on Water-Use Estimates 121
Water Supply and Demand: National View 122
Water Supply and Demand: Regional View 127
Water Supply and Demand: State View 131
Total Withdrawals (Surface and Ground) 131
Surface Water 135
Groundwater 135
Geographic Information Systems (GIS) 136
Types of GIS 137

5

ECONOMICS AND WATER RESOURCES

Price Theory: Supply and Demand 145
Water Markets 148
Market versus Government Allocation 150
Welfare Economics 151
The Federal Subsidy to Water Users 152
Cost–Benefit Analysis 153
Structuring the Cost–Benefit Analysis 153
Cash Flow Diagram 155
Discount Factors 155
Hypothetical Cost–Benefit Analysis 160
Optimality Criteria 162
Identification of Benefits and Costs 163
Problems with Cost–Benefit Analysis 163
Random Nature of Future Payments and "Expected Value" 165
Evolution of Water Resource Planning Objectives and Procedures 166
Financial Analysis 168
Resource and Environmental Economics 168

6

WATER DEMAND AND SUPPLY: MANAGEMENT AND PLANNING

Water Demand 174
Urban Water Demand Planning 176

Water Supply 177
Dams and Reservoirs 178
Interbasin Transfers 191
Planning and Decision Making 200
Planning Entities 201
The Planning Process 203
Public Participation 206
Constraints on Water Management Decision Making 208

7

OFFSTREAM WATER USE: URBAN AND AGRICULTURAL USES

Urban Water Systems 212
Urban Water Supply 212
Geographic Pattern of Public Water Supply 219
Drinking Water Quality 220
Wellhead Protection and Management Programs 225
Irrigation 227
Historical Development 227
Irrigation in the World Today 229
Irrigation in the United States 229
Water Application Methods 231
Estimating the Evapotranspiration Requirement (E_{tp}) *234*
Estimating the Irrigation Water Requirement (IRR) 238
Irrigation Efficiency 238
Economically Efficient Water Application 239
Water Conservation in Agriculture 241
Environmental Impacts of Irrigation 242

8

INSTREAM WATER USE: HYDROELECTRIC POWER AND RECREATION

Hydropower 246
Concepts and Definitions 246
Types of Hydroelectric Power Plants 247
Categories of Electrical Power 249
Hydroelectric Power Potential 251
Regional Water Use for Hydroelectric Power Generation 253

State Water Use for Hydroelectric Power Generation 254
Hydroelectric Dams 255
Environmental Challenges for Hydroelectric Power 257
Water for Recreation 260
Recreation on Federal Public Land 262
Federal Land Resource Agencies 263
Federal Water Resource Agencies 267
Environmental Values and Instream Flows 270

9

WATER QUALITY AND ECOSYSTEM HEALTH

The Water Molecule 274
Acidity/Basicity 274
Pollution Concentration 276
Categories of Pollutants 276
Disease-Causing Organisms 276
Oxygen-Demanding Wastes 277
Sediment 281
Nutrients 283
Organic Chemicals 285
Inorganic Chemicals 288
Heat 289
Radioactive Material 290
Water Quality Management 292
Point Source Pollution 292
Nonpoint Source Pollution 295
Water Quality Monitoring 296
The Quality of Our Nation's Water 298
Rivers and Streams 299
Lakes, Ponds, and Reservoirs 300
Estuaries 301
The Watershed Approach to Water Quality 301
Ecosystems and Water Quality 303

10

FLOODS AND DROUGHTS

The Human-Ecological Model of Hazards 308
Floods 309
The Physical System 309

Streamflow Analysis *311*
The Human System *318*
A Brief Review of Flood Management in the United States *320*
Typology of Human Adjustments to Reduce Flood Losses *321*
The Evolving Approach to Floods Mitigation *324*
Drought 326
Drought Definition *326*
Drought Analysis *328*
Human Adjustment to Drought in the United States *336*

APPENDIX 1
Unit Conversions and Equivalents 341

APPENDIX 2
Sources of Water-Related Information 345

APPENDIX 3
Investment and Discount Factors 347

BIBLIOGRAPHY 357
INDEX 367

PREFACE

BACKGROUND AND APPROACH

This book is about how we use water in the United States. Water is one of the most abundant, naturally occurring substances on Earth, and it is one substance we cannot live without. This life-or-death relationship is true for air as well as for water, but unlike air, water is unevenly distributed in space and time. Water is not always available where it is needed, when it is needed, or in the quantity or at the level of quality that is required. The spatial and temporal variabilities in both water supply and demand provide the most fundamental reasons for water planning and management. However, among the many other reasons for planning and managing water resources are minimizing conflicts between existing water users, accommodating new water uses or the expansion (contraction) of traditional uses, anticipating or responding to the changing values and priorities people place on different water uses, minimizing the environmental impacts associated with using water, and minimizing the impacts to society and the environment from the hydrologic hazards of floods and droughts. As more and more people make increasing demands on the limited water supplies, the importance of effective planning and management becomes ever more critical.

The principal goal of *Water Use, Management, and Planning in the United States* is to convey to students the breadth of the subject we call "water resources" and the depth of society's involvement in manipulating (managing) the hydrologic cycle. The only reasonable way to study water resources today is to use an interdisciplinary approach. Humans are now an important, if not the most important, agent of environmental change. The growing diversity in the type and scale of human impacts on water, land, and even the atmosphere is manifested as complex cascades of changes in the physical environment and human systems. Ideally, an interdisciplinary study incorporates the perspectives and knowledge of individual scientific disciplines and systematically integrates them to develop a "holistic" understanding of the topic. The more thoughtful the integration, the more successful the interdisciplinary approach. But the most successful interdisciplinary analysis goes beyond integration and achieves synergism among the individual elements.

What are the individual disciplines with a stake in water resources? There are many and they come from both the social sciences and the nonsocial (natural) sciences. From the social sciences, history, economics, geography, planning, law, and political science all contribute knowledge and a unique perspective about water as a natural resource. Among the natural (earth and life) sciences, hydrology, meteorology, geology, geomorphology, biology, and chemistry are providing distinct insights into the movement of water through the environment, how its physical and chemical properties are changed by human activities, and how those changes influence the structure and functioning of other natural systems, particularly ecosystems. Mathematics must also be included in our list because quantitative tools and analytical methods are fundamental to both social and natural scientists. As a geographer interested in water resources, I am accustomed to looking at the interaction between different factors in attempting to understand the patterns of water use and management. It would be presumptuous to say that geographers as a group are necessarily skilled in integrative analysis, but it is true that geographers are trained, and are therefore predisposed, to consider the interplay between the physical environment and human socioeconomic systems when trying to understand the use of natural resources. For me, an interdisciplinary approach to the topic of water seems only natural.

I thus wrote *Water Use, Management, and Planning in the United States* to satisfy my need for an interdisciplinary textbook, to use in an introductory, one-semester course on water resources. The presentation and material are targeted to an upper-division undergraduate and first-year graduate audience of students from a variety of disciplines within the natural and social sciences.

ORGANIZATION

The first problem encountered in using an interdisciplinary approach is knowing where to begin. I start with hydrology in Chapter 1. This chapter is divided into several parts. One section discusses the components and processes that form the hydrologic cycle, or as I prefer to call it, the hydrologic system. Another section introduces and explores the concept of a water balance. A water balance is founded on the conservation of mass—water cannot be created or destroyed—so input to some "subsystem" must equal the output plus any change in water storage. This physical continuity makes a water balance one of the most important tools for planning and managing water. The final section considers the issue of global climate change and some potential impacts on hydrology and water resources. In addition to discussing hydrology, I use this chapter to define many of the terms encountered in the field of water resources. Like any other subject, water resources has distinctive, even unique, terms and concepts, and a challenge for any introductory book is to carefully define the terms that professionals in the field take for granted.

Chapter 2 is a review of the history of water resources in the United States. While tracing the historical developments in water resources, the chapter shows that our attitudes about water reflect broader, more general values in society. This chapter also provides a context and foundation for understanding many issues discussed in later chapters. The historical analysis begins in the early 19th century, when the only philosophy guiding the use of natural resources was that of exploitation, and water management meant water development. By the 20th century, the new philosophies of utilitarian conservation and natural resource preservation had emerged. As we move through the 20th century, we see that increases in environmental degradation coupled with fundamental changes in values gave rise to the environmental movement by the 1960s. In the past few decades, the emphasis on water development has largely given way to a focus on improved water management and "sustainable" water use. What is meant by sustainable use, however, is not always clear. Much of the history of water resources is a history of the activities of the federal government, largely because the federal government had the money to invest in large-scale water development, had a variety of incentives to do so, and possessed most of the technical expertise.

In Chapter 3 we turn our attention to water (quantity) law, which has been influenced by history and hydrology as well as several social institutions. This chapter begins with an examination of state laws for surface and groundwater use, followed by a look at federal laws affecting and controlling water use. The basic questions that underlie this chapter are who has the right to use water and how much are they allowed to use? The evolution of water law, and its reliance on precedent, gives this chapter a strong historical flavor.

Patterns of water use by source, water-use sector, and geographic region are the topics of the first part of Chapter 4. Because of the conventions and units used to report water data, it was logical to position this chapter after Chapter 1, where these terms are defined. The second part of this chapter covers geographic information systems (GIS). GIS technology is much more than methods and technologies for data storage, and I have provided a brief discussion of the basic types of GIS and their potential applications to resource analysis.

Chapter 5 deals with the discipline of economics. Because economic principles and analytical techniques are vitally important to resource decision making, this material is presented before the material on planning and decision making in Chapter 6. A number of the economic topics encountered throughout the book tend to be quantitative. Quantitative analyses in this and other chapters can be omitted if desired. However, quantitative material is always followed with a practical example to demonstrate its application. The major economic topics discussed include price theory, supply and demand, water markets, welfare economics, cost–benefit analysis, and environmental economics.

Chapter 6 addresses planning and management for water supply. The focus here is on water supply for the urban/domestic sector, but many of the principles can be applied to other water-use sectors. Where water supply planning was once limited almost exclusively to finding new sources of water, planners and managers are now turning to managing the demand. The second half of the chapter describes the rational planning model, some suggested (nonrational) modifications to the model of rational decision making, and the role of public participation in resource decision making.

Chapter 7 is a detailed examination of urban and agricultural water use—how water is used, how much is used, the spatial pattern of use by state, and the increasing role of conservation in balancing supply and demand. The section on urban water supply includes a discussion of drinking water quality and the 1996 amendments to the Safe Drinking Water Act and their influence on public water supply planning and management. Industrial water use is not covered, an omission necessary to design the text for a one-semester course.

Chapter 8 covers three instream water uses—hydropower, recreational uses, and instream uses for environmental purposes. Emphasis is primarily on the first two because environmental issues are considered again in Chapter 9 under the topic of water quality. Hydropower is a traditional water use that came of age during the first half of the 20th century. In contrast, the trend toward valuing water primarily for recreational uses emerged after World War II. The discussion of recreation focuses on the types of uses and levels of participation at facilities on federal public lands.

Chapter 9 discusses water quality and ecosystems. Some of the main topics addressed include water properties, a categorical treatment of water pollutants, point versus nonpoint sources of pollution and their control, water quality management under the Clean Water Act, and the status of aquatic species in the United States. Related resource issues are discussed within the framework of each pollutant category. A section on water quality management under the Clean Water Act considers the basic requirements and administration of the act, the emergence of nonpoint sources of pollution as the leading cause of water quality degradation, and the reemergence of the concept of basin-wide planning in the 1990s.

The final chapter examines the two hydrometeorological hazards—floods and droughts. This chapter uses the geographer's "human ecological" paradigm of natural hazards to explore how hazards are created through the interaction of a natural (physical) system and a socioeconomic system for resource use. Some of the more popular methods used to study the physical phenomena of these hazards and how society copes are discussed. As with water quality management, flood hazard mitigation strategies are evolving from the limited purpose of protecting people and property to a more holistic concern for natural floodplain uses and environmental protection.

FEATURES OF THE BOOK

The book includes a wide variety of pedagogical features to enhance its usefulness. Important terms and concepts are italicized for emphasis. Numerous illustrations, 67 of which are original, and tables and many line, bar, and pie graphs present supporting details on water and related resources. I have also included 20 quantitative worked problems. The large diversity in the types of problems—from cost–benefit analysis in Chapter 5, to reservoir storage analysis and streamflow simulation in Chapter 6, to flood frequency analysis in Chapter 10—is yet another example of the interdisciplinary character of the subject. Each problem is worked in a step-by-step fashion that shows the types of data required and which is the appropriate equation(s) for solving the problem. I realize that quantitative analysis can be intimidating to some students and even some instructors, but in my experience even the most adamant "mathphobe" can gain additional insights by following the examples and doing similar types of exercises.

Water Use, Management, and Planning in the United States includes numerous focus boxes to present a variety of material. Within some boxes I have consolidated and highlighted important points and themes being discussed at length in the adjacent text. Some boxes provide examples of real water resource activities and programs relevant to that particular topic. Other boxes are used to frame water resource case studies, seven of which are included in the book. Some case studies are relatively brief and cover only one or two pages. Others, such as the story of how Los Angeles developed its water supply, are necessarily a bit longer. A unique feature of the water law chapter (Chapter 3) is the three exerpted court cases featuring important legal principles and precedents. Reading these decisions provides a taste of the nuances of legal analysis and judicial decision making.

Finally, I have included three appendixes containing common formulas for converting between different systems of units, a list of government sources of water data along with World Wide Web (WWW) addresses, and discount rate tables for cost–benefit analysis. Students are thus encouraged to browse the Web and discover important water resource sites.

Water resource management is in a state of change. Gone are the days when we could assume that there would always be a plentiful supply of water to meet our growing demands. Gone are the days when we sought to control rivers with impunity and by the force of our technology. Gone, too, are the days when we manipulated water resources for the benefit of "man" without regard to the impacts on natural systems and other species of life. This book is about how we use water in the United States, but every author must choose what to include and what to omit because it is impossible to cover everything. Every person has his or her own set of priority issues; mine are laid out here. This book owes its inspiration to all of the wonderful teachers I have had along the way. Anything that is commendable about the book, I owe to them. I alone take responsibility for the book's shortcomings.

Stephen A. Thompson

1

THE PHYSICAL SYSTEM

Early Ideas about Hydrology
Water Measurement
The Hydrologic Cycle
Water Storage
Water Cycling Processes
Precipitation
Infiltration and Soil Moisture
Runoff
Groundwater
Water Balance
Climate Change
Computer Simulation of Climate Change
Evidence for Climate Change

Water management involves manipulating the hydrologic cycle. This chapter focuses on the physical side of water management and is divided into five sections. The first section reviews early ideas about the hydrologic cycle. The second section covers water measurement including dimensions and units. The third section discusses the major components and processes of the hydrologic cycle. The fourth introduces the concept of a water balance. And the final section examines some water–related issues associated with the potential human alteration of the Earth's climate. Changes in established climate patterns could have profound impacts on water availability and established socioeconomic systems.

EARLY IDEAS ABOUT HYDROLOGY

Every society has had to face the problems of water management, though some civilizations more than others. The Egyptians are credited with irrigation and flood control projects dating back to 5400 B.P. The ancient cities of Jericho, Babylon, and Carthage built water systems to provide drinking water to residents. The ancient Romans constructed water supply aqueducts, and they also built a closed sewage system which was more advanced than facilities available to millions of people in developing countries today (Gleick, 1993). Early societies accomplished amazing feats of water engineering even though they did not understand the hydrologic cycle or basic hydraulic principles. The Greek philosopher Aristotle thought water was held in a great underground sea and flowed up against gravity into the mountains, thence to flow back to the ocean. Plato's explanation for why water flowed down from the mountains was that cold temperatures converted air into water. These interesting but erroneous concepts about the hydrologic cycle persisted for centuries.

It was not until the Renaissance that accurate ideas about the hydrologic cycle emerged. Bernard Palissy (1510–1590) is credited with being the first person to state unequivocally that rivers have no source other than precipitation. Pierre Perault (1608–1680), a disbarred French lawyer turned natural scientist, undertook a field experiment to prove Palissy's assertion. Perault calculated a water balance for the Seine River basin and concluded that precipitation supplied more than enough water to the river. The noted English astronomer Sir Edmond Halley (1656–1742) was prompted to conduct studies on evaporation after condensation interfered with his telescopic observations. He determined that ocean evaporation was sufficient to replenish the rivers flowing back to the ocean.

The work of Renaissance scholars ushered in an age of hydrologic experimentation and quantification. But misperceptions about the hydrologic cycle persisted. In the United States the notion that local evaporation was primarily responsible for local precipitation was embodied in federal legislation. The Timber Culture Act of 1872 gave homesteaders in certain western states an additional quarter–section of land (160 acres) if they planted trees on 40 acres. It was thought the trees would increase transpiration and create more precipitation in the immediate area. It was also widely believed by the pioneers moving into the Great Plains in the second half of the 19th century that plowing the natural grassland vegetation caused the climate to become wetter. They were convinced that plowing increased evaporation from the soil, which in turn led to increased local rainfall. The fact was these settlers simply drew an incorrect conclusion from the cues offered by the naturally variable environment. Explorers to the western United States in the first half of the 18th century proclaimed it the "Great American Desert." This description is accurate for the extreme southwest, but the region of the central plains has a transitional climate fluctuating between semiarid and subhumid. It is characteristic in this type of climate to experience periods of abundant rainfall alternating with times of desiccating droughts. The early explorers saw this region during drought, while homesteaders in the 1870s and 1880s came during a wet spell. The homesteaders concluded that their action (plowing) modified the climate, causing it to become wetter. This conclusion was consonant with a prevailing cultural attitude that human transformations of natural landscapes were "improvements" over nature. By the 1890s drought returned to the region, forcing widespread abandonment of the land (Warrick and Bowden, 1981). The myth that "rain follows the plow" died in the dust of the deserted farms.

While the 19th–century notion that plowing caused the regional climate to become wetter is considered naive today, recent research has indicated there is a relationship, though poorly understood, between the persistence of droughts in the central United States and soil moisture levels (Oglesby and Erickson, 1989). It is unlikely that the early settlers influenced weather patterns in any significant way, but we should be careful in saying that their actions had absolutely no influence. The possible effect of land use on regional climate reemerged in the 1970s in the debate over the causes of *desertification* in the Sahel region of sub–Saharan Africa. Desertification is a complex process of vegetation and soil degradation believed to result through the interaction between land use and natural climate variation. One theory on desertification in the Sahel was that overgrazing had changed the reflectivity of the land surface, which through complex feedback processes changed local energy budgets and ultimately the amount of regional precipitation (Charney, 1975; Jackson and Idso, 1975). In other words it was thought that people in the Sahel had helped change the regional climate through abusive land use practices. Land–atmosphere feedback theories have been postulated to explain desertification in the overgrazed lands of the southwest United States as well (Schlesinger *et al.,* 1990). One of the most important questions facing scientists and decision makers today is whether human activities are changing climate on a global scale. We will return to this topic at the end of the chapter. It was not until the 1930s that the age of "modern" hydrology emerged based on scientific observation, experimentation, and theory. In the United States one of the leaders in the field was Robert Horton, who investigated a wide variety of hydrologic and geomorphic phenomena.

WATER MEASUREMENT

The United States has resisted adopting the metric system of measurement so working with hydrologic and meteorologic data in the United States requires converting between conventional English and metric unit systems. This situation will continue into the foreseeable future. Even if the country adopted the metric system tomorrow the conversion problem would still exist because all of our historical data are in English units. Appendix 1 contains some useful conversion formulas.

Water has mass, it occupies space, it moves, it changes state, and it changes temperature. All of these physical properties are *dimensional quantities* that are measured using a system of units. Length (L) is a spatial dimension and is measured using units such as inches, feet, millimeters, centimeters, and meters. Volume is a three–dimensional quantity, $V = (L \times L \times L) = L^3$. Discharge ($Q$) is volume per

EXAMPLE 1.1

Assume 0.5 inch of rain fell uniformly over an area of one square mile. Convert this into the equivalent volume of water. The 0.5 inch is a length (L) measurement. Area is (length × length) = L^2. The first step is to get all the data into the same units. A convenient English unit of length for this problem is feet.

Convert 0.5 inch into feet:

(0.5 in)(1 ft/12 in) = 0.0416 ft.

Convert square miles into square feet:

(1 mi)(5280 ft/mi) = 5280 ft.

Thus one square mile is

(5280 ft)(5280 ft) = 27,878,400 ft^2.

Multiply the precipitation in feet by the area in square feet to get the volume of water in cubic feet:

(27,878,400 ft^2)(0.0416 ft) = 1,161,600 ft^3.

In other words, 0.5 inch of rain over one square mile equals more than 1.16 *million* cubic feet of water. Since few people are familiar with cubic feet, convert cubic feet into gallons:

(1,161,600 ft^3)(7.48 gal/ft^3) = 8,688,768 gal of water.

A small amount of rain can become a large volume of water when multiplied by area.

---

time, $Q = L^3/T$. Discharge is measured for streamflow and for wells. Example 1.1 demonstrates the relationship between precipitation measured in length and the equivalent volume of water over an area. Example 1.1 shows that relatively small amounts of rain can produce very large volumes of water. This is why storms that drop large amounts of rain over large areas produce devastating floods.

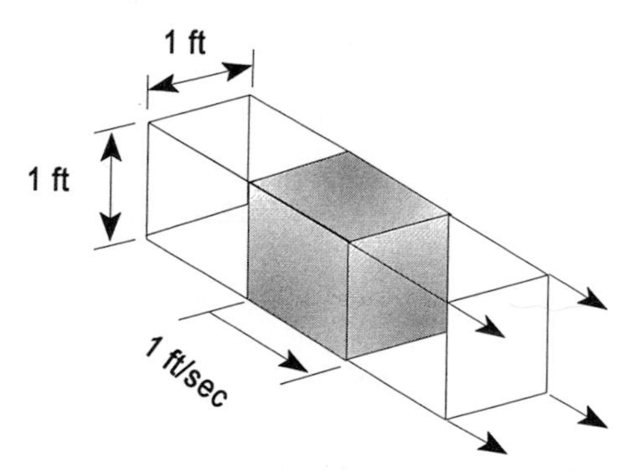

FIGURE 1.1 This figure shows that one cfs is produced by water moving at a velocity of one foot per second through a cross-sectional area of one square foot.

Discharge is a volume of water per unit of time (L^3/T). Discharge is calculated by multiplying water velocity (L/T) by the cross–sectional area (L^2) through

EXAMPLE 1.2

In this example a stream flows into an empty reservoir. The reservoir is a square 600 feet on a side with vertical walls. The average discharge into the reservoir for a 24–hour period is 86 cfs. How deep is the water in the reservoir at the end of 24 hours?

First find the volume of water by multiplying average discharge by the time period. Since discharge is in ft^3/s, convert time (24 hours) into seconds:

$$(86\ ft^3/s)(60\ s/min)(60\ min/hr)(24\ hr) = 7{,}430{,}400\ ft^3.$$

This again is the principle of getting all quantities into the same units. Next, find the area of the reservoir:

$$(600\ ft)(600\ ft) = 360{,}000\ ft^2.$$

To find the water depth (L) divide the water volume (L^3) by reservoir area (L^2):

$$7{,}430{,}400\ ft^3 / 360{,}000\ ft^2 = 20.64\ ft.$$

The depth of water is 20.64 feet after 24 hours.

---

which the water flows. In the United States stream discharge is measured in cubic feet per second (ft^3/s or cfs). Again, most other countries use cubic meters per second (cms). Discharge from a pumping well is usually reported in gallons per minute. One cfs equals 7.48 gallons of water flowing past a reference point each second. Figure 1.1 is a sketch of discharge calculated as the water velocity times cross–sectional area. Discharge is converted into volume by multiplying the discharge by the time period over which it occurred. Example 1.2 demonstrates some relationships between discharge and volume.

From Examples 1.1 and 1.2 it is apparent that measuring water in cubic feet quickly produces very large numbers. It is cumbersome to work with so many digits so a larger unit for measuring water volume is frequently used. This unit is the *acre–foot.* An acre–foot is 1 acre covered to a depth of 1 foot. In case you did not learn this before, 1 acre equals 43,560 ft^2. An acre–foot thus equals 43,560 ft^3 of water. How many gallons of water are there in 1 acre–foot?

THE HYDROLOGIC CYCLE

WATER STORAGE

The hydrologic cycle is the continuous cycling of water from the oceans, to the atmosphere, down to the land surface, and back again to the oceans (Figure 1.2). Water is stored in the oceans, on the continents in lakes and rivers, underground in the soil and as groundwater, and in the atmosphere as water vapor and clouds. Approximately 96.5 percent of all the water on Earth is found in the oceans (Table 1.1). This water is too saline to be used directly for water supply. Desalinating seawater is possible but it is energy intensive and extremely expensive. Of the remaining water on Earth only 2.6 percent is freshwater on land. Most of this, about 1.74 percent, is stored as ice in continental glaciers and ice caps. The next

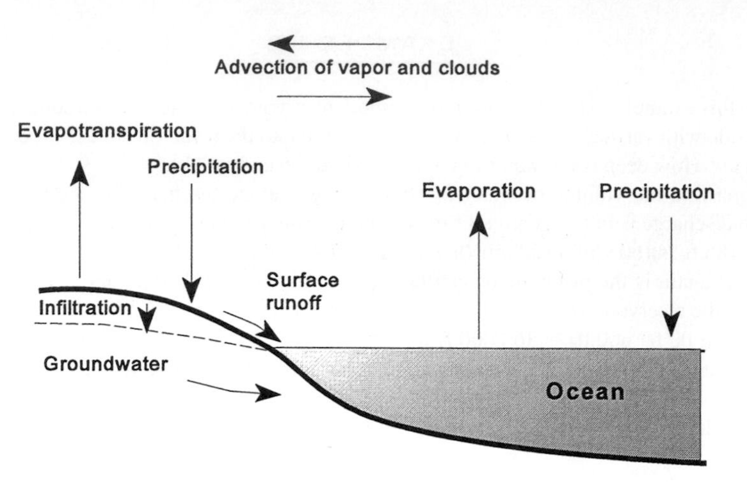

FIGURE 1.2 A simplified diagram of the global hydrologic cycle. (Source: Thompson, 1998)

largest store of freshwater (0.76 percent) is groundwater. Only about 0.27 percent of all the freshwater on Earth is found in lakes and rivers.

WATER CYCLING PROCESSES

Hydrologic processes that transfer water from one storage location to another include *evaporation* and *evapotranspiration, precipitation, infiltration,* and *surface*

TABLE 1.1 Water Storage on Earth (Shiklomanov, 1993; Gardner, 1977)

Location	Volume (10^3 kilometers)	Percentage of global reserves: Total	Freshwater	Average residence time
World oceans	1,338,000	96.5	—	2600 years
Glaciers and permanent snow	24,064	1.74	68.7	100–100,000 years
Groundwater	23,400	1.70		1–50,000 years
Fresh	10,530	0.76	30.1	
Saline	12,870	0.94	—	
Lakes	176.4	0.013	—	100 years
Fresh	91.0	0.007	0.26	
Saline	85.4	0.006	—	
Soil	16.5	0.001	0.05	3 months
Atmosphere	12.9	0.001	0.04	10 days
Rivers	2.12	0.0002	0.006	20 days

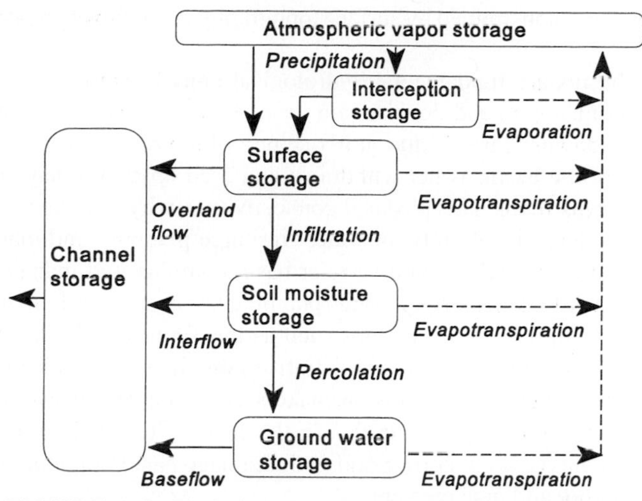

FIGURE 1.3 The basin-scale hydrologic system (cycle) showing water storages and processes. (Source: Thompson, 1998)

and *subsurface* runoff. Figure 1.3 shows some important hydrologic storages and processes at the scale of an individual *drainage basin.* A drainage basin is an area of land that drains water to a common outlet (Figure 1.4). In the United States drainage basins are also called *watersheds.* The line delineating a drainage basin is the drainage *divide.* A common simplification is to consider the drainage basin only in terms of surface runoff and to neglect groundwater flow. Groundwater usually flows into the stream channels of the basin but not necessarily. Since

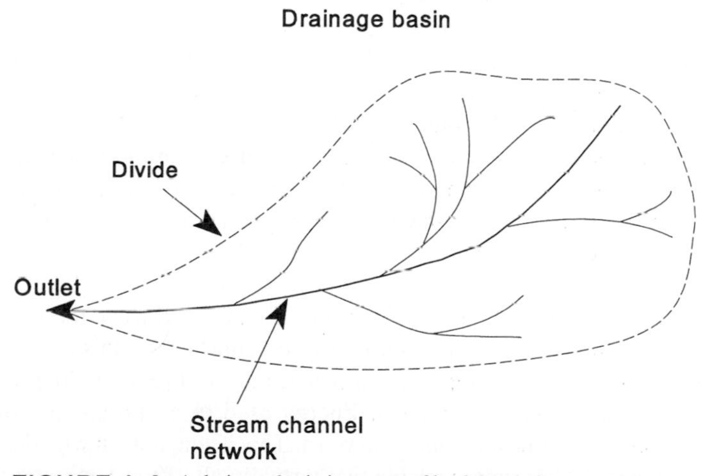

FIGURE 1.4 A drainage basin is an area of land that drains water to a common outlet. The drainage divide separates basins.

groundwater is not constrained by surface topography it can flow beneath surface–water divides.

Drainage basins are fundamental hydrological units because hydrologic processes connect upstream and downstream areas. Land–use change upstream can affect flooding downstream. Pollution in one part of the basin is carried by water to other parts of the basin. Water withdrawn and used upstream may be unavailable for use downstream. This physical connectivity is why drainage basins have long been considered ideal units for water–resource planning and management. Throughout history interest in unified river basin planning and management has waxed and waned. In general basin–wide planning has been more talk than reality, though there are significant exceptions such as the Tennessee Valley Authority. The reason is that political and administrative boundaries rarely coincide with natural basin boundaries, and decision makers are more responsive to political boundaries. It is interesting to note that in the 1960s Great Britain reorganized local political boundaries to correspond with drainage basin boundaries to facilitate water planning and management.

Evaporation and Evapotranspiration

Evaporation is when energy is used to change the state of water from the liquid to the gaseous state, and the water molecules drift up into the atmosphere as water vapor. *Sublimation* is when water molecules change from the solid state (ice) directly into gas. Evaporation occurs from the surface of lakes, rivers, oceans, and wet soil. Since 71 percent of the Earth's surface is covered by ocean, most of the world's evaporation (about 86 percent of the total) occurs from the oceans. Transpiration is the same liquid–to–gas state change but it occurs as part of plant respiration, and the water molecules escape to the atmosphere through stomata on the leaves. On land, evaporation from wet soil and transpiration from plants are collectively called evapotranspiration. The hydrologic cycle starts with evaporation and evapotranspiration. The Sun, either directly as radiant energy or indirectly by warming the air, the water, or the soil, provides the energy for the change of state. When the Sun burns itself out some 12 billion years in the future, the hydrologic cycle will quickly come to a halt.

Three meteorological factors control the rate of evaporation—energy availability, wind speed and the humidity level of the overlying air. The more energy available, the faster the wind, or the drier the air, the greater the evaporation. You probably recognize these variables as the essential components of an electric hair drier. In addition to the meteorological factors, vegetative and soil factors affect the rate of evapotranspiration. In moving water from the surface to the atmosphere evapotranspiration plays an important role in cooling the Earth's surface. Solar energy basically does two things when it reaches the surface—it either heats the surface, or it evaportranspirates water. Energy used to evaporate and transpire water is unavailable for heating and vice versa. This is one reason why deserts are so hot; there is so little water available that virtually all of the Sun's energy goes into heating the land and the overlying air. (Ah, but it is a dry heat.) In humid climates with abundant vegetation and wet soil a significant amount of the Sun's energy is used for evapotranspiration. This keeps air temperatures lower but of course raises the humidity level of the air.

Measurement of Evaporation and Evapotranspiration

Evaporation and evapotranspiration are measured or estimated using evaporation pans, lysimeters, and various formulas. Like precipitation, evaporation and evapotranspiration are measured as a depth of water, i.e., inches or millimeters. The most common type of evaporation pan used in the United States is the Class A Pan. The Class A Pan is galvanized steel with a diameter of 4 feet and a depth of 10 inches (Figure 1.5). One method of operating the pan is to fill it to a depth of 8 inches and then refill it when the water level drops to 7 inches. Measuring the water level in the pan on a daily basis gives an estimate of daily pan evaporation. Measurement errors can occur if birds drink from the pan, debris falls in, or water splashes out. The biggest problem with pans is that they evaporate more water than would a shallow lake located in the same area. This is because the pan sits above the ground and additional energy conducts through the walls of the pan into the water. This added energy increases pan evaporation relative to evaporation from a nearby lake. To correct for this upward bias a *pan coefficient* is used to lower the pan evaporation value so that it more closely represents evaporation from the lake. The most common value for the pan coefficient is 0.7. In other words, multiply the measured pan evaporation by 0.7 to get the estimated lake

FIGURE 1.5 A Class A evaporation pan. The pan is exposed on a platform to allow air to flow under the pan. In the background is an instrument shelter and a nonrecording precipitation gauge.

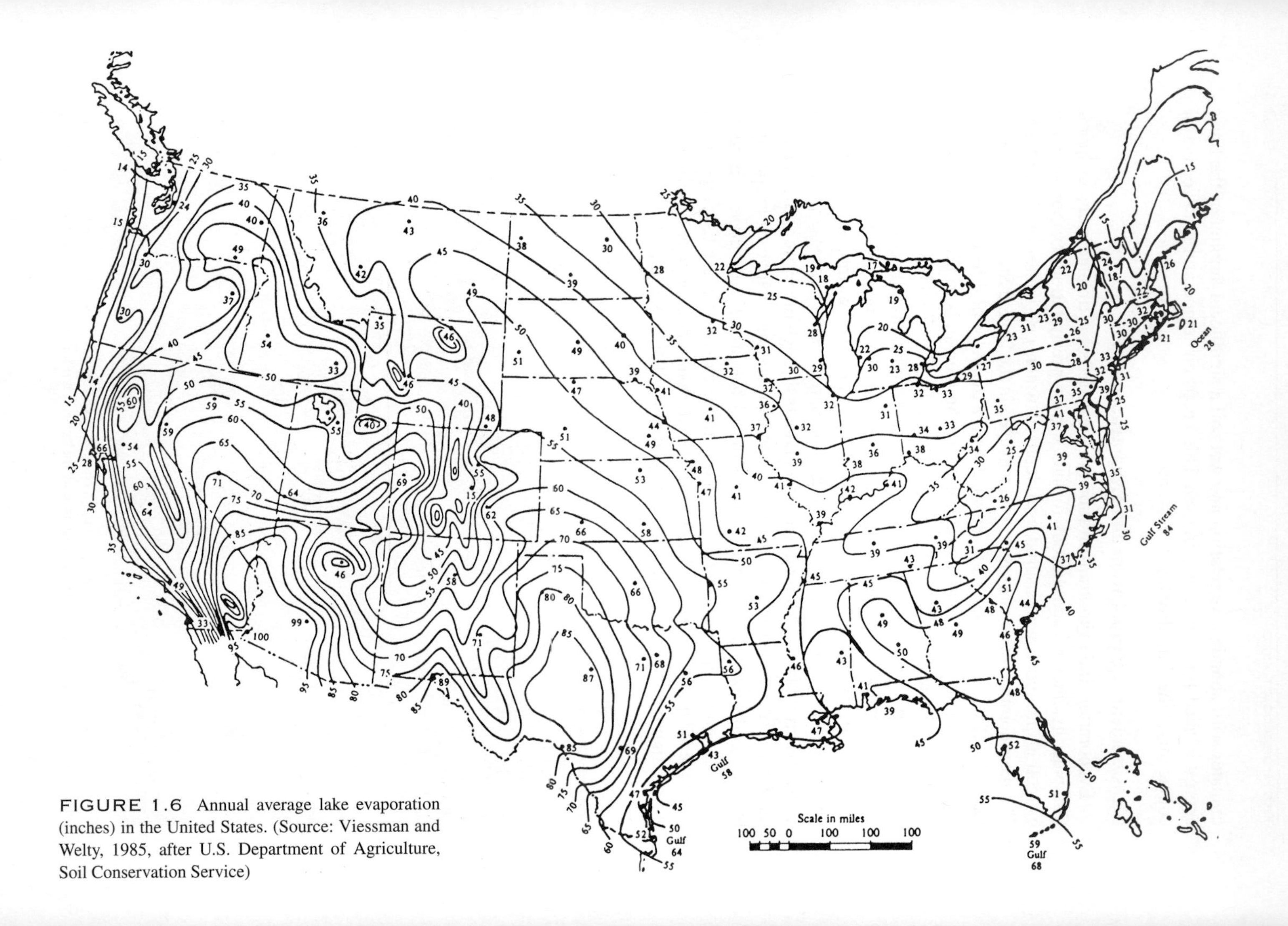

FIGURE 1.6 Annual average lake evaporation (inches) in the United States. (Source: Viessman and Welty, 1985, after U.S. Department of Agriculture, Soil Conservation Service)

evaporation. Typical values for daily evaporation are around one–tenth of an inch on a cool day, to more than a half an inch on a hot, windy day in the summer. Figure 1.6 shows isolines of annual average lake evaporation in the United States. Annual evaporation ranges from 20 inches per year in the Northeast and Northwest, to more than 100 inches per year along the lower Colorado River.

Lake evaporation is a major water management issue in the southwest United States. Water evaporated from reservoirs is lost from a water management point of view. In fact the single largest "use" of water in the entire state of New Mexico is evaporation from Elephant Butte reservoir on the Rio Grande. The only way to control evaporation is by covering the reservoir, which is feasible only for small municipal water supply reservoirs. No one has yet devised a way to reduce evaporation from large reservoirs.

Lysimeters are soil–filled tanks used to measure evapotranspiration. They are buried flush with the ground and planted to the same type of vegetation as that found in the surrounding area (Figure 1.7). There are two basic types of lysimeters—the nonweighing and the weighing lysimeter. The nonweighing lysimeter calculates evapotranspiration as the difference between the inputs of precipitation and irrigation water and any water draining out the bottom. This device does not measure soil moisture and should only be used in situations where the change in soil moisture is negligible. The weighing lysimeter uses a scale that allows the change in soil moisture storage to be determined from the change in the weight of the device.

A third way to estimate evaporation and evapotranspiration is by mathematical equations using meteorological data. The most sophisticated equations use solar energy, wind speed, and humidity as input variables. Penman's (1948) equation is

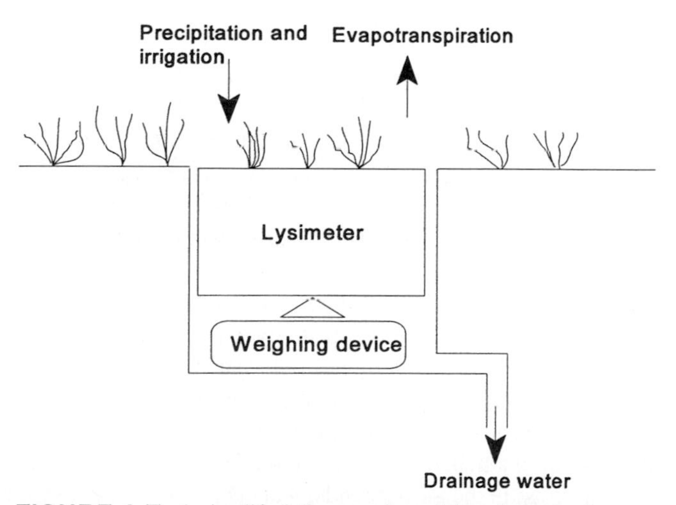

FIGURE 1.7 A simplified diagram of a weighing lysimeter for measuring evapotranspiration. (Source: Thompson, 1998)

of this type and is called a *combination equation* because it combines all three meteorological variables into one equation. The simplest equations are empirically derived and use only air temperature. The two most popular temperature–based evapotranspiration equations are Thornthwaite's equation (1948) and the Blaney–Criddle (Blaney, 1955) method. The trade–off is between the more accurate, but more data–intensive, combination equations versus the less accurate, but more data–forgiving temperature equations. Which equation you use depends upon the accuracy required and availability of meteorological data.

Potential versus Actual Evapotranspiration

In hydrology and water resources the terms *potential* and *actual evapotranspiration* are frequently used. Potential evapotranspiration is the amount of evapotranspiration that would occur from a fully vegetated surface with adequate soil moisture at all times. In other words it is the rate of evapotranspiration limited only by the meteorological conditions and not limited in any way by vegetation or soil factors. Potential evapotranspiration is sometimes assumed to be the same as the rate of evaporation from a free water surface. As the soil becomes drier plants have a harder time extracting moisture, and their transpiration decreases. Thus the actual evapotranspiration may be lower than the potential value.

In the water–resources literature the term *consumptive use* is used as a synonym for either evaporation or evapotranspiration. Water that is consumed is no longer available for use by others. The distinction between the amount of water withdrawn for use and the amount consumed during use is important for water planning and management. For example, thermal (steam) electric power generation and irrigation agriculture both withdraw about the same amount of water on an annual basis in the United States. Using water to cool a power plant consumes about 3 percent of the water withdrawn, whereas irrigation agriculture consumes about 55 percent of the water withdrawn.

PRECIPITATION

Precipitation is when water—in either liquid (rain) or solid form (snow)—returns from the atmosphere to the Earth's surface. There are other forms of precipitation including freezing rain, sleet, and hail, but for water management the most important forms are rain and snow. Precipitation is highly variable in space and time. Figure 1.8 shows *isohyets* of annual average precipitation in the continental United States. The spatial variability of precipitation is most extreme in the West, where annual precipitation varies from over 100 inches per year in parts of the Pacific Northwest to less than 4 inches per year in the desert Southwest. The extreme variability is due in large part to the mountainous topography. Westward flowing air masses are forced to rise up the windward (western) slopes of mountain ranges, resulting in adiabatic cooling, cloud formation, and precipitation. On the leeward (eastern) side the air descends, warms adiabatically by compression, and results in clear skies. The west–facing slopes receive more precipitation and are the source of most of the major streams and rivers in the western United States.

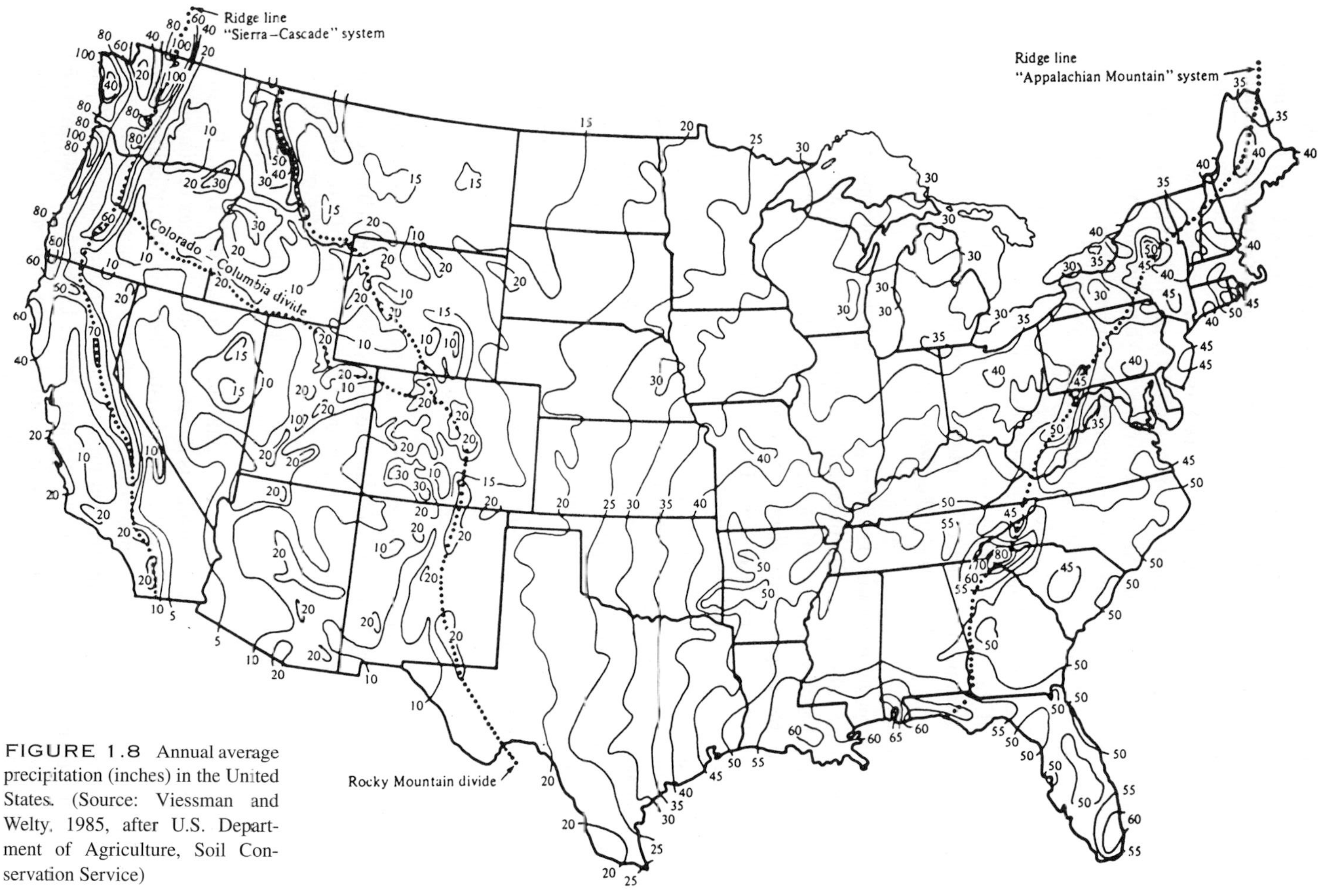

FIGURE 1.8 Annual average precipitation (inches) in the United States. (Source: Viessman and Welty, 1985, after U.S. Department of Agriculture, Soil Conservation Service)

Throughout the eastern United States annual precipitation is more spatially uniform. Annual values range from 60 to 70 inches per year in the South to around 40 inches per year in the Northeast. Again the more uniform pattern is largely because of the more gentle topography. The dividing line between the arid and semiarid West and the subhumid to humid East is conventionally taken as the 20–inch isohyet, because 20 inches of precipitation per year in the midlatitudes is about the minimum required to support trees. Where annual precipitation is between 10 and 20 inches per year, natural grassland vegetation dominates, and with less than 10 inches per year desert scrub prevails. The 20–inch isohyet is usually found near 98° W longitude. Looking at Figure 1.8 it is apparent that many locations in the western United States could have a water *quantity* problem. On the other hand the East has ample precipitation and water quantity has, at least until recently, not been as much of a problem. Many places in the East have long had water *quality* problems. We will address issues of water quantity and quality in later chapters.

Precipitation varies in time as well as space, though this is not apparent in Figure 1.8. Figure 1.9 shows graphs of monthly average precipitation at five locations in the United States. In southern New York precipitation is fairly evenly distributed throughout the year with every month having between 2.5 and 4 inches. In the winter much of it comes as snow. In south Florida precipitation is more seasonal with the wettest months in the late summer and fall. Wet summers and dry winters are typical of subtropical climates around the world. The fall maximum also reflects the importance of hurricanes and tropical storms. To the west in northcentral Colorado annual precipitation is less than half that of southern New York and Florida, and has a spring maximum. Much of the spring precipitation comes as snow. In the southwest desert of New Mexico the annual average precipitation is barely 10 inches per year, with half the total falling in just 3 months from July to September. This is the so–called summer monsoon season in the Southwest. Finally, the central coast of California is an example of a subtropical dry summer (also called mediterranean) climate where winters are wet and the summers are extremely dry. The water management problem here is that the time of greatest water demand (summer) is out–of–phase with the time of greatest water supply (winter). Hundreds of dams and reservoirs have been built to balance the timing of supply and demand by capturing the spring runoff for use later in the summer.

Precipitation Measurement

Precipitation is measured using precipitation gauges (Figure 1.10). The standard National Weather Service gauge is a metal cylinder 8 inches in diameter. A removable funnel channels the water into a smaller inner cylinder. Precipitation is measured by inserting a graduated measuring stick into the inner cylinder. This is a *nonrecording gauge* because it measures only the total precipitation. Precipitation is recorded as a depth (length) of water. In the United States we measure precipitation in inches while most other countries use millimeters. *Recording*

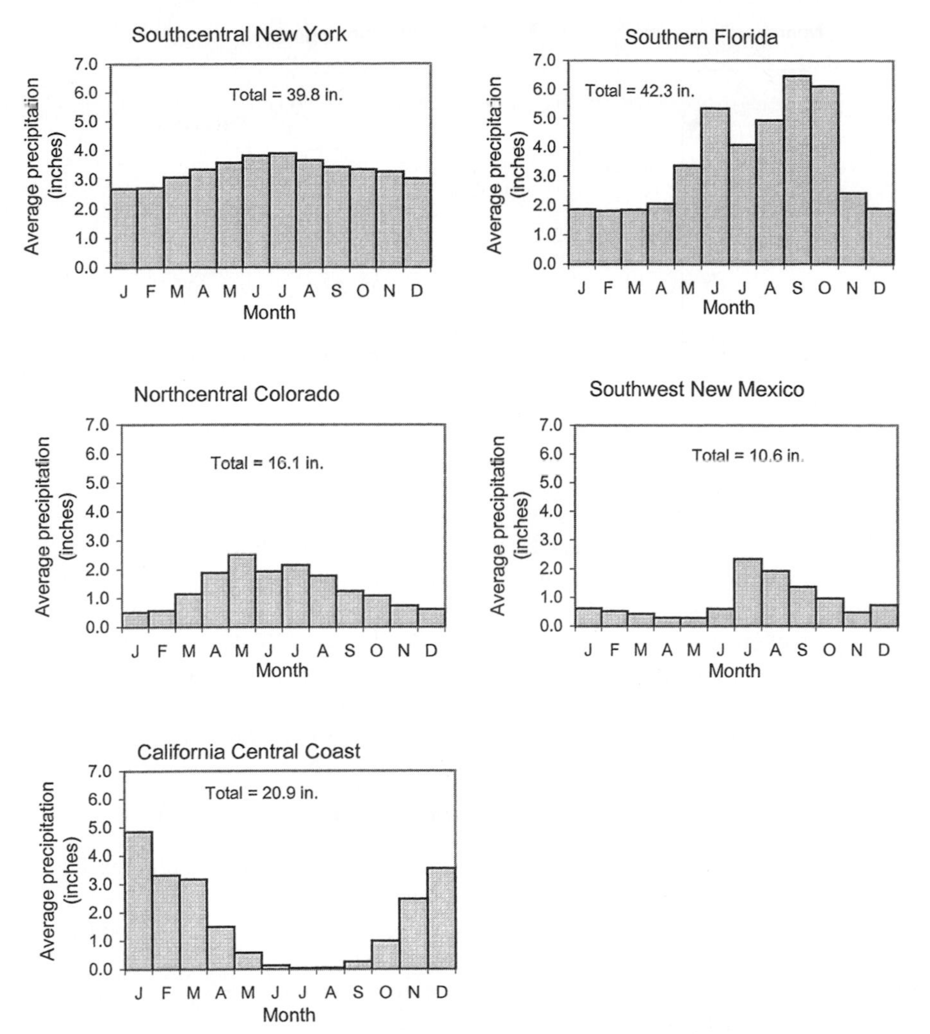

FIGURE 1.9 Monthly average precipitation for five locations in the United States.

gauges incrementally record precipitation during a storm. One type of recording gauge is the tipping–bucket gauge. Rain is directed into a small bucket having a capacity of 0.01 inch. When the bucket fills it overtips and brings a second bucket under the funnel. The tipping action triggers a data–recording device indicating that 0.01 inch of rain has fallen. The recording gauge generates a record of the time distribution of rainfall through a storm. A recording gauge thus gives values of rainfall *intensity,* e.g., inches per hour. Recording gauges are used in rainfall–runoff studies. A graph of precipitation or precipitation intensity versus time is called a *hyetograph* (Figure 1.11).

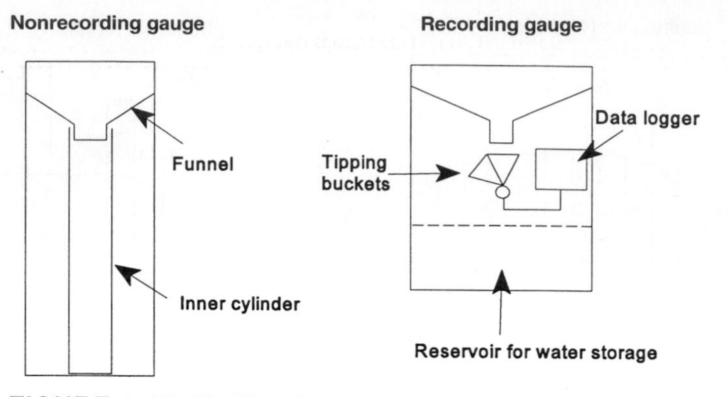

FIGURE 1.10 Simplified sketches of a standard nonrecording precipitation gauge (left) and a tipping-bucket recording gauge (right). (Source: Thompson, 1998)

Remote sensing is also used to measure precipitation. Satellite imagery is combined with ground–based measurements to monitor snow depth and water content in the mountains of the western states. Snowpack conditions are used to generate water supply forecasts. The National Weather Service's NEXRAD doppler radars measure precipitation as a function of the returned power of the radar echo. The radar integrates precipitation measurements over time to give estimates of total precipitation. This is useful for severe storm analysis and flood forecasting.

INFILTRATION AND SOIL MOISTURE

Infiltration is water soaking into the ground. It is important in hydrology because water that infiltrates replenishes soil moisture and deeper groundwater resources. Water that does not infiltrate quickly runs off and can cause flooding along streams and rivers. Infiltration is affected by the type of soil and the type of land cover or use. Water infiltrates sandy soils faster than clayey soils. Water

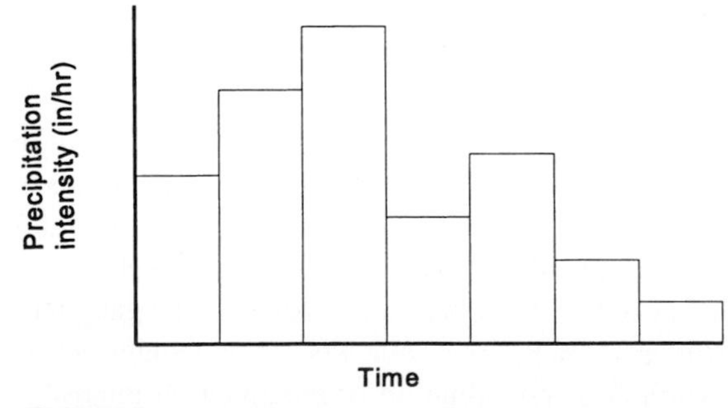

FIGURE 1.11 Hyetograph showing precipitation intensity over time.

TABLE 1.2 Classification of Soils into Hydrologic Soil Groups by Infiltration Capacity after Prolonged Wetting when Planted to Clean–Tilled Crops (Musgrave, 1955)

Soil group	Characteristics	Minimum infiltration capacity (in/hr)	(mm/hr)
A	Soils that are sandy, deep, and well drained.	0.3–0.5	8–12
B	Sandy loams, and moderately deep and moderately well-drained soils.	0.1–0.3	4–8
C	Clay loam soils, shallow sandy loams often with a low-permeability horizon impeding drainage.	0.05–0.1	1–4
D	Heavy clay soils with a high swelling potential, water-logged soils, or shallow soils over an essentially impermeable layer.	0–0.05	0–1

infiltrates soils with a good cover of vegetation, like an undisturbed forest or pasture, faster than bare soil, or even soils with row crops such as corn. Infiltration is also affected by the *antecedent moisture* condition of the soil. Infiltration is usually greater for dry soil than wet soil. Infiltration is difficult to quantify for large areas because the controlling factors vary over short distances. In many planning studies soils are classified into one of four Hydrologic Soil Groups (HSG). Hydrologic Soil Groups are soil texture–based classifications with corresponding infiltration rates (Table 1.2).

RUNOFF

Runoff from land occurs as both surface runoff and subsurface flow (Figure 1.3). Surface runoff occurs either when the precipitation rate exceeds the infiltration rate or when underground water reemerges at the surface and flows overland toward stream channels. When it rains so hard that infiltration cannot absorb all of the water, the excess flows over the surface as *hortonian overland flow*. This runoff process is named for Robert Hortan who first described it in the 1930s. Hortonian overland flow is common in urban areas where rain falls on impermeable roads, sidewalks, and parking lots. It occurs from natural surfaces where bare rock is exposed, vegetation is sparse, or the soil is thin, clayey, or compacted. The second type of surface runoff process is called *saturation overland flow* and can occur after prolonged rainfall or snowmelt completely saturates the ground. The areas most prone to saturation overland flow are the bottom of concave hillslopes and low–lying land adjacent to stream channels. These areas can have high infiltration rates and normally do not generate overland flow. But once the ground becomes saturated, additional rain or meltwater cannot infiltrate and runs off over the surface. Water flowing as *interflow* from further upslope can also be forced to return to the surface upon reaching the saturated area.

Once the surface or subsurface runoff reaches a stream it is channelized flow. The study of water flowing in open channels is one of the most extensively studied fields of hydrology. Since streamflow is on a direct path out of the drainage basin it is common to consider and refer to it as runoff as well. Chapter 10 includes a more extensive discussion of streamflow analysis.

Water that infiltrates and percolates to groundwater flows slowly toward stream channels and the ocean. Groundwater flow velocities are much lower than the velocities for surface water, and it could takes months, years, decades, or even centuries for groundwater to reach a stream. Groundwater flow provides the *base-flow* that keeps perennial streams flowing between storms. Figure 1.12 shows annual average runoff for the continental United States. The greatest runoff occurs in the mountainous Pacific Northwest, while the desert Southwest has the least amount of runoff.

Human Impact on Infiltration and Runoff

Changing land use and land cover changes infiltration rates. Land development for houses, roads, and shopping malls increases the amount of impermeable surface. This in turn increases the amount of precipitation that becomes surface runoff, and can exacerbate flood problems downstream. In the language of economics, land development upstream creates a negative *externality* on people downstream. Development produces an external cost (excess water and increased flood hazard) that is borne by the people downstream. Since the 1970s state and local governments have passed subdivision regulations requiring developers to control any increased runoff. In most cases this means setting aside a portion of the development for use as a detention basin to temporarily store the runoff. These ordinances internalize the externality because the developer pays to build the detention basin. Figures 1.13a and 1.13b are based on Tourbier and Westmacott (1981, 1992). The figures shows how land–use change alters the amount of water following different hydrologic pathways. The values in Figure 1.13 are unique to the particular geographic combinations of soil and vegetation but they are indicative of the types and magnitudes of the changes that occur.

The complementary problem stemming from reduced infiltration is the reduction in groundwater recharge. Groundwater is an important source of water supply in many areas and the spread of impermeable surfaces potentially threatens the long–term viability of the groundwater supply. While it may be possible in some cases to divert surface runoff into "recharge areas," this creates another problem. Surface runoff from streets and parking lots carries a variety of pollutants and these should not be introduced into the groundwater. Surface runoff rarely flows more than a few hundred feet before reaching a channel. The hydraulics of open–channel flow are different from those of overland flow and there is an extensive body of literature devoted to analyzing open–channel flows. We will discuss

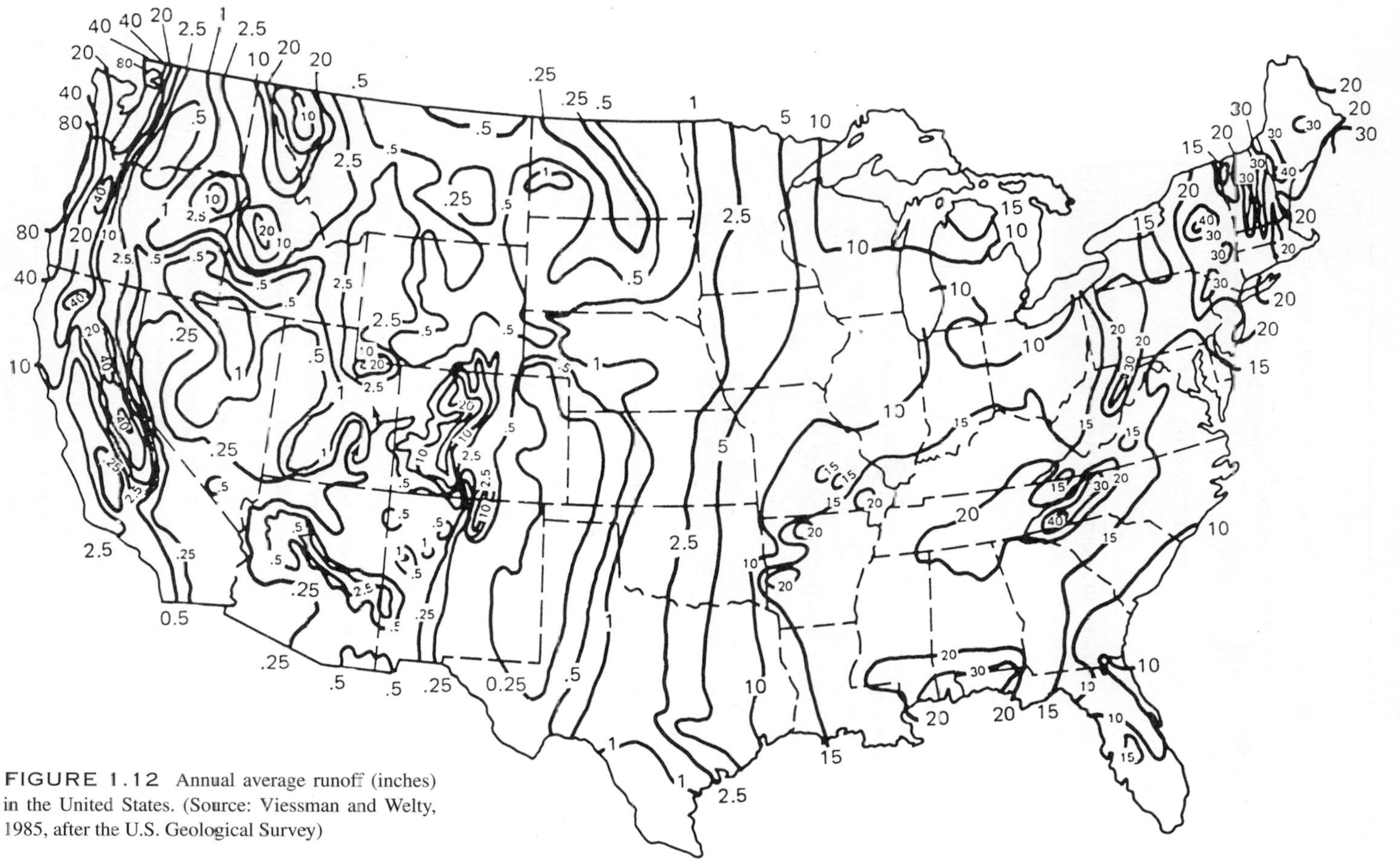

FIGURE 1.12 Annual average runoff (inches) in the United States. (Source: Viessman and Welty, 1985, after the U.S. Geological Survey)

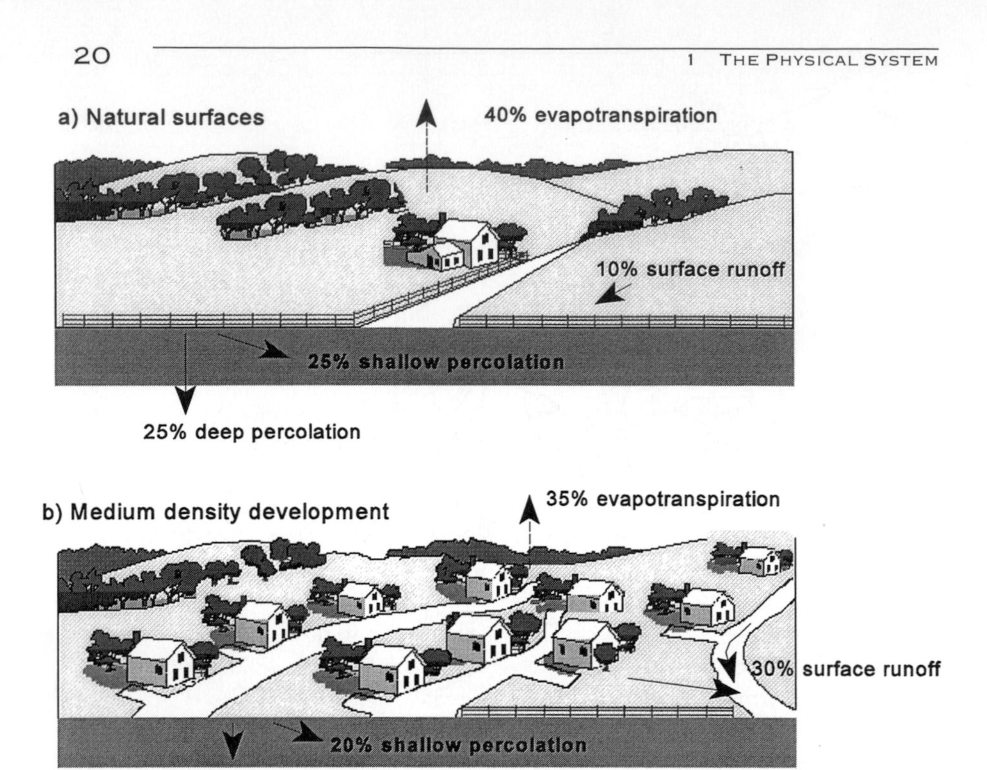

FIGURE 1.13 Hydrologic changes following land development. (Based on data from Tourbier and Westmacott, 1981, 1992).

stream discharge when we consider water supply in Chapter 6 and floodplain management in Chapter 10.

GROUNDWATER

The last major component of the hydrologic cycle depicted in Figure 1.14 is groundwater. Groundwater is stored in the interparticle pore spaces and cracks within rock material. The difference between groundwater and other types of underground water, e.g., soil moisture, is that with groundwater the interparticle spaces are saturated. In the soil the interparticle spaces are usually an unsaturated mixture of water and air. Groundwater represents a massive storage for water on Earth (Table 1.1) and in the United States. About 45 percent of the groundwater on Earth is fresh; the remainder is brackish. Groundwater moves at very low velocities, on the order of a few feet per day or less. Surface water, on the other hand, flows at velocities of a few feet per second. When groundwater flows in underground crevasses or solution channels it flows like channelized surface water, but this type of groundwater flow is relatively rare. Most groundwater actually flows *through* the porous rock material. The lowest groundwater velocities are found in clays and solid crystalline rock where velocities may only be a fraction of an inch per year or per decade.

From a water–resource point of view an *aquifer* is a water–bearing rock formation that yields water in economically usable quantities. Water stored in rock formations, but which cannot be used economically, is still groundwater to the hydrologist but it is not a resource. This definition of an aquifer depends on the intended use of the water. If you drill and pump a well and it produces 10 gallons per minute, is this an aquifer? The answer is yes if the water is intended for in–house domestic uses. A 10–gallon–per–minute well is a good well for a single–family household. The answer is no if you had intended to irrigate 500 acres of corn. A good agricultural well produces hundreds to thousands of gallons of water per minute.

The two basic types of aquifers are the *unconfined* aquifer and the *confined* aquifer (Figure 1.14). An unconfined aquifer has an extensive water table open to recharge by precipitation. Unconfined aquifers are also called water table aquifers. When a well pumps water from an unconfined aquifer the water is released through drainage of the pore spaces. In order for water to flow to a well a decreasing pressure gradient must exist in the direction of the well, which is why a *cone of depression* forms around a pumped well in an unconfined aquifer. Water flows downslope toward the pumping well.

A confined aquifer does not have an extensive water table; the water table exists only in a spatially restricted recharge area. The confined aquifer is sandwiched between relatively impermeable layers of rock that limit recharge from above and below. These layers are called *aquicludes* if they are impermeable and *aquitards* if they are semipermeable and slowly transmit (leak) water into the aquifer. One defining characteristic of confined aquifers is that the water is pressurized. If the

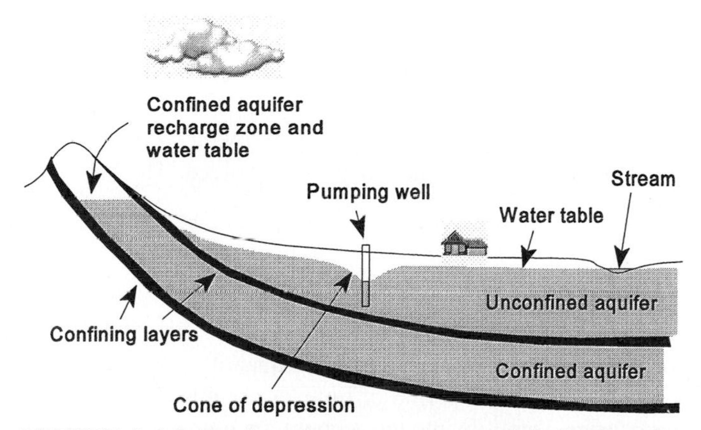

FIGURE 1.14 This figure is a cross section through the Earth's surface showing an unconfined aquifer and a confined aquifer. The unconfined aquifer has an extensive water table open to recharge. The confined aquifer is sandwiched between impermeable layers and is only recharged in the spatially restricted recharge zone. (Source: Thompson, 1998)

internal pressure is high enough, water will flow from the well without pumping. This is called a flowing *artesian well.* Water is released from a confined aquifer through compression of the aquifer, much like squeezing a saturated sponge. Unlike the unconfined aquifer, a cone of depression does not form in a confined aquifer because it remains saturated as it compresses. Compression may lead to subsidence of the overlying rock and eventually the land surface. In many places around the world, including London, Beijing, Mexico City, and the San Joaquin Valley in central California, surface subsidence has exceeded 25 feet.

Two major issues for groundwater management are overdrafting of aquifers and groundwater contamination. Overdrafting or "mining" occurs when the groundwater is pumped from the aquifer faster than it is recharged. This is a problem in arid and semiarid environments where groundwater is the primary source of water supply and recharge is limited. Overdrafting may lead to subsidence as mentioned above and/or the intrusion of salt water if the aquifer is located along a coast. Groundwater contamination can happen anywhere when pollutants are accidentally or intentionally released on or below the land surface.

From a global perspective there is plenty of water on Earth, and the hydrologic cycle continuously replenishes the sources of water on which society depends. But at regional and local scales there are critical imbalances between the demand for water and the available supply. The hydrologic cycle ensures we will never run out of water, though we could run short of water in certain areas, and we could run short of clean water anywhere.

WATER BALANCE

Planning and managing water requires that we measure the various components of the hydrologic cycle. With the exception of chemical reactions, water is neither created nor destroyed; it must go somewhere and be accounted for. It is this simple fact—the conservation of mass—that allows us to use water balances for water management. A water balance is a type of mass balance. Theoretically, a mass balance can be done for any physical quantity; however, in practice it may prove difficult to measure all of the quantities involved. The basic equation for any mass balance is

$$I = O + \Delta S, \tag{1.1}$$

where I is the inputs, O is the outputs, and ΔS is the change in the amount of mass in storage. (The Greek symbol Δ means change.) Equation (1.1) says the inputs minus the outputs must equal the change in the quantity stored. All of these quantities are measured over some period of time. A checking account at a bank is a type of mass balance. Deposits into the account are inputs, checks written against the account are outputs, and storage is the amount of money in the bank. If inputs (deposits) are greater than outputs (value of the checks), then ΔS is positive and the amount of money in the bank increases. If inputs are less than outputs the account balance goes down. For a water balance what are considered inputs and

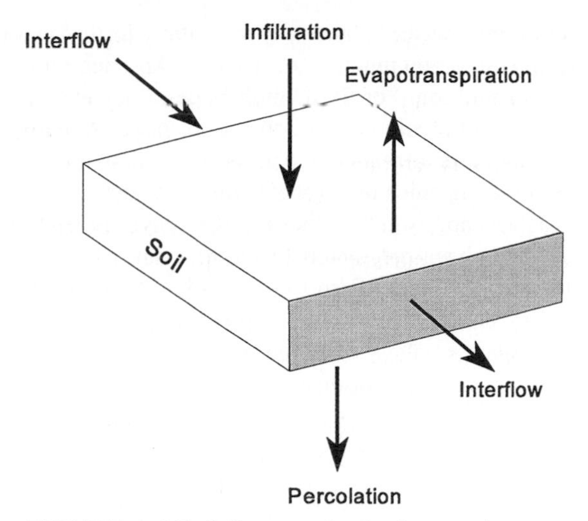

FIGURE 1.15 Soil system showing inputs and outputs. (Source: Thompson, 1998)

what are considered outputs depends upon the particular water store we are analyzing. If we are analyzing soil moisture the inputs are infiltration and lateral subsurface interflow (Figure 1.15). Outputs from the soil are evapotranspiration, downward percolation to groundwater, and interflow exiting the soil volume downslope. Just as with a checking account, if the inputs exceed the outputs, the amount of water in storage increases. Unlike a checking account there is a minimum and a maximum amount of water that can be stored in the soil. When outputs exceed inputs, soil moisture storage goes down, but it can never be less than zero. Likewise, when the soil becomes saturated no additional water can be stored. If inputs and outputs happen to be equal over some time period, then $\Delta S = 0$ and the soil moisture content remains constant. This condition of inputs equaling outputs, with storage constant, is called *dynamic equilibrium* or *steady state,* and occurs in many natural systems. A water balance can be done for any part of the hydrologic cycle, or any human water–resource system. For example, Table 1.3

TABLE 1.3 Annual Average Water Balance for North America

	km^3/yr^a	$(mm/yr)^b$	$(in/yr)^b$
Precipitation	15,561	645	25.4
Runoff	5,840	242	9.5
Evapotranspiration	9,721	403	15.9

[a] Data taken from Speidel and Agnew (1988).

[b] Depths are approximate and were calculated by dividing water volumes in km^3 by the area 24,120,000 km^2.

gives the annual average water balance for the entire North American continent. Fully 62.5 percent of the precipitation over North America returns to the atmosphere as evapotranspiration. For the United States alone evapotranspiration returns about 70 percent of precipitation to the atmosphere. Water balances are routinely calculated for water management purposes in individual drainage basins. Public water supply companies use water balances to detect leaks in their water system. Mass balances are used for other types of environmental analyses. Mass balances for salt have been calculated to monitor salt transport and storage in irrigated farming areas (Gomez–Ferrer *et al.,* 1983). Mass balances for nitrogen and phosphorous are done as part of nonpoint source pollution management (Hall and Risser, 1993). On a smaller scale, fuel mass balances are used by retail gas stations to monitor underground storage tanks for leaks that might lead to groundwater contamination. The lysimeter and evaporation pan mentioned earlier also work on the principle of a mass balance.

CLIMATE CHANGE

Climate is the average weather at a location. Climate is what you expect in terms of precipitation, temperature, sunshine, wind, etc.; weather is what you get, that is, the day–to–day variation in these elements. Figure 1.6 (evaporation) and Figure 1.8 (precipitation) show climate in terms of those two elements. Climate includes the seasonal changes as well the annual averages, which is why the graphs in Figure 1.9 are more informative than Figure 1.8. It is one thing to know that the central coast of California gets about 21 inches of rainfall per year, but from a water management viewpoint it is much more useful to know that virtually no rain comes during the summer.

One of the most hotly debated areas of science today is the issue of human–forced climate change, the so–called global warming problem. Balling (1993) rather sarcastically refers to global warming as "the mother of all environmental scares." Other scientists feel we are performing a global–scale experiment on the atmosphere and, of course, ourselves. There is no question that human activities are changing the chemical composition of Earth's atmosphere. We are increasing the concentration of carbon dioxide by burning carbon–based (fossil) fuels and by deforesting large areas. Methane is added to the atmosphere from rice–paddy agriculture and landfills, nitrous oxide comes from fossil fuel burning, and chloroflourocarbons are released through uses as diverse as air conditioners, plastic foams, and aerosol propellants. All of these gases have one thing in common, they are all greenhouse gases. This means they absorb longwave infrared energy radiating from the Earth's surface toward space. The physics are quite simple—the more greenhouse gases in the atmosphere, the more infrared energy absorbed by the atmosphere, and the higher the temperature. The theory is simple; the reality is much more complex. The atmosphere is an incredibly complex interactive system responding to both internal changes in chemical composition and external influences from the ocean, continental land surfaces, and even variations in solar energy from the Sun. For those who followed the quasi–scientific topic of chaos, the atmosphere is an example of a nonlinear, dynamical system. In this chaotic atmosphere a butterfly flapping its wings over a field in Switzerland could conceivably set in motion nonlinear interactions that could eventually change the weather in Kansas. Quite frankly we simply do not know what will happen 10, 50, or 100 years down the road as a result of our climate experiment today. To make matters even more interesting, the Earth's climate in the geologic past has been both warmer and colder than it is today due to natural forces unrelated to any anthropogenic influence. Is the Earth's climate changing due to natural process? We do not know. Are human activities changing climate, and if so are we causing changes that add to, or counteract, possible changes driven by natural processes? Again, we simply do not know the answer with certainty.

COMPUTER SIMULATION OF CLIMATE CHANGE

Despite our profound ignorance scientists have glimpsed possible scenarios of a greenhouse future. They have done this using computer–based mathematical models called global circulation models (GCMs). "Climate models are based on physical laws, represented by mathematical equations, that are solved using numerical methods" (Trenberth, 1997). In the grid type of model, the Earth is subdivided horizontally into a grid perhaps 3° latitude by 3° longitude. The atmosphere is further divided into as many as 15 layers vertically (Figure 1.16). There are a half–dozen GCMs at major atmospheric research centers and universities

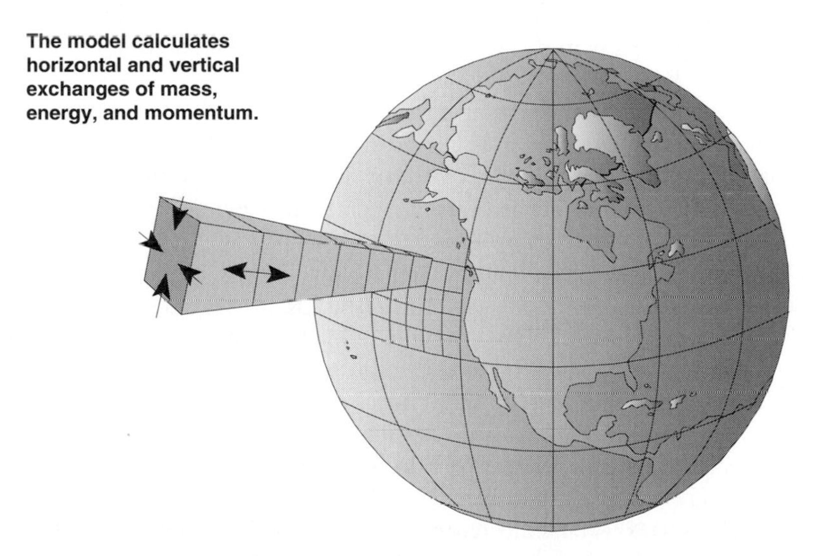

FIGURE 1.16 Representation of a grid-based GCM. The Earth is gridded horizontally and the atmosphere is divided into a number of layers vertically. (Henderson-Sellers and McGuffie, 1987)

worldwide. In these models the components and processes of the climate system are described by mathematical equations. When the model is run, equations describing the flux and conservation of mass, energy, and momentum are repeatedly solved in all three dimensions. For each grid polygon and atmospheric layer the model calculates values for atmospheric variables like air temperature, precipitation, and wind speed. The most advanced GCMs are coupled ocean–atmosphere models. Once the model is calibrated and verified, different scenarios can be tested. The classic "equilibrium experiment" is to double the effective carbon dioxide (CO_2) concentration in the model and run the model until it come into dynamic equilibrium with the new conditions. The effective CO_2 is the radiative forcing from all the greenhouse gases expressed as an equivalent concentration of CO_2. An alternative experiment is to run the model and change the effective CO_2 concentration incrementally over time. This is a more realistic simulation of how greenhouse gases are released, but it is a more difficult experiment to run. The results from the various models and experiments indicate that a doubling of the effective CO_2 might cause the average temperature of the Earth to increase anywhere from 1.5 to 4.5°C. The spatial pattern of temperature change is not uniform. Temperatures in high–latitude zones could increase twice as much as the global average, while the equatorial region might see little or no change.

The potential changes in precipitation are more complex. Some models show certain areas, like monsoon Asia, receiving more precipitation while others receive less. All the models predict an increase in the global average precipitation because a warmer atmosphere holds more water vapor; hence, there is a more vigorous hydrologic cycle. Beyond this there is considerable disagreement between the models on potential precipitation changes.

Another important hydrologic variable to consider is soil moisture. Soil moisture storage is the net of the changes in inputs and outputs (Figure 1.15). Precipitation may go up, but if higher temperatures mean greater evapotranspiration, the net effect could be a drier soil. It is interesting that one of the few areas of agreement between the different GCMs is that they all show soil moisture decreasing in the central United States. This is significant since the central United States is the agricultural breadbasket of the world.

The author simulated potential hydrological impacts from a hypothetical change in climate along a transect from semiarid western Kansas through humid eastern Missouri (Thompson, 1992). The particular climate–change scenario was a 3°C increase in temperature and a 10 percent increase in precipitation over a 50–year period. The results showed very little impact on springtime soil moisture in the humid area (Missouri), but spring soil moisture decreased significantly in the drier region (Kansas). Summer soil moisture decreased at all locations, and annual runoff decreased around 25 percent at most locations.

Researchers have taken temperature and precipitation results from GCMs and used them as inputs to drive smaller–scale hydrologic models of individual drainage basins. One such exercise looked at what would happen to runoff along the western Sierra Nevada mountains in California if climate changed according to the prediction of one GCM (Lettenmaier *et al.,* 1988). The simulation showed that overall there would be more precipitation, which is good news for water supply planners. However, there was a change in the timing of runoff. The higher air temperatures in the spring caused the snowpack to melt earlier and more of the annual precipitation came in the form of rain rather than snow. This change in the timing of runoff could produce new challenges for water supply management in this area.

EVIDENCE FOR CLIMATE CHANGE

Careful examination of climatological records indicates that the global average temperature for the Earth has increased about 0.5°C over the last 100 years (Figure 1.17). This is only about one–half the temperature change predicted by the GCMs given the amount of greenhouse gases released into the atmosphere. The temperature change has been greater in the southern hemisphere than in the northern hemisphere. The average temperature for the United States has increased about 0.3°C during this century, with most of this increase occurring in just the last 30 years (Figure 1.18). The increase in temperature has been due mainly to an increase in the nighttime minimum temperatures during the winter and spring, and most of the change has occurred in the West. Is this evidence of climate change? Perhaps, but this magnitude of temperature increase could be expected to occur naturally 1 time in 20 (Karl *et al.,* 1995).

The precipitation record for the United States over the last century is less clear. Since 1970 average precipitation has increased about 5 percent above the previous 70 years (Figure 1.19). The increase has been mainly in the summer and fall

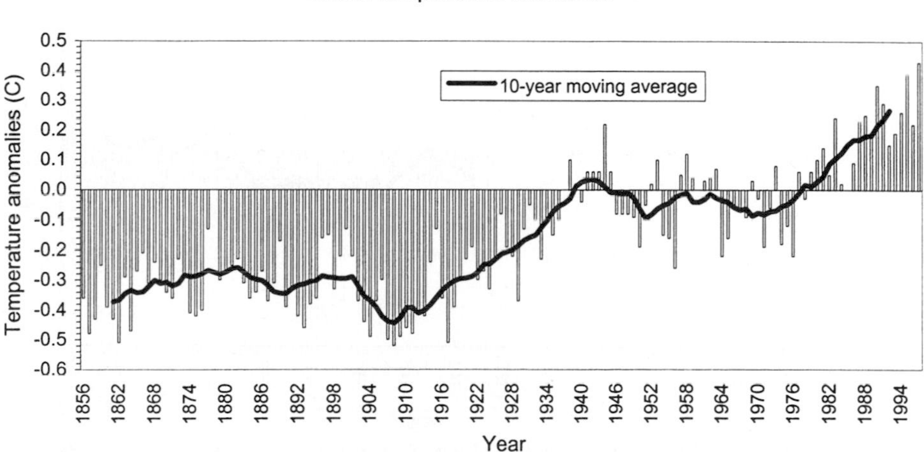

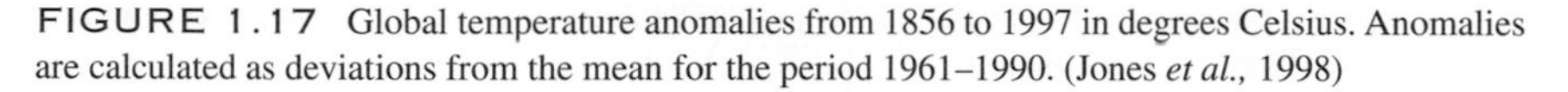
FIGURE 1.17 Global temperature anomalies from 1856 to 1997 in degrees Celsius. Anomalies are calculated as deviations from the mean for the period 1961–1990. (Jones *et al.,* 1998)

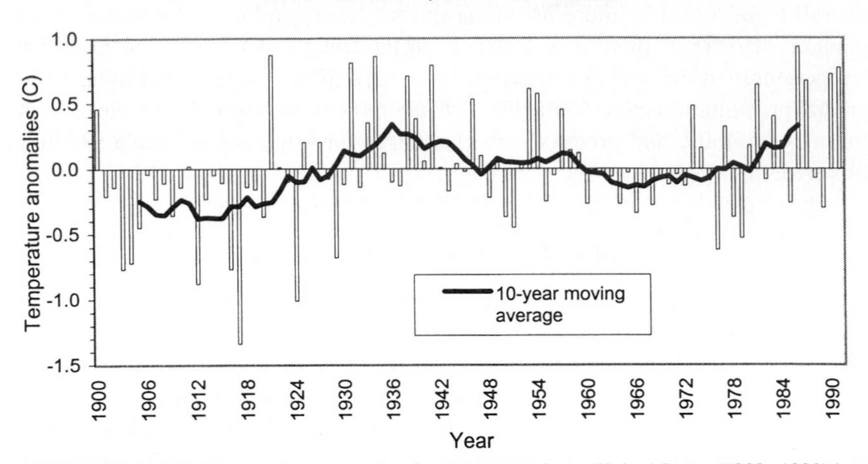

FIGURE 1.18 Temperature anomalies for the conterminous United States (1900–1992) in degrees Celsius. Anomalies are calculated as deviations from the mean for the period 1951–1980. (Karl *et al.*, 1994a)

seasons. Unlike temperatures, the increase in precipitation has been widespread across the country. Regions where precipitation *decreased* significantly include most of California, Colorado, Wyoming, Montana, and northern New England. Another change that shows up in the precipitation data is that extreme precipitation events have become more frequent. This increase in the frequency of intense

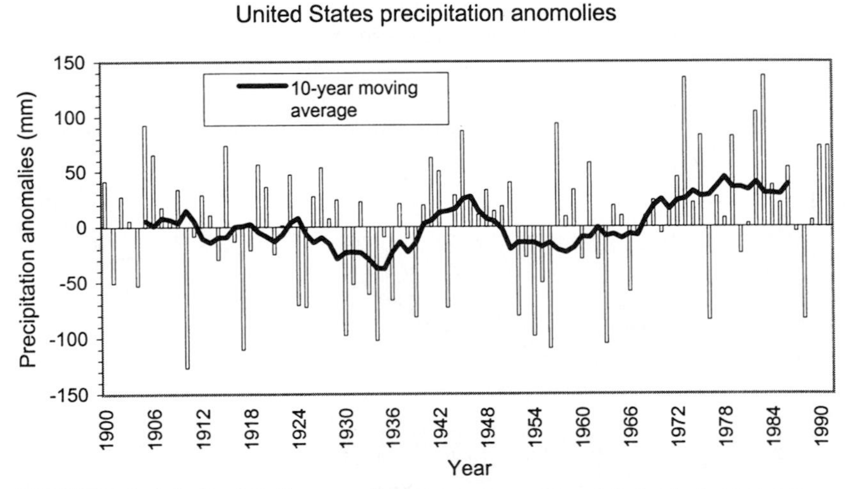

FIGURE 1.19 Precipitation anomalies of area–averaged precipitation in the United States (1900–1992). Anomalies are calculated as deviations from the mean for the period 1951–1980. (Karl *et al.*, 1994b)

rainfall events is highly significant, and would be expected to occur naturally only 1 time in 1000 (Karl *et al.,* 1995).

Lins and Michaels (1994) analyzed streamflow records for the United States from 1941 to 1988 and found that streamflows across most of the conterminous United States have been increasing. The increase is most apparent in the fall and winter seasons. The increase in winter season streamflow is interesting in light of the fact that there has been no apparent increase in winter season precipitation. We do not know the answer but it could be a result of a decrease in evaporation and evapotranspiration.

One piece of evidence that seems to indicate a warming of the Earth is the recently observed melting of ice around the margins of Antarctica. Satellite images conclusively show the reduction in ice–covered area (*Scientific American,* 1995a). Recent research also seems to point to past climate change that was much faster than previously thought possible. Conventional wisdom holds that sea–level fluctuations of 50 feet or more would take thousands of years. New evidence from the Bahamas Islands seems to show that 120,000 years ago sea levels rose 20 feet above current levels and then plunged 30 feet below modern levels, all in less than 100 years (*Scientific American,* 1995b).

We cannot answer the question with confidence as to whether climate is changing, or, if it is, whether the change is natural or driven by human action. The data do not allow us to draw such unequivocal conclusions. There appears to be some changes that are consistent with GCM predictions, but other data are inconsistent with model results. One thing we can say for sure is that, if climate does change it will be the changes in the hydrologic cycle that will be the most significant for society.

2

A History of Water Development in the United States

1800–1900: A Time of Resource Exploitation
1900–1921: The Progressive Period
1921–1933: Post–World War I
1933–1943: The New Deal
Other New Deal Activities
1943–1960: Congressional Control of Water Resource Activities
New Water–Resource Issues
Other Water–Resource Issues
1960–1980: The Environmental Period
Water Quality
The National Water Commission
Traditional Water Agency Activities
1980 to the Present: A Time of Changing Focus

History provides a context for understanding present and future actions. History is also a teacher as we can learn from past experience and hopefully avoid some of the more monumental mistakes. Fully appreciating where we are today in managing water requires understanding the path by which we arrived. As Socrates was purported to have said, "No wind is favorable if one does not know from what port he has departed and to what port he is heading" (Zinser, 1995). A number of reasonably distinct periods can be identified in the history of water–resource management in the United States. These periods are not defined by the water–related activities; rather the attitudes and methods applied to water management followed larger political, social, and economic currents moving through society. Different areas of water–resources management have been dominated by different actors and decision makers at different times. The federal government's interest in navigation via the commerce clause of the Constitution clearly dominated planning and development early on. Urban water supply planning has always been predominately a local activity. In other water–resource fields like irrigation and flood control, early private initiatives were later subsidized by large–scale federal investment. With the exception of the canal–building period in the early 1800s, state governments were not a major force in water–resource activities until the 1960s when concern for the environment began to shift water–resource activities away from a dominant focus on water development to a concern with environmental issues.

1800–1900: A TIME OF RESOURCE EXPLOITATION

At the beginning of the 19th century the federal and state governments were trying to define their respective authority, and their relationship to the private sector. This was especially true in water resources because waterways were the primary arteries of commerce. No one knew the limits of the federal government's power. Could the federal government tax and spend to improve navigation on rivers, or was this something that should be done by the states or private enterprise? Everyone agreed that the country needed to grow, but should growth be with direct aid from the federal government or by a more laissez–faire approach?

The link between the army and water resources was forged early. Ever since the Revolutionary War the army had wanted to emulate the French model of the "scientific builder"—an army officer with a scientific degree (Shallat, 1992). Congress authorized a French–style Corps of Engineers with the establishment of the Military Academy at West Point in 1802. In 1808 Treasury Secretary Albert Gallatin submitted the Gallatin Report to Congress. It was a comprehensive plan for a federally sponsored program of road and waterway development. Even though Gallatin spoke of direct government aid the plan was couched in the best Jeffersonian terms stressing the virtues of farming and how internal improvements would allow every industrious citizen to become a freeholder yeomen. The War of 1812 sparked new interest in Gallatin's proposals as some congressmen felt federal spending for military roads, navigable waterways, and protected harbors was perfectly justified as a matter of national defense. Some said the constitutional power to declare war and finance an army gave Congress the power to finance public works (Shallat, 1992). Others felt that such projects were too overtly commercial and not properly the role of government.

The year 1824 was an important year for water resources in the young country. First, the Erie Canal, begun in 1817, was nearly complete, and the vision it promised sent engineers to the drawing boards to plan other canals. The states of Virginia, Ohio, New Jersey, Indiana, Pennsylvania, and Illinois all started planning or building canals. Some canals were never completed and all were made obsolete by the railroads within a few decades. The second event was the Supreme Court decision *Gibbons v. Ogden* (1824). New York had given monopoly rights to Ogden for steamship operations between New Jersey and Manhattan. Thomas Gibbons, a competitor, sued and the case was appealed to the U.S. Supreme Court. The court found for Gibbons, saying no state had the power to interfere with interstate commerce. The decision declared federal supremacy over all navigable waters by authority of the commerce clause. This meant the interests of state government, local government, and private individuals were subservient to the interest of the federal government when it came to navigable waters. A number of court decisions would follow that defined just exactly what navigable meant.

The last event was the General Survey Act of 1824. The act established a Board of Engineers for Internal Improvements. The Board of Engineers would plan public projects while the Corps of Engineers and the Topographical Bureau would carry out the work. In the act, Congress appropriated $75,000 for navigational improvements on the Ohio and Mississippi Rivers. The appropriation was the first ever for army–directed water projects (Shallat, 1992).

Francis Grund (1837) wrote that "Americans love their country, not, indeed, as it is, but as it will be." The heady optimism of the American people in their perceived ability to successfully bend nature to their will was captured by Mark Twain and Charles Warner, as recounted by White (1969):

> The approach of the American people to the development of their natural waterways was caught by Mark Twain and Charles Warner in *The Gilded Age* when they had the eager young engineer take the visitor out to the banks of the local stream, locally known as Goose Run but recently endowed with the name of the Columbus River. With a glow in his eye the engineer says, "The Columbus River, if deepened, and widened, and straightened, and made a little longer, would be one of the greatest rivers in our western country." (pp. 24–25)

Planning and feasibility studies for canals established early on a distinct culture toward water projects within the Corps of Engineers. General Simon Bernard, head of the Board for Engineers, in defending before Congress the astronomical cost projections for the Chesapeake and Ohio Canal, said, "When a nation undertakes a work of great public utility, the revenue is not the essential object to take into consideration: its views are of a more elevated order" (Shallat, 1992). Bernard saw intangible benefits associated with national pride and satisfaction in accomplishing such a great public project. Bernard was performing his own personal version of cost–benefit analysis; the costs were fairly well specified in monetary terms, but the benefits went far beyond mere dollars and cents.

By midcentury flooding on the Mississippi River pushed flood control into the potential orbit of federal authority. Flooding had always been considered a local concern; it was not an issue broad enough in scope to warrant federal involvement. Along the Lower Mississippi local governments had built a patchwork of small levees to protect individual towns. Major floods repeatedly overwhelmed this disjointed and poorly constructed levee system. In 1850 the federal government authorized a survey of the Mississippi River and the investigation of a flood control plan. The report was submitted to Congress in 1861 and it considered the two issues of flood control and navigation as interrelated water management problems (Petersen, 1984). In 1879 Congress created the Mississippi River Commission. The commission was charged with developing plans for navigation and flood control, but only on the Mississippi River. In establishing the commission Congress recognized that flooding along the Mississippi River had national repercussions, and that the problem was too difficult for local governments. As with the first efforts in navigation the government's involvement in flood control proceeded incrementally. First there was no involvement because the authority of the federal government was unclear. As federal authority was recognized, activities were geographically restricted to the Mississippi River. The Mississippi was seen as a special case; only the Mississippi was big enough to justify a federal presence. It would be 1917 before the federal government appropriated money for flood control outside of the Mississippi Basin. In that year Congress appropriated money for flood control on the Sacramento River in California. The flood control plan for the Mississippi was based on levees and diversion channels. The Corps of Engineers was not authorized to build dams, and, in any event, the technological advancements needed for constructing large dams were still decades away. This levee–based approach was at the heart of the flood control controversy that erupted more than a century later. In the summer of 1993, and again in the spring of 1995, floods ravaged towns along the Mississippi, causing $15 billion in damages in 1993 alone. The efficacy of levees, the concept of flood *control* itself, and the wisdom of river engineering in general were all called into question.

Another activity that was enthusiastically engaged during the 1800s was the draining of swamps. Today they are wetlands, and are recognized as critical components of the environment. Wetlands provide lush feeding grounds and nesting sites for wildlife, they act as natural reservoirs to detain and store excess flood waters, and they provide natural water filtration and purification through biological processes. But in the 19th century they were swamps, and the only good swamp was one that had been drained. The "Swamp Acts" of 1849 and 1850 ceded large areas of overflow lands unfit for cultivation to the states, which in turn sold the land to individuals for reclamation. The formation of drainage districts in the Mississippi Basin and in the Sacramento–San Joaquin Delta led directly to the formation of levee and flood control districts later on (Petersen, 1984).

California was in the midst of a gold rush in 1849. We will see in the next chapter that the miners created a new water law doctrine called *prior appropriation,* which dominates throughout the western states today. Many would–be prospectors realized they could make a better living as farmers down in the Central Valley growing the food and fiber needed by the miners. Farming in California meant irrigating because, with the exception of the high mountains, the climate was either arid, semiarid, or dry–summer subtropical (mediterranean). The development of irrigation agriculture followed a similar sequence in many western states. First, individual farmers dug ditches and diverted water to their farms near the river. As farms were developed farther from the river the task of supplying water became more difficult, so farmers formed irrigation companies and districts. These organizations could undertake larger projects by mobilizing a larger labor force, raising revenue through levies on developed land, and by securing larger loans from banks. In joining the organization, farmers pooled their individual water rights in exchange for shares of stock in the new company. Shares were sometimes given as payment for work on company projects. Each share was worth a certain amount of water. In some states private investors backed the development of irrigation agriculture. This happened in northeast Colorado where English investors financed the development of irrigation corporations in the late 1800s. All of these investment corporations eventually failed, but the irrigation facilities were reorganized into successful quasi–governmental irrigation companies.

But even local irrigation districts were not up to the task of undertaking truly big projects, such as building a dam to control the Colorado River. By the late 1800s the third phase of irrigation development emerged—petitioning the federal government for direct aid in irrigation development. At first the government responded only by revising the homestead laws in the four driest states and eight territories. The Desert Land Act (1877) gave homesteaders title to an entire section of land (640 acres, or one square mile) if the land were irrigated within three years. The government expected the irrigation infrastructure to be privately developed. In the humid eastern states individuals could only homestead a quarter section (160 acres). As with most of the homesteading acts, the Desert Land Act was rife with fraud. According to Reisner (1986).

> As for the Desert Land Act and the Timber and Stone Act, they could not have promoted land monopoly and corruption more efficiently. A typical irrigation scene under the Desert Land Act went as follows: A beneficiary hauled a hogshead of water and a witness to his barren land, dumped the water on the land, paid the witness $20, and brought him to the land office, where the witness swore he had seen the land irrigated. Then, with borrowed identification and different names, another land application was filed, and the scene was repeated. If you could pull it off six or seven times you had yourself a ranch. (p. 45)

The Desert Land Act is an example of a federal statute that was passed for one reason (to further land occupation) but was later interpreted by the courts as establishing an important rule in water law. The act was interpreted at one time by the U.S. Supreme Court as having given the states ownership of all water not specifically reserved by the federal government. Later rulings contradicted this interpretation, further fueling the debate over who really owns the waters of the United States. A number of national irrigation congresses were held beginning in 1891 to promote federal aid for irrigation development. As we see in the next historical period, the government finally responded when it passed the Reclamation Act in 1902.

There was small but significant controversy over how the driest lands of the United States should be developed. Starting in Ohio in the 1790s the federal government began surveying and selling public land based on the township and range system. In the original colonies, property boundaries were set by "metes and bounds," and it was common for property lines to follow natural features, e.g., streams, ridges, and coast lines. The township and range system divided the landscape into nice neat squares (sections). Townships are numbered north and south, while ranges run east and west. The rectangular pattern paid absolutely no attention to natural features. In the humid states a quarter section was usually adequate for a family farm. Out West 160 acres was wholly inadequate unless the farm was irrigated. The famed explorer of the Colorado River, Major John Wesley Powell, submitted his recommendations for developing the arid West to Congress in his *Report on the Lands of the Arid Region of the United States* (1879). The report was a sensitive appraisal of the resource limitations of the western United States. By Powell's estimate only 1 to 3 percent of the arid region could be successfully reclaimed by irrigation. Among Powell's main recommendations were abandoning the township and range system, developing the region on a watershed basis where water and land would be developed as a unit, and abandoning the rigid 160–acre homestead rule. He reasoned that in areas with good soil and water supply 160 acres may not be necessary for a successful farm. In other areas farms might need to be much larger to be viable. His vision of development was out–of–sync with that of Congress and his report was largely ignored, but it makes great reading for any student of natural–resource history. Two other resource–related controversies were smoldering in the latter half of the 19th century—one environmental and the other sociological.

The 1800s were a time of resource exploitation with little concern for environmental degradation. The country was thought to hold an inexhaustible supply of natural resources. If in exploiting those resources the environment was damaged, so be it, there was unlimited virgin land over the horizon. Born partly of the experience in Europe, but mainly from the accumulating legacy of environmental degradation left in the wake of westward migration, there was a small but growing concern about resource conservation. Even President Washington had rebuked some fellow farmers for their lack concern over soil conservation. One of the constants of westward expansion was the logging of forests. As soon as one virgin stand was cleared, loggers moved on to next. The stump–dotted hillsides left behind were stripped by erosion. There were even instances where sediment from the logged–over hillslopes clogged eastern harbors. In California, sediment, this time a legacy of the gold rush, washed down from the foothills, choked the stream channels, and covered farm fields. The hydraulic mining techniques literally washed away hillsides, carrying sediment downstream. The bed of Bear Creek was raised 90 feet in nine years by sediment deposition, and Sleepy Hollow Creek was raised 136 feet. Along the Yuba River 16,000 acres of farmland were covered with sediment (Teclaff and Teclaff, 1973). Since channel sedimentation threatened to impair navigation on larger rivers downstream this prompted action by the federal government. In 1893 Congress created the California Debris Commission to restore the Sacramento and San Joaquin Rivers to their 1860 condition.

In and around major cities the water pollution in some rivers must have been staggering as indicated by nuisance lawsuits filed at the time. Teclaff and Teclaff (1973) give the following examples. From *Clark v. Peckham* (1871),

> A municipal corporation cannot turn its sewage into a navigable water way in such a way as to *fill it up* [emphasis added] to the injury of navigation.

Or, from *New York v. Baumberger* (1867),

> A city is entitled to an injunction restraining the discharge of mash from a brewery through the sewer into a navigable river the free use of which for purposes of navigation is impeded by diminishing the depths of water so that vessels will be prevented from coming to the city's wharves, thereby depriving it of dockage and wharfage.

Section 13 of the River and Harbors Act of 1899 was titled the Refuse Act. The Refuse Act prohibited the discharge of any nonliquid material into navigable waterways. This act became the foundation for federal water quality regulations later on.

Concern about the transformation of wild nature was expressed by Thoreau in his manuscript *Walden* published in 1854. But one of the most significant events in support of the cause of conservation was the publication of George Perkins Marsh's book *Man and Nature or Physical Geography as Modified by Human Action* (1864). Marsh's treatise is considered the first scientific assessment of human–caused environmental degradation. Drawing largely on European examples Marsh detailed how abuse of natural resources led to environmental problems. Much of his focus was on erosion, sedimentation, and flooding that followed deforestation. Marsh provided examples and documented them with references. His analysis stands in contrast to earlier works which raised similar concerns, but were poorly documented and were based more on hearsay than verifiable evidence.

The emerging social controversy in the latter half of the 19th century was the growing resentment of a wealthy upper class that had gained monopolistic control over natural resources, the means of production, and transportation systems. One industry that figured prominently as a focus of the public's ire was the railroads (see Figure 2.1). Farmers in the West relied on rail to get their products to market; they had no real alternative. Good road transportation was nonexistent and, unlike their eastern counterparts who had a choice between rail and barge, there were few navigable waterways. The railroads treated the farmers poorly. They gave rate breaks to large shippers and inflated rates for farmers and manufacturers who shipped small quantities and/or short distances. The railroads also controlled marketing and prices since they owned the grain elevators. The railroad tycoons were getting rich off the "peoples land" (Petulla, 1988). In fact, well into the 20th century one of the major justifications for federal investment in waterway improvements was to control freight rates on the railroads. This was "reminiscent

FIGURE 2.1 "The Grange Awakening the Sleepers." This 1873 political cartoon shows a farmer warning ordinary citizens of the West of the oncoming "Consolidation Train." The train bears a variety of unpleasant social burdens associated with the growing monopolies. The citizens ignore the farmers warning. (The log ties upon which the rails were laid were called "sleepers.") (Source: Culver Pictures, New York)

of Charles Lamb's arrival at roast pig by burning the entire house: construction investment in the waterway is sacrificed in order to obtain a change in freight rates" (White, 1969, p. 25).

Farmers were not alone in their struggle; millions of wage workers in eastern factories saw the ever–widening division of society into the haves and have nots. It was not just the railroads, but bankers, steel barons, equipment manufacturers, "middlemen" in general, and corrupt public officials who were all increasingly unpopular. They were all seen to be getting rich and powerful at the expense of the everyman. The growing resentment of the wealthy upper class was evident by the emergence of a number of independent, antimonopoly political parties in the

1870s. Congress eventually responded with the Sherman Antitrust Act in 1890. This growing populist movement helped define the next historical period.

A final indicator of the changing attitude toward natural resources was the passage of federal legislation reserving portions of the public domain for national parks and forests. Yellowstone became the nation's first national park in 1879 through the combined effort of various groups who wanted to preserve it from miners, timber cutters, and industrialists. The General Revision Act of 1891 allowed the President to set aside public reservations "wholly or in part covered with timber or undergrowth." Passage of the Organic Act in 1897 gave the government authority to reserve national forests and created the Forest Service. By 1901, 50 million acres had been withdrawn from the public domain, and President Roosevelt withdrew another 150 million acres in the next decade. Controversy still rages today over these reserved lands, but there was no doubt by the end of the century that a new age for resource management had dawned.

1900–1921: THE PROGRESSIVE PERIOD

The populist movement that began in the late 1800s became a dominant political theme when Theodore Roosevelt became president in 1900. Popular discontent with the existing social order, an economic depression, and the growing awareness of natural–resource abuse had brought the issue of resource conservation to the highest levels of the federal government. The conservation movement benefitted enormously from the psychological impact of the 1890 Census announcement, echoed by historian Fredrick Jackson Turner, that the western frontier was closed. For 250 years the frontier had been synonymous with unlimited opportunity and abundance. The country suddenly seemed middle–aged and conservation had a new appeal (Nash, 1973). Unlike McKinley before him, Roosevelt was antimonopoly and supported breaking up big business trusts. Viessman and Welty (1985) identify five elements of the progressive period that in one way or another influenced water–resources planning:

- The conservation of natural resources
- A desire to encourage small, independent enterprises, especially family farms
- Belief in a strong federal government having the ability to affect the nation's economic life
- Protecting equal opportunity and promoting the well–being of the people
- Guarding the public domain from giveaways to special interests

The term conservation was coined in a conversation between Gifford Pinchot and fellow forester Overton Price. Price noted that government forests in India were called conservancies. The two men liked the term, and thus a concept that had originated in the seminal thinking of men like Thoreau and Marsh now had a name (Coggins and Wilkinson, 1981). But just exactly what conservation meant was unclear. Two different interpretations of the concept had emerged. One view was that of Gifford Pinchot, the first Chief Forester of the United States. To Pinchot conservation meant the "wise use" of resources. Resources, he reasoned, should be *used* to provide the greatest benefit to the greatest number of people. The wise use philosophy was the essence of progressivism; it was antimonopoly at its core, and implied scientific resource management. The other view of conservation was held by John Muir, the founder of the Sierra Club. To Muir conservation meant preservation—non use—of natural resources. While the philosophical differences between Pinchot and Muir were debated more over wilderness than water, it was a water–resource battle in the Hetch–Hetchy Valley that irreconcilably galvanized the two groups. Hetch–Hetchy was a scenic valley adjacent to Yosemite National Park in the foothills of the Sierra Nevada Mountains in California. Hetch–Hetchy Valley had been reserved by the act creating Yosemite in 1890. The city of San Francisco purposed damming the Tuolumne River in the valley to create a water supply reservoir. San Francisco needed to improve its water supply because its population was increasing, and the city had recently been devastated by the fire that followed the earthquake in 1906. Muir and his followers vehemently opposed the dam, arguing the fundamental necessity for wilderness in a civilized society. Pinchot supported the dam as a much better use of the Tuolumne River. In 1913 Congress agreed with San Francisco, the preservationists lost, and the dam was built. Today Hetch–Hetchy Dam and reservoir sit in the Sierra Nevada foothills, and water is carried to San Francisco via an aqueduct of the same name. Now there were three resource–use philosophies applied to natural resources in the United States—unbridled exploitation, wise use, and preservation.

Two federal water–related agencies were created at this time. The Reclamation Service (renamed the Bureau of Reclamation in 1923) was created in the Department of Interior by the Reclamation Act of 1902. The Reclamation Service was authorized to build single–purpose irrigation water supply facilities in the 16 western states. Its mission was soon expanded to multipurpose irrigation and hydropower facilities, and Texas was added in 1906, making the service area 17 western states. By 1906 the Reclamation Service had started projects in 15 states for the irrigation of 2.5 million acres. The money to build reclamation projects initially was to come from the sale of public land; later it came from the sale of hydroelectricity generated at the federal dams. Irrigation water from federal projects was to be used only on small family farms, i.e., not exceeding 160 acres, and was not to be delivered to absentee land owners. In theory the Reclamation Act embodied all five of the progressive–period elements listed above.

The second federal agency created at this time was the Federal Power Commission (FPC) created by the Federal Water Power Act of 1920. The act gave the government control over the construction of hydroelectric dams on navigable rivers. The federal government became concerned with hydropower development for two reasons. One was the need to protect the federal navigation servitude. Second was the concern with economic development. The most important aspect of water conservation, as far as Gifford Pinchot was concerned, was preventing prime dam sites from falling into the hands of private developers. Pinchot feared multipurpose development would be stifled and the lucky entrepreneurs might extend their control from hydropower to all of industry (Nash, 1973). The government feared private monopoly control over electricity because electricity was the key to modernization. The government felt it needed to regulate such a vital resource. Now the government could regulate private electric utilities two ways: first by granting operating licenses through the FPC, and second by direct competition. By building its own hydroelectric facilities the government was able to sell cheap power, thus forcing private power companies to lower their rates.

Roosevelt appointed the Inland Waterways Commission (IWC) in 1907. The commission's report in 1908 emphasized the interrelated problems of natural–resources management, for example, how water management could conserve coal, iron, soil, and forests resources. The IWC recognized the need for *multipurpose* water planning, and specifically how federal planning for navigation should account for other uses. It emphasized cooperation between the various levels of government and the private sector, and in populist tones spoke of the equitable distribution of the benefits from water resources. The report also recognized the growing problem of water pollution, and that the problem of floods could be addressed through both structural and nonstructural means. In transmitting the report to Congress, Roosevelt talked of the fundamental unity of resource management in a river basin. Finally, the report recommended that Roosevelt convene a conference of all the state governors to discuss the issues surrounding the conservation of natural resources. The IWC addressed issues which are still relevant today—multipurpose water–resource planning, integrated river basin management, and federal, state, local, and private cooperation. That we still discuss these issues today shows the prescience of the IWC, a frustrating lack of progress in the last century, or an element of both.

Taking his cue from the IWC Roosevelt convened the first White House Conference of Governors in 1908. The conference's theme was conservation of the Nation's natural resources. Papers were presented on water, soil, and mineral resources. At its conclusion the conferees called for the creation of a National Conservation Commission (NCC), which Roosevelt created by executive order the very same year (1908). The NCC was organized into four sections—minerals, water, forests, and land—with Gifford Pinchot as Chairman. The NCC carried out the first formal inventory of the Nation's natural resources. The final report of the NCC (1909) recommended a program of scientific study and called for a unified plan for multipurpose water uses including navigation, flood control, irrigation, hydropower, water supply, and pollution control.

The third water commission created in the first decade of the 20th century was the National Waterways Commission (NWC). The NWC's report called for federal construction of certain waterways and hydropower facilities, but it still thought flood control was primarily a local responsibility. The report recommended legislation to protect waterways from railroad competition. The report acknowledged that deforestation affected streamflow, but hydrologic science was still too uncertain to recommend specific land management practices (White, 1969). The most directly ascertainable and verifiable impacts of forest management on water resources were on soil erosion and sedimentation.

An important piece of federal legislation at this time was the Weeks Act (1911). The Weeks Act authorized federal purchase of *private* land in headwaters regions to protect the navigability of rivers. While most of the national forests in the western states came about through the reservation of public land, most of the national forests in the eastern United States owe their origin to purchases under the Weeks Act.

The progressive period was characterized by an unprecedented rise in the concern for the conservation of natural resources thanks to the support of President Roosevelt. But the operative word in the progressive philosophy of conservation was resource use. The best water–resource plan was the one that squeezed the maximum number of uses out of the river. While the preservationist philosophy of nonuse of wilderness was tentatively accepted, in principle at least, nonuse of water was not (Teclaff and Teclaff, 1973). Through the various commissions and conferences, two concepts evolved that have endured throughout the 20th century. One concept is the multipurpose project; the other is basin–wide planning. The multipurpose project became a reality in the next historical period, while basin–wide planning has had a mixed reception. The advent of the multipurpose project marks an important change in water management in the United States. Up until this time single–purpose construction by both public agencies and private individuals was the predominant strategy. As for single–purpose public construction, White (1969, p. 32) observed,

> Single–purpose public construction . . . is remarkably free from experimentation with alternative means. It is largely impervious to doubts as to economic justification. One type of construction came to be associated with one aim by one form of public agency—municipal, district, or federal. It is a ponderous strategy using a limited number of blunt instruments, insensitive to economic indicators, and highly conservative in dealing with risk and uncertainty.

On the other hand White found single–purpose construction by private individuals, e.g.,water supply for the home, to be more sensitive to a wider range of alternatives means, more technologically innovative, and more attuned to marginal economics.

1921–1933: POST–WORLD WAR I

During the progressive period the federal government was called upon to reign in private enterprise, protect natural resources from monopolies, and promote regional economic development. The government responded, and to a large extent the promise of America had been restored. Warren G. Harding, a Republican, was elected President in 1920, which also happened to be the first year that America's urban population outnumbered its farm population. Harding was backed by a solid Republican majority in Congress. But even with the ascendency of Republican control, some progressive–period concepts continued to influence water–resources development. The country was sliding into an economic recession, and it was estimated that the number of wage earners in manufacturing had been reduced 25 percent between 1919 and 1921. Government statistics indicated that almost 5 million workers were unemployed and 453,000 farmers had lost their homesteads (Chronicle of America, 1993). By 1921 the recession ended and the country was prospering once again. Predictably, the mood in Washington was to remove as much competition between the government and the private sector as possible. President Coolidge said, "America wouldn't be America if the people were shackled with government monopolies." Coolidge was a man for the times and the times were decidedly laissez–faire once again, that is, with the exception of water–resources development. Herbert Hoover, first as Secretary of Commerce and later as President, supported the progressive–era concepts of multipurpose projects and basin–wide planning.

The planning functions of the Corps of Engineers and the Bureau of Reclamation continued to expand. First in 1925 in conjunction with the Federal Power Commission, and then again in 1927, Congress authorized the Corps to undertake a comprehensive study of the Nation's rivers with the view to coordinating development for navigation, flood control, irrigation, and power production. The original list of surveyed rivers was published in House Document 308; hence, these basin surveys came to called "308 reports." In the following 20 years the Corps made over 200 separate 308 basin investigations. Many 308 reports were used as blueprints for river development decades later. The one basin the Corps was not allowed to study was the Colorado; the Colorado River was the exclusive domain of the Bureau of Reclamation. Already by the 1920s federal agencies were skirmishing over water–development turf. The nascent fractionation of federal water development did not go unnoticed. A report of a joint committee on reorganization of executive departments in 1924 recommended transferring the nonmilitary engineering activities of the Corps of Engineers from the War Department to the Department of the Interior. Later President Herbert Hoover supported a similar recommendation as a way to reduce duplication, but Congress voted against it.

The year 1928 was another important year for water resources. Floods on the Mississippi River once again prompted Congress to react and pass flood–related legislation. This is a familiar scenario—federal legislation followed a major natural disaster. Congress has always been better at reacting to catastrophe than proactively planning for such a contingency. The Flood Control Act of 1928 authorized the Corps of Engineers to develop a comprehensive flood control strategy for the Lower Mississippi. While the act continued to emphasize levees and diversion channels on the main channel, for the first time it gave the Corps permission to study flood control reservoirs on the tributaries. What is truly astounding was that Congress declared the federal government would pay the entire cost of flood control on the Mississippi. In just a few decades the government went from a policy of viewing floods as a local problem inappropriate for federal involvement to a policy of multimillion dollar construction with the government picking up the entire tab! As expected this new policy strongly influenced local decision making. When the federal government declared it would build flood control structures at virtually no cost, local decision makers understandably saw this as the best solution to their flood problem. The 1928 act further solidified the federal government's structural approach to dealing with floods, an approach that would dominate for the next 40 years.

The Boulder Canyon Project Act passed Congress in 1928. This act set in motion a host of activities in the Colorado basin. The act consummated the Colorado River Compact of 1922; it authorized the building of Boulder Canyon Dam (renamed Hoover Dam) and the All American Canal; and eventually it was used to divide the waters of the Colorado River between the states of California, Arizona, and Nevada. At the end of Chapter 3 we examine the management of the Colorado River in detail. By this time the technology to build large dams existed. Earth–moving equipment, reinforced concrete, advanced dam design, long–distance electrical transmission technology, and a willing federal government all combined to make the multipurpose dam a reality. Hoover Dam on the Colorado River was the prototype (Figure 2.2). Authorized in 1928 and completed in under four years,

FIGURE 2.2 Hoover Dam, the architectural masterpiece among all dams, rises 70 stories from the bed of the Colorado River. (Photo by Andrew Pernick, U.S. Bureau of Reclamation, 1996)

it stands 726 feet high with a mass of 4.4 million cubic yards of concrete. The dam created Lake Mead, which extends 110 miles upstream. Hoover Dam was the latest symbol of our ability to harness and transform the natural world. In 1993 the American Society of Civil Engineers proclaimed it one of the "seven wonders of the United States."

Hoover Dam was planned and authorized in the "Roaring Twenties," a time of prosperity throughout the country. That all changed abruptly when the New York stock market crashed in 1929, and the next year drought began squeezing the Great Plains dry. As topsoil from Kansas settled on the shoulders of unemployed factory workers queued at the soup kitchen doors on the streets of New York, they could not help but think that somehow, some way, abuse of the land had contributed to the economic catastrophe. The country was headed into an unprecedented economic depression and water resources would play a pivotal role in the government's plan for recovery.

1933–1943: THE NEW DEAL

President Franklin Roosevelt took office in 1932 as both the economic depression and the drought intensified. The jobless rates in some cities exceeded 50 percent; nationally unemployment was above 25 percent. More than 5000 banks had collapsed and with them went more than 9 million savings accounts. Two million people wandered the country as vagrants. In his inaugural address Roosevelt vilified the bankers and financiers, saying the "moneylenders have fled from their high seats in the temple of our civilization." Roosevelt said if the traditional executive–legislative measures did not quickly bring about change he would ask Congress for unprecedented executive power to wage war on the depression.

One of Roosevelt's primary strategies for economic recovery was to use federal public works projects to create jobs and lift the country out of the depression. Water projects provided many of those jobs. Many water projects that had been planned in the 1920s and early 1930s got under way, including projects in the Columbia River basin, the Central Valley of California, Hoover Dam, and eventually the St. Lawrence Seaway. To avoid the appearance of pork barrel politics the administration wanted public works projects tied to comprehensive river basin plans. In some cases these were the 308 plans produced by the Corps. The New Deal was about more than just economic recovery, it was about social reform. Part and parcel of social reform was the notion of planning for "national resources." National resources included the traditional land, water, and minerals, but the New Deal's concept of national resources included people too. Ultimately this proved a politically fatal mistake.

To accomplish his goals Roosevelt experimented with new institutions. Roosevelt created the Public Works Administration (PWA) with Secretary of the Interior Harold Ickes in charge. Ickes established the National Planning Board (NPB) in

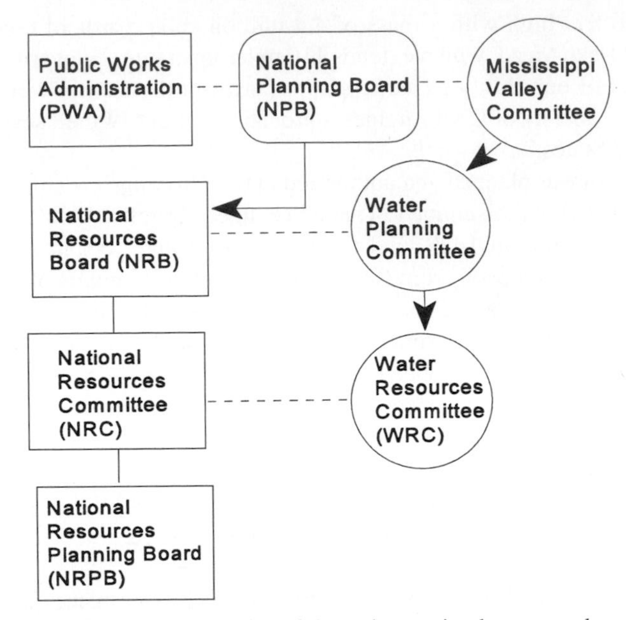

FIGURE 2.3 Evolution of the various national resource planning committees under Roosevelt in the 1930s.

1933 as an advisory board to the PWA (Figure 2.3). The President's uncle Frederic A. Delano was chairman of the NPB. Delano and the other two members of the board, Wesley Mitchell and Charles Merriam, were strong supporters of urban park and city planning. Mitchell and Merriam came from academia—Mitchell was an economics professor at Columbia University, and Merriam was a political science professor at the University of Chicago. This was a significant time because it marked the entrance of social scientists into the national water–planning arena. Up until then water projects were both planned and constructed by engineers. The engineers planned and analyzed projects for their ability to produce specific outputs such as electricity, water supply, or flood control. Engineers were the technological magicians transforming nature into projects producing goods and services. Very little attention was given to the social, and even less to the ecological, impacts associated with these projects. Now social scientists—geographers, economists, and political scientists—as well as biologists had a voice in national–resource planning. Merriam recruited other social scientists from the University of Chicago (Reuss, 1992). Harlan H. Barrows, Chairman of the Geography Department at the University of Chicago, was one of these who followed Merriam to Washington. A year later Barrows brought along his graduate student Gilbert F. White. White would play an influential role in national water policy for decades to come.

In terms of water–related activities the NPB's most important contributions were to coordinate the President's Committee on Water Flow and establish the Mississippi Valley Committee (MVC). The committee's report, *Development of Rivers in the United States* (House Document 395), recommended multipurpose plans for 10 drainage basins. One of those basins was the Tennessee Valley, which is discussed in more detail below.

In June 1934 Roosevelt reorganized and renamed the NPB the National Resources Board (NRB) as a separate agency outside the PWA. The MVC was renamed the Water Planning Committee of the new NRB. The NRB's major effort was *A Report on National Planning and Public Works in Relation to Natural Resources and Including Land Use and Water Resources* published in December of 1934. The report identified 17 drainage basins for detailed engineering, social, financial, and legal study, and it recommended that the studies analyze the basins as comprehensive units.

The next metamorphosis came just a year later in June 1935. Roosevelt changed the name of the NRB to the National Resources Committee (NRC). The Water Planning Committee of the NRB now became the Water Resources Committee (WRC) under the NRC. The WRC was again predominantly composed of engineers, but Barrows was one of two original nonengineer members, and White served as secretary (Clawson, 1981). The WRC was instructed to undertake a nationwide drainage basin study with three objectives: determine the principle problems in each basin; outline an integrated plan for development to solve those problems; and identify specific projects as elements of a plan for each basin and prioritize them. The NRC issued its report, *Drainage Basin Problems and Programs* (1935, revised 1936). The report supplemented and extended the work of many earlier committees. *Drainage Basin Problems and Programs* is still impressive today in its scope and detail. Initially 115 basins were studied (revised to 45), major problems were identified, and specific projects and their construction priority were recommended. The report represented the apex of New Deal multipurpose, basin–wide planning for water resources.

In its final incarnation the NRC was reorganized and renamed the National Resources Planning Board (NRPB) in 1939. The NRPB continued to carry out studies on national resources and worked with the Bureau of the Budget to review the agencies' construction plans. This marked the tentative beginning of separate economic evaluation of agency projects. Through the 1930s Roosevelt paid close attention to the activities of the various national planning boards, but this all changed with the onset of World War II. Congress had always been wary of including human resources under the rubric of national–resource planning; it sounded a little too much like socialism. With Roosevelt distracted by the war Congress zero–budgeted the NRPB in 1943, effectively killing it. And so ended the Executive Branch's 10–year experiment with national–resource planning. This was a truly unique time as scientists rallied to the Nation's capital under the banner of national planning; it had never happened before, and it has not happened since.

OTHER NEW DEAL ACTIVITIES

The drought and dust storms in the Great Plains led to the creation of the Soil Erosion Service within the Department of Interior in 1933. In 1935 it was renamed the Soil Conservation Service (SCS) and moved to the Department of Agriculture under the direction of Hugh Hammond Bennett, the "father" of soil conservation in the United States. (The SCS was renamed the Natural Resources Conservation Service in 1995.) Bennett wasted no time in setting up demonstration projects, research stations, soil conservation nurseries, and the Civilian Conservation Corps. Bennett's cadre of soil conservationists taught farmers techniques of contour plowing, terracing, crop rotation, and the use of cover crops. In the years to come the SCS would play an increasingly important role in hydrologic investigations and small–scale flood control activities.

One of Roosevelt's most ambitious New Deal experiments was the Tennessee Valley Authority (TVA). In the 1930s the Tennessee Valley was poverty stricken and economically depressed, and had suffered decades of clear–cut logging and soil erosion. Congress granted Roosevelt's request and created the TVA as a government corporation in 1933. Roosevelt saw in the TVA the opportunity to undertake comprehensive, integrated, natural–resources planning for an entire drainage basin with the explicit goal of promoting social change. The TVA employed local labor to construct multipurpose dams, improve hundreds of miles of channels, teach soil conservation, and undertake watershed reforestation projects. The dams provided cheap electricity, flood control, and improved navigation. Electricity from government facilities was sold at wholesale prices to nonprofit public distribution systems and was used for making fertilizer. The TVA also began setting up regional programs in everything from public health planning to adult education (Petulla, 1988). TVA was a vast social experiment where social scientists tested hypotheses against reality. TVA research "became a small industry within the social science community" (Reuss, 1992).

While Congress passed many important pieces of legislation in the 1930s, two are especially noteworthy—the Taylor Grazing Act of 1934 and the Flood Control Act of 1936. The Taylor Grazing Act ended homesteading in the United State. For more than a century Congress had been selling or giving away public land to private individuals for family farms, to the railroads as inducements to extend their lines, and to the states for various purposes. The federal policy was to transfer the natural resources of the public domain into the hands of the states and private individuals. Starting with the reservations for national parks and forests in the late 1800s, that policy began to change. The Taylor Grazing Act marked the end of the era of federal resource transfers, and the beginning of federal retention and management of resources. The act created the Grazing Service, with its charge being to develop regulations for private grazing on federal land. As long as the government's policy was to give away resources there was relatively little conflict; once the government began reserving those resources, private and state interests began to howl.

The Flood Control Act of 1936 declared once and for all that federally sponsored flood control promoted the general welfare and was a proper activity for the federal government. What makes the 1936 act important was that Congress stated that the government should engage in flood control and navigational improvements "if the benefits to whomsoever they may accrue are in excess of the estimated costs, and if the lives and social security of people are otherwise adversely affected" (49 Stat. 1570, 33 U.S.C. 701a). In this act the federal government had qualitatively stated for the first time the principle of *cost–benefit analysis.* Congress gave no guidance on what were costs, what were benefits, or how either should be measured, but this act is generally acknowledged as the beginning of cost–benefit analysis. The details of cost–benefit analysis are discussed in Chapter 5. The 1936 act also established local cooperation as a requirement for federal participation. Local cooperation meant donating land for right–of–ways, holding the government free of liability from damages due to construction, and operating and maintaining the project after construction. In 1938 these local contributions were eliminated and then partially reinstated in 1941. Future legislation increased local participation by requiring locals to share some of the costs.

A third event brought about by the 1936 Flood Control Act was that it institutionalized the "upstream–downstream" conflict between the Corps of Engineers and the Department of Agriculture regarding flood control strategies. The Department of Agriculture, through the Forest Service and now the SCS, maintained that upstream land treatment to reduce surface runoff was the best way to prevent floods. The Corps of Engineers, which of course had many more years of experience and considerable turf to protect in any discussion of flood control, maintained that structural engineering works downstream on the main channel were the superior approach. The controversy was difficult to resolve because the downstream hydrologic response to upstream land use was still uncertain, even after years of watershed experimentation (White, 1969). The 1936 act gave the Department of Agriculture authority to explore upstream land treatment and the Corps of Engineers the authority to investigate building more dams. The Department of Agriculture initially pursued a primarily nonstructural approach, but over time it asked for and received authority to construct small dams and reservoirs on upstream tributaries.

An alternative approach to dealing with floods was formulated at this time by geographers Barrows and White in their work on the Water Resources Committee. Barrows had stated in a subcommittee document for the *Drainage Basin Problems and Programs* report that "if it would cost more to build reservoir storage than to prevent floodplain encroachment, all relevant factors considered, the latter procedure would appear to be the better solution" (Reuss, 1992). Barrows' subcommittee endorsed the idea of nonstructural floodplain management, e.g., zoning and land–use regulations to keep people off the floodplains. The subcommittee warned that structural works (levees, dams, etc.) encouraged new floodplain development, which inevitably produced demands for increased protection (Reuss, 1992). Not everyone on the subcommittee agreed, especially the Chief of Engineers for the Corps. Abel Wolman, chairman of the WRC, was sympathetic to Barrows' viewpoint and personally brought it to Roosevelt's attention. Gilbert F. White took up the issue of nonstructural floodplain management in his doctoral dissertation, *Human Adjustment to Floods* (1945). This became the seminal work on flood hazard adjustment in the United States, and it continues to influence discussions on how society adjusts to natural hazards, and the use of natural resources generally (Platt *et al.,* 1997).

The New Deal period saw the twin concepts of multipurpose projects and basin–wide planning reach their zenith through the coordinating efforts of the Executive Branch. Among the social sciences the influence of geographers and economists was particularly noteworthy. The role of economists increased as agency plans were reviewed by other agencies, including the Bureau of the Budget. In addition to these tentative evaluations of costs and benefits, economic theories were developed to help understand patterns of resource production and consumption, and their effects on social welfare. Geographers made their mark in two areas. They championed the concept of basin–wide planning, drawing on their tradition of regional analysis. Some economists would argue later that the basin–wide perspective was most persuasive when it focused on resource production, e.g., hydroelectric power, timber, and water supply, but was of little utility when considering questions of resource consumption (Ciracy–Wantrup, 1961). Geographers also advocated broadening the range of choice in how floods were dealt with by considering nonstructural flood hazard adjustments.

1943–1960: CONGRESSIONAL CONTROL OF WATER RESOURCE ACTIVITIES

The demise of the NRPB meant the Executive Branch lost control over federal water–resource planning. With the exception of the Bureau of the Budget's review process, which was still not formalized and mandatory, there were few mechanisms for coordinating and reviewing agencies' plans. And there was a tremendous amount of planning and construction going on at this time. The 308 reports and the detailed listing of construction projects in *Drainage Basin Problems and Programs* provided the agenda, the new era of prosperity following the war provided the money, and the previous decade of public works activity provided the momentum. Thus the Corps of Engineers, Bureau of Reclamation, and the Soil Conservation Service pushed full speed ahead with their own separate water–development agendas. The agendas were set more by local interest and the narrow mission of the agency than by any national policy or plan. The typical planning process began with local and/or state interests contacting their congressman and water–development agency field office about a building a water project. The congressman would negotiate with his colleagues back in Washington to get the project authorized and an appropriation placed in the agency's budget. It was classic

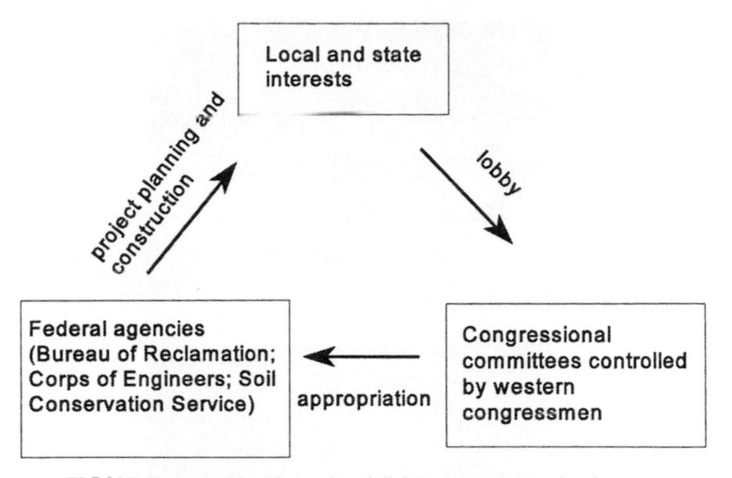

FIGURE 2.4 The "iron triangle" for western water development.

pork barrel politics where congressmen swapped support for each other's pet projects. The agency would do the planning and design, and, once the money was appropriated, it would build the project. In the West this triumvirate became known as the "iron triangle" and was immensely successful because western congressmen held key positions on congressional committees (Figure 2.4).

What coordination existed between agencies at this time was carried out through the informal Federal Inter–Agency River Basin Committee (FIARBC or "firebrick"). The origins of FIARBC went back to 1939 and an executive order requiring the agencies to share information and review each other's plans prior to submitting them to Congress. While FIARBC promoted dialogue, it had no power to alter the activities of the participating agencies. One of FIARBC's most important contributions was in advancing formal economic evaluation procedures for water projects. In 1946 FIARBC appointed a Subcommittee on Benefits and Costs to formulate standardized principles for project planning and evaluation, especially cost–benefit analyses. The subcommittee report, *Proposed Practices for Economic Analysis of River Basin Projects* (FIARBC, 1950), became known as the Green Book (Figure 2.5).

Floods in 1943 and 1944 led to the Flood Control Act of 1944. The act stipulated coordination of basin reports among the different federal agencies. Hence, we see at this time the creation of interagency basin committees for the Missouri (1945), Columbia (1946), Pacific Southwest (1947), Arkansas–White–Red (1950), and New England–New York (1950). The 1944 Flood Control Act also authorized the Pick–Sloan plan for the Missouri River. The development of the Missouri River is a stellar example of the lack of coordination between different federal water–development agencies (see Box 2.1). Two other important changes brought about by this act were Section 8, which provided that Corps

FIGURE 2.5 The "Green Book" which significantly advanced cost-benefit analysis of water projects. (The cover really is green.)

reservoirs in the 17 western states could now include irrigation purposes, and Section 4, which gave the Corps specific authority to provide outdoor recreation facilities at its projects.

NEW WATER RESOURCE ISSUES

By the 1940s a number of new water–resource issues were emerging alongside the traditional purposes of water development. Water pollution had been recognized as a problem as early as the report of the Inland Waterways Commission in 1908. The lead federal agency on water pollution was the Public Health Service, which was interested in drinking water quality and disease prevention. During the New Deal period the NRPB raised questions about water pollution that went beyond the threats to human health to include effects on wildlife and recreation. Congress took its first tentative step in recognizing water pollution as a national problem when it passed the Federal Water Pollution Control Act (FWPCA) of 1948. The act deferred to the primacy of the states because as with flood control before it, pollution control was originally considered a state and local issue (White, 1969). The act authorized grants and low–interest loans for planning studies and wastewater treatment plant construction; however, no funds were ever

BOX 2.1 THE MISSOURI RIVER AND THE PICK–SLOAN PLAN

Developing the Missouri River was not a particularly high priority for either the Bureau of Reclamation or the Corps of Engineers. The Bureau had not been very enthusiastic because irrigation development in Montana and the Dakotas, with their cool temperatures and short growing season, was generally not economical. The Corp had been decidedly unmotivated because the Missouri was barely navigable. It was as someone said, "too thick to drink and too thin to plow." The Bureau had built a number of economically marginal irrigation projects in the decades after the Reclamation Act of 1902. In the late 1930s Glenn Sloan, an engineer in the Bureau's Billings, Montana, office began working on a basin–wide plan for irrigation development. Sloan's plan hinged on building multipurpose dams high in the basin on the upstream tributaries. The dams would generate electricity which would be used to help pay for the otherwise uneconomical irrigation facilities. The "Sloan Plan" was nearly complete by 1943 when the Bureau found out that the Corp was working on its own plan for the Missouri River.

The Corps of Engineers completed Fort Peck Dam in the early 1940s. Fort Peck was started during the New Deal and its construction was justified largely on the basis of regional economic stabilization rather than any water–related purpose. Fort Peck, is enormous and at the time was the world's largest dam. It is still the sixth largest dam in the world today. But floods along the Missouri in the spring and early summer of 1943 inundated Omaha, Nebraska, which also happened to be the regional headquarters of the Corps of Engineers. Colonel Lewis Pick was the director of the Omaha office at the time and later became Chief of Engineers. Perhaps taking the flood as a personal affront, Pick prepared and sent a basin–wide flood control plan for the Missouri to Washington by October 1943. This was the "Pick Plan."

The Sloan Plan emphasized reservoirs high in the basin so as to maximize hydroelectric power generation to help subsidize irrigation. The Pick Plan emphasized reservoirs low on the main stem for flood control and navigation. Thus Congress in 1943 was confronted with two entirely different plans for the same river. Roosevelt suggested creating a TVA–like authority for the Missouri and taking the development of the basin out of the hands of both agencies. This was the impetus to get the two agencies negotiating. In a two–day meeting they reconciled their plans and agreed on the compromise "Pick–Sloan Plan." The Pick–Sloan Plan included every dam and project proposed in the two separate plans, and then some. According to Reisner (1986, p. 193), Henry Hart, a journalist and historian who covered the Pick–Sloan controversy at the time, observed that reconciliation of the two plans "meant chiefly that each agency became reconciled to the works of the other."

appropriated. Because the 1948 act did little to stem the deterioration of the Nation's waters, Congress took a bolder step with the Federal Water Pollution Control Act of 1956. The 1956 act continued the program of grants for planning and construction of municipal wastewater treatment plants, but provided for federal enforcement on interstate streams. The federal government began asserting its authority in the area of water quality because, with few exceptions, the states refused or were incapable of taking action. Some states did negotiate interstate pollution control compacts (Table 3.2). Amendments to the FWPCA in 1961 authorized water storage in federal reservoirs for pollution control; reservoir water would be released to dilute pollution—dilution was a solution to pollution. The amendments clearly stated that storage should not be a substitute for treatment or control of pollution at the source. Water pollution was seen mainly as a coming from *point sources* (pipes, tanks, etc.) in and around metropolitan areas. *Nonpoint pollution sources* such as runoff from farms and logged–over forests were recognized, but were not addressed by the legislation. The pollution cleanup approach that was evolving emphasized the construction of wastewater treatment plants and pollution dilution. Even though Congress recognized that pollution prevention could be an alternative means of pollution control, reducing the release of pollutants at their source would not be seriously explored until the 1980s. By the 1960s water quality had become a major water–resource issue, though just exactly what "clean water" meant was unclear. Some argued that clean water meant keeping all pollution out of the waters. Others argued it meant using the water to assimilate waste, consistent with other water uses (White, 1969).

Other new water issues that were materializing included an increasing demand for outdoor recreation, and concern for fish and wildlife. Ownership of private automobiles increased dramatically after World War II. This greater mobility made it possible for families to visit the many new reservoirs, and quite unintentionally reservoirs gained yet another purpose—water–based recreation. The Fish and Wildlife Coordination Act of 1948 provided for cooperation among agencies to minimize the harmful impacts to wildlife from project construction. The act was amended in 1958, adding wildlife enhancement as a planning purpose, and required the water–development agencies to consult with the Fish and Wildlife Service during the planning phase.

The President's Water Resources Policy Commission, chaired by Gilbert F. White, was created and charged with identifying and assessing major issues in the development of the Nation's water resources. The commission's report (PWRPC, 1950) endorsed drainage basin planning based on a set of clearly specified national objectives, the creation of river basin commissions to coordinate agency programs, a program of basic data collection, and standardized procedures for economic evaluation of projects. The commission made other recommendations on project financing and reimbursement, reclamation, water pollution, and wildlife. The report lists a variety of objectives, seven to be exact, to be pursued by water–resource development. These objectives either explicitly or implicitly dealt with national economic development, regional economic development, the enhancement of social welfare, and environmental improvement. This marks the beginning, or at least the recognition, of *multiobjective planning*. Multiobjective planning is quite different from multipurpose planning. Until this time water-resources planning was driven by the single objective of economic development. With a few exceptions, most notably the TVA, the objectives of social welfare and environmental improvement were acknowledged, but were not considered sufficient reasons for undertaking water development. The final report of the commission was released in three volumes. Volume 1 dealt with general policy issues and recommendations, volume 2 was a geographic analysis of 10 major drainage basins in the United, and the third volume was devoted to water law.

A number of other federal commissions were formed in the early 1950s. The primary focus of the various commissions was the reorganization of the Executive Branch to reduce duplication between the water-development agencies. These commissions were a direct reaction to the decentralized and uncoordinated water-development process that had evolved.

Finally, in 1955 the Presidential Advisory Committee on Water Resources released its report. In the letter establishing this committee President Eisenhower noted that the country had an agency-based policy for water development, but not a coordinated policy for the federal government. The committee reiterated many of the same recommendations offered by earlier commissions and committees, even as far back as the Inland Waterways Commission (1908). The committee again recommended a plan for collecting basic water data, and that water planning should be a coordinated effort between the different levels of government. This committee was less taken with the drainage basin as the fundamental planning unit, but it supported the idea of a national project review board, the creation of regional river basin commissions, a uniform set of principles and standards for project planning and evaluation, cost sharing for projects, and the notion that each project should stand on its own merits and be authorized separately. The committee also noted how important water law was to water management and planning.

OTHER WATER RESOURCE ISSUES

Advances in well drilling and pump technology led to massive groundwater exploitation at this time. In basins where groundwater and surface water were hydrologically interconnected, this created major conflicts. The pumping of groundwater reduced the amount of water in the rivers, which were already being fully used. On the High Plains, development of the center pivot sprinkler system meant farmers could tap the vast groundwater reserves of the Ogallala Aquifer. Up until the 1950s farmers on the semiarid plains grew wheat and other grains by relying exclusively on the variable natural precipitation. A dependable supply of irrigation water meant farmers could switch to more profitable crops like cotton and corn. The adoption of the new irrigation technologies was primarily a private-sector decision. The Bureau of Reclamation was reluctant to investigate or

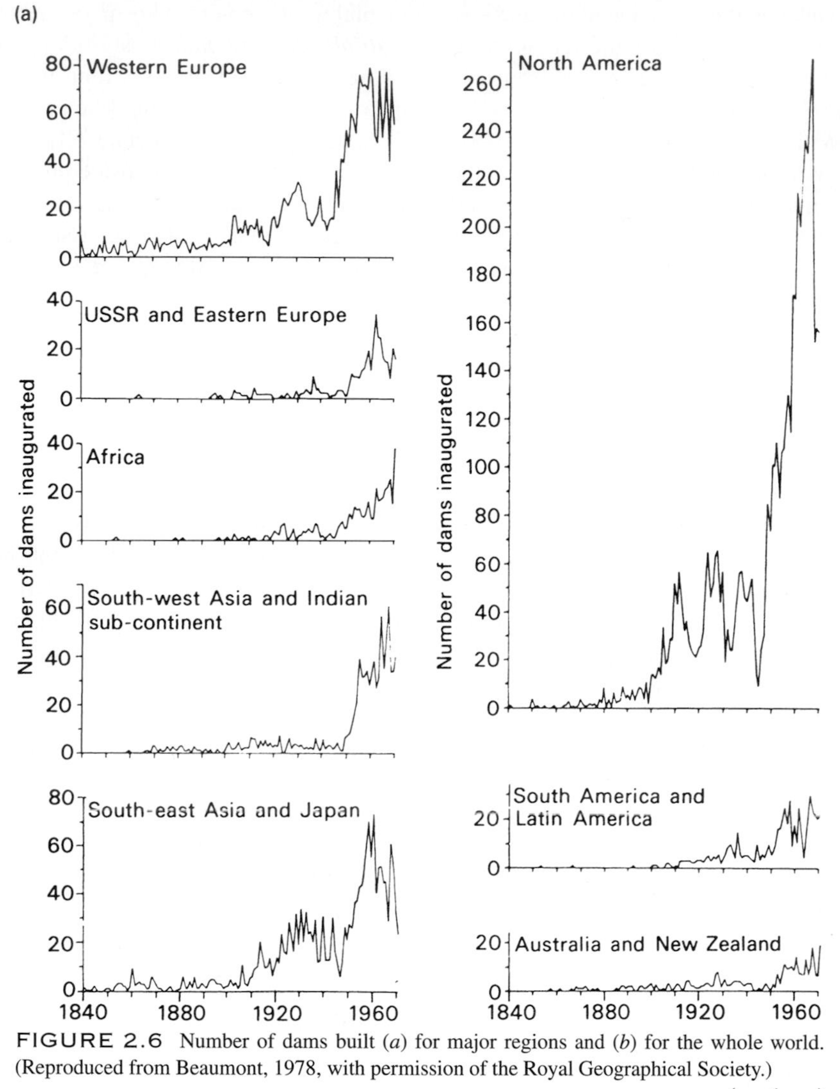

FIGURE 2.6 Number of dams built (*a*) for major regions and (*b*) for the whole world. (Reproduced from Beaumont, 1978, with permission of the Royal Geographical Society.)

(*continues*)

promote new sprinkler technologies. It is rather remarkable how inflexible the federal agencies were with regard to exploring alternative means of reaching their particular objectives. In the area of flood control the Corps of Engineers refused for years to give serious consideration to nonstructural alternatives and clung tenaciously to structural works even in the face of mounting flood damages. The TVA was the first government entity to broaden the range of choice in dealing with floods when it initiated a community floodplain management program in 1953.

The period from 1940 to 1960 was the "go–go" years for dam construction. Between 1945 and 1969 35,000 dams were built (Figure 2.6). Today there are

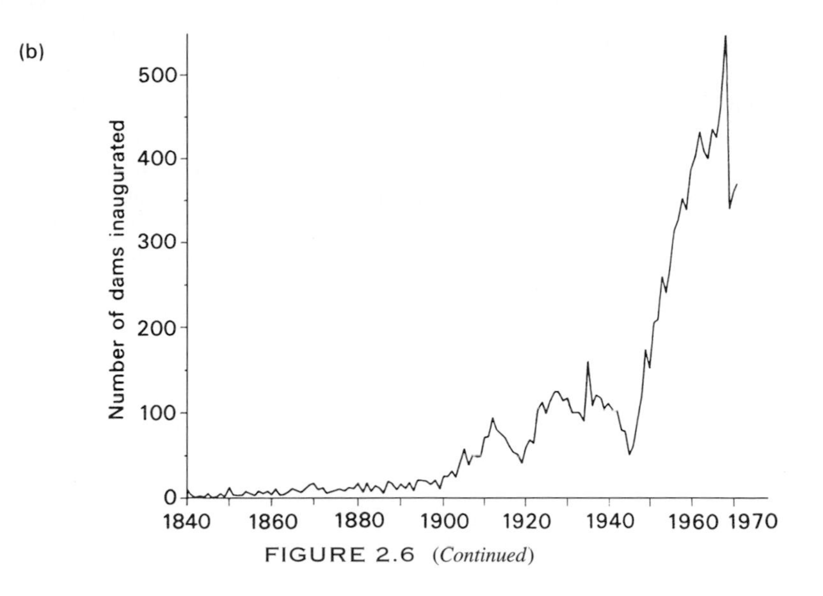

FIGURE 2.6 *(Continued)*

something like 75,000 dams throughout the country. But even in the midst of this frenzy of construction a new era was dawning, an era that would be driven by a concern for the environment. Preservationists had taken a backseat to utilitarian conservation for the last 50 years. People began to want more from natural resources than just economic production—they wanted recreation and aesthetic pleasures. A new confrontation was building in the early 1950s, and as with Hetch–Hetchy half a century earlier, it centered on a dam in the West. With the passage of the Upper Colorado River Basin Compact in 1948, the Bureau of Reclamation put together a comprehensive plan for development of the upper basin composed of a series of multipurpose dams and nearly a dozen new irrigation projects. The plan was the Colorado River Storage Project. Echo Park Dam was to be built on the Green River, and its reservoir would inundate part of Dinosaur National Monument in Utah and Colorado. Public opposition to Echo Park Dam was organized by the Sierra Club and the Wilderness Society. The public outcry so overwhelmed Congress that it postponed consideration of the entire project. Ultimately the preservationists prevailed and Echo Park was dropped from the project. This was a critical turning point for the new environmental movement because it was the first time preservationists had won a battle against the utilitarian conservationists. It also signaled the coming of age of public interest groups in environmental decision making. However, the price that preservationist paid in defeating Echo Park became apparent when Congress finally passed the Colorado River Storage Project Act in 1956. The bill included Glen Canyon Dam on the main stem of the Colorado River, which eventually backed water up into Canyonlands National Park. Glen Canyon Dam was the cash register for further development in the Colorado basin, as revenues from hydroelectric power would finance future construction throughout the basin.

In the 1960s the Bureau of Reclamation proposed building two dams on the Colorado River that would flood parts of the Grand Canyon. When the Sierra Club led the challenge to save the Grand Canyon it came under the scrutiny of the Internal Revenue Service. This infuriated the public and catapulted the Grand Canyon dams issue into the national spotlight. By now the public's opinion of large dams had changed and the necessity of dams for economic development was being questioned. While the two Colorado River dams made it to the drawing boards, that was as far as they got. The Colorado River Basin Project Act (1968), which authorized the Central Arizona Project, sounded the death knell for the Grand Canyon dams by stating "nothing in . . . this Act shall be construed to authorize the study or construction of any dams on the main stem of the Colorado River between Hoover Dam and Glen Canyon Dam."

Congress passed the Central and Southern Florida Project Act in 1948. The "River of Grass" that was the Everglades was viewed as a wasteland, not a valuable wetland ecosystem. The Corps of Engineers started a multimillion dollar project of diking, draining, and channeling excess water to open up the potentially rich agricultural land around Lake Okeechobee. Ninety–eight miles of the meandering Kissimmee River were channelized to control water flow. Hundreds of miles of canals and levees now regulate the water that used to drain naturally from Lake Okeechobee and flow south through the Everglades during the wet season. The result of the Corps' projects, as well as many private drainage and irrigation projects, has been the creation of a highly productive agricultural economy in southcentral Florida—and the starvation of the remaining wetlands in Everglades National Park. By the early 1980s the prospect of the ecological collapse of the Everglades led to proposals to dismantle some of the Corps' projects. Today wading bird populations are down 90 percent and the populations of various vertebrates are down anywhere from 75 to 90 percent (*Environment,* 1995). In 1995 the Corps allocated $370 million to return the Kissimmee River to its original meandering condition. The total cost from both state and federal sources will probably exceed $2 billion. Whether ecological restoration will be successful, or is even possible, remains to be seen. Another problem in the Everglades is water quality. Phosphorous from agricultural runoff and mercury from an unknown source are both significant pollutants in the Everglades. An important message from the Everglades experience is that water–resource planning needs to be more flexible. Water planning and development tacitly assumed that society's institutions and values would not change over the 50– to 100–year lifetimes of these projects. But after just 40 years in south Florida, costs and benefits have completely reversed. The benefits of wetlands today exceed the costs of tearing out the water control structures; in the 1940s the wetlands had no value whatsoever.

In the years from 1900 to 1960 public perceptions and values underwent a dramatic shift. Traditional water uses—irrigation, flood control, and navigation—were losing their appeal, while recreation, water quality, and environmental issues were on the rise. To environmentalists the federal government had gone from being the protector of our natural resources at the turn of the century to the perpetrator of environmental destruction by midcentury. Among the more radical environmentalists in the early 1960s, water projects, especially dams, were regarded as quintessential monuments to environmental destruction. But the traditional water–development lobby, e.g., the iron triangle, was still largely intact. The increasingly active environmental movement on the one hand, and the entrenched water–development culture on the other, meant that conditions were ripe for social conflict in the next period.

1960–1980: THE ENVIRONMENTAL PERIOD

On April 20, 1959, Congress created the Senate Select Committee on Water Resources and initiated yet another era for water management in the United States. The report of the Senate Select Committee (U.S. Congress, 1961) contained some old and familiar recommendations regarding the need for better state and federal cooperation and the desirability of comprehensive river basin planning, but there were new recommendations acknowledging water as a scarce resource, the serious deterioration in the nation's environment, and the water–related needs of a changing society. The committee called for a periodic assessment of the Nation's water resources on a regional basis, a federally sponsored program of scientific research on water, and programs to encourage water–use efficiency and water conservation. Improved water–use efficiency and water conservation meant using less water to do the same job or to generate the same benefits. The utilitarian conservation philosophy that had dominated water development for so long emphasized getting as many benefits from the water as possible. It was a philosophy that clearly intended that all water should be put to work. The concept of water–use efficiency had not played an important role in utilitarian water management. One reason was that water was viewed as a free resource. Water was just there, in the streams and underground, free to use. Another reason was that water was always considered a special resource, different from other natural resources because it was essential for life. The cold calculus of economic efficiency, where price mediates supply and demand, did not, or should not, apply to water.

The Senate Select Committee encouraged broadening traditional agency responsibilities to incorporate emerging issues related to water quality, municipal and industrial needs, recreation, fish and wildlife, and floodplain regulation. What really makes the Senate Select Committee's report unique is that this time Congress actually listened (to itself) and enacted some, but not all, of the recommendations. In 1962 Senate Document 97 was released. Senate Document 97 contained updated procedures for water project planning and economic evaluation. The document embraced multiobjective planning and declared three national objectives for federal water–resource planning: national and regional economic development; preservation of resources, including rehabilitation of damaged resources; and social well–being. While the well–being of people was supposed to be the overriding determinant, economic development was probably more important. Social well–being was more intangible and difficult to measure; economic objectives, on the other hand, could be analyzed quantitatively. Naturally

TABLE 2.1 Annual Appropriations (in millions of dollars) for Water Resource Programs, 1966–1971[a]

Year	Corps of Engineers	Bureau of Reclamation	Soil Conservation Service	Total
1966	1330	295.4	102.6	1728.0
1967	1293	247.5	106.9	1647.4
1968	1305	225.0	100.1	1630.1
1969	1246	204.4	104.0	1554.4
1970	1157	180.8	117.5	1455.3
1971	1310	232.7	136.6	1674.3

[a] Source: Holmes, 1979.

the agencies were inclined to pursue those objectives that could be measured more easily.

Of the federal water agencies the Corps of Engineers dominated water development as measured by appropriations from Congress. Table 2.1 shows annual appropriations to the Corps of Engineers, the Bureau of Reclamation, and the Soil Conservation Service for the years 1966 to 1971. The Corps of Engineers commanded fully 80 percent of the federal water–development appropriations at this time.

Congress passed the Water Resources Research Act in 1964, establishing a water research institute at the land–grant university in each state and a research grant program. In 1965 Congress passed the Water Resources Planning Act. The act authorized the creation of federal–state river basin commissions and created the Water Resources Council (WRC). The WRC was composed of the Secretaries of Agriculture, Army, Health, Education and Welfare, and Interior, as well as the Chairman of the Federal Power Commission. The council had a director, staff, and a real budget. The WRC was directed to coordinate, review, and evaluate agency plans; conduct periodic assessments of the Nation's water resources; develop and refine the principles and standards for water–resource planning and evaluation; conduct symposia, sponsor research; and distribute information to the public. The Water Resources Council was the most important administrative entity for federal water–resource planning since the Water Resources Committee of the 1930s, and it had much more responsibility.

By the 1960s the failings of the Nation's structural flood control strategy were painfully evident. Gilbert F. White and his students at the University of Chicago pointed out that despite billions of dollars spent in trying to control nature, the annual damages from floods rose every time the waters did. The more money we spent on flood control, the greater the damages, which was not exactly the result we were trying to achieve. This was dubbed the "flood control paradox." Apparently dams and levees gave the impression that floodplains were now protected and safe from future inundation. Levees were particularly effective at fostering this misperception in the public's mind. The more money the Nation spent on structural flood control, the more floodplain development that occurred, and the greater the damage potential. When the next flood came along that exceeded the project's design, the damages were greater than if no protection had been provided in the first place. In Section 206 of the Flood Control Act of 1960 Congress recognized the need for greater state and local floodplain management and directed the Corps of Engineers to prepare floodplain information studies similar to the program developed by the TVA in the early 1950s. The Bureau of the Budget created a Task Force on Federal Flood Control Policy chaired by White, while at the same time the federal Department of Housing and Urban Development explored the feasibility of a federal flood insurance program (Burby and French, 1985). In 1968 Congress created the National Flood Insurance Program (NFIP). The NFIP provided federally subsidized flood insurance to communities that regulated floodplain land use. In theory, having occupants purchase flood insurance means they assume more of the risk of living in the floodplain. The NFIP was a significant step in the evolution of nonstructural flood hazard mitigation, but the program had to be overhauled in 1973 because very few communities participated voluntarily.

In 1964 and 1965 Congress continued to respond to the public's increasing demand for recreation and passed legislation to promote outdoor recreation and fish and wildlife enhancement at federal water projects. In 1968 Congress enacted the Wild and Scenic Rivers Act. The Wild and Scenic Rivers Act was designed to preserve, in a free–flowing condition, certain waters possessing outstanding scenic, recreational, fish and wildlife, geologic, historic, or other cultural characteristics. However, a wild and scenic designation means only that a relatively narrow corridor—one–quarter mile on each side of the channel—is protected and preserved. There are about 150 rivers or segments of rivers, totaling approximately 10,000 miles, in the Wild and Scenic River System (Table 2.2).

TABLE 2.2 Annual Total River Miles in the National Wild and Scenic Rivers System in the United States[a]

Year	Total miles
1968	773
1972	895
1976	1610
1980	5662
1984	7217
1988	9264
1992	9972
1993	10,764

[a] Sources: Gleick, 1993 and Zinser, 1995.

The most far–reaching piece of environmental legislation ever passed in the United States, and perhaps the entire world, was the National Environmental Policy Act (NEPA) of 1969. The NEPA required that the federal government be cognizant of how its activities affect the environment. Section 102 of the NEPA required any proposed federal action that might have a significant impact on the environment to be preceded by an *Environmental Impact Statement* (EIS). Water–development projects would almost certainly require an EIS. The NEPA created the Council on Environmental Quality (CEQ) to advise the President on environmental matters and to prepare an annual report on environmental conditions and trends.

Another significant piece of environmental legislation was the Endangered Species Act (1973). The power of the Endangered Species Act to influence water development was glimpsed early when the discovery of a fish species called the "snail darter" threatened to halt construction of the TVA's Telico Dam on the Little Tennessee River. In the 1990s the Endangered Species Act has threatened to disrupt the entire water–resource management system in the Columbia and Snake River basins in the Pacific Northwest. These rivers have been managed primarily for power generation, irrigation water supply, and navigation. The listing of certain species of salmon as endangered has meant traditional water management priorities must change. The changes are intended to make the rivers more hospitable to salmon migration, and could cost hundreds of millions or even billions of dollars in lost power revenues, structural modifications to dams, and reduced agricultural production. There has even been talk of removing two dams on the Elwha River in western Washington state to improve fish migration.

WATER QUALITY

Continued deterioration of the nation's waters combined with a lack of progress by the states forced Congress to pass the Water Quality Act of 1965 as amendments to the FWPCA. The two most important changes were the creation of the Federal Water Pollution Control Administration (FWPCA) and the provision for establishing and enforcing water quality standards on interstate waters. The FWPCA took over most of the responsibilities previously under the Public Health Service. The act recognized that water quality had become much more than a public health issue. The act provided that all states establish enforceable water quality standards for interstate streams, and if they did not, or if their standards were not acceptable to the Secretary of the Interior, he would establish standards himself. The standards were intended to enhance water quality, protect public health and welfare, and give consideration to the use of water for public water supply, fish and wildlife, recreation, agriculture, and industry. The standards were criteria for the receiving waters; they were not intended to be limits on pollutant concentrations in effluent. However, it did not take long before the standards were interpreted to mean that all municipal wastes must receive secondary (biological) waste treatment.

The "ancient authorities" of the Corps of Engineers, i.e., the codification of various early regulations in the 1899 River and Harbors Act, were being reinterpreted largely as a result of pressure from the environmental movement. The Refuse Act (Section 13) originally prohibited discharge of any refuse into navigable waters "except that flowing from streets and sewers and passing therefrom in a liquid state." A number of decisions by the U.S. Supreme Court in the late 1960s made it apparent that Section 13 was now being interpreted to mean any pollutant except municipal sewage (Holmes, 1979).

The Environmental Protection Agency (EPA) was created in 1970 and assumed responsibility for the disparate collection of federal programs related to air quality, solid waste, and water quality. While including all environmental programs under one agency had obvious benefits in terms of coordination, one drawback was that it separated the government's traditional water quantity activities from the emerging water quality activities.

The turning point in federal water quality legislation came with the passage of Public Law 92–500, the Federal Water Pollution Control Act Amendments (FWPCAAs) of 1972. The FWPCAAs set as a national goal that the waters of the United States be "fishable and swimmable" by 1983. The amendments intended to reach this goal through a two–pronged approach. First, they authorized a massive construction grant program for municipal wastewater treatment plants. Unlike earlier legislation, this time Congress was forthcoming with the funds. The federal government would eventually spend over $20 billion constructing municipal wastewater treatment plants between 1972 and 1992. With the FWPCAAs the federal government agreed to pay up to 75 percent of construction costs; back in the 1950s and 1960s the federal share varied between 20 and 55 percent. The second component was the National Pollution Discharge Elimination System (NPDES) permit program. It was now illegal for anyone to discharge any pollutants into the surface waters of the United States without a NPDES permit. The NPDES permit program was founded on the old permit requirement of the Refuse Act. In effect the federal government had extended the navigation servitude up the smallest nonnavigable tributaries. Section 401 established the NPDES program within the EPA; Section 404 established a separate permit program for dredge and fill material within the Corps of Engineers. Section 404 permits are used to regulate the draining and filling of wetlands, and the 404 permit process incorporates all of the applicable federal environmental laws. For example, the Corps of Engineers once denied a 404 permit for the construction of a reservoir not because of any detrimental effect of the reservoir on the surrounding area, but because of the detrimental environmental effects it would have on the migratory whooping crane (an endangered species) some 300 miles downstream (Dzurik, 1990).

Another significant component of the 1972 FWPCAAs was Section 208. Section 208 recognized pollution coming from "area–wide" sources. This was the first statutory recognition of nonpoint source pollution by any level of government. Even though nonpoint sources were recognized as a problem, Section 208 was really toothless. Controlling nonpoint source pollution meant controlling land use and the federal government was not prepared to open the pandora's box of "regulatory taking" of private property—at least not yet. The term "taking" refers to the constitutional provision that says the government cannot take private property without just compensation. The 1972 FWPCAAs were amended and renamed the Clean Water Act in 1977. The 1977 amendments corrected some deficiencies and added language to deal with yet another emerging environmental problem—toxic waste. In 1987 the government was forced to revisit the nonpoint source pollution problem.

Congress gave EPA additional responsibilities when it passed the Safe Drinking Water Act (SDWA) in 1974. Prior to the SDWA the only enforceable federal drinking water standards were the Public Health Service's standards dating from 1912 for infectious water–borne diseases, such as cholera, dysentery, and typhoid. The SDWA required public water supplies to meet minimum standards. The SDWA has no effect on private wells or small water systems serving fewer than 25 people or 15 hookups. This means something like 43 million people are beyond the protective reach of the SDWA. Congress directed the EPA to develop *primary* and *secondary drinking water standards.* Primary standards are enforceable and are meant to protect public health. Primary standards are set for constituents like toxic substances, infectious organisms, and inorganic and organic chemicals that might pose a threat to human health. Secondary standards are not enforceable and are simply aesthetic guidelines for such things as water color, odor, hardness, and iron content. Drinking water standards and the SDWA are discussed in Chapter 7.

The increasing concern with toxic and hazardous wastes was evident with the passage of the Resource Conservation and Recovery Act (RCRA) and the Toxic Substances Control Act, both of which became law in 1976. The regulations implementing RCRA created a "cradle to grave" tracking procedure for hazardous waste. Ostensibly from the moment a hazardous substance is created until it is ultimately disposed of, its whereabouts are known and documented. RCRA was amended in 1984, tightening the controls on land disposal of hazardous waste and adding a regulatory program for controlling underground storage tanks. The technological disaster at Love Canal, New York, in 1978 was a rude wake–up call to the threat hazardous wastes posed to groundwater resources, the environment, and people.

The 1960s and 1970s saw state and local governments broaden their water–resource planning and management expertise. Local and state government expertise was well developed in some fields but not in others. The administration of state water rights, urban water supply planning and development, and state regulation of water quality for public health are examples of where state and local governments had considerable expertise. But state and local governments were generally not prepared to handle the new federal initiatives related to wastewater treatment plant construction, floodplain management, hazardous waste regulation, and river basins commissions.

THE NATIONAL WATER COMMISSION

The furor over the Bureau of Reclamation's proposal for two dams on the Colorado River in the mid–1960s was the original impetus for the creation of

Box 2.2 THEMES OF THE NATIONAL WATER COMMISSION

1. The level of future water demand is not inevitable, but derives in large part from policy decisions within the control of society.
2. There has been a shift in national priorities from development of water resources to enhancement and restoration of water quality.
3. Water planning should be tied more closely to land–use planning.
4. Sound economic principles should be applied in evaluating water projects and programs.
5. Water conservation needs to be actively promoted through policy.
6. Laws and legal institutions should be reevaluated in light of contemporary issues.
7. Development and management of water resources should be at the level of government nearest the problem and most capable of effectively representing the vital interests involved.

one final commission to study the Nation's water resources. The National Water Commission (NWC) was created in late 1968 to study and report on virtually all water problems, programs, and policies. The NWC commenced more than 60 background studies on 22 subject areas (Holmes, 1979). The final report (NWC, 1973) contained literally hundreds of separate recommendations. Here we mention seven themes from the final report (see Box 2.2). Arguably the NWC was the most penetrating and insightful analysis of the Nation's water resources. The themes are as relevant today as they were in 1973. The first theme listed in Box 2.2 recognizes that we control our future water use by the planning decisions we make today. Planners had assumed that water use would inevitably increase, as though it was somehow beyond our control. Significantly, total water use in the United States actually decreased in the mid–1980s. We examine water–use trends in Chapter 4. Themes 2 and 6 recognize that as society changes, existing institutions need to change as well. Theme 7 encourages the primacy of state and local control. While this is a good principle, it also reveals that even an independent commission responds to political pressure. The NWC was operating during a Republican administration, which favors state and local control.

TRADITIONAL WATER AGENCY ACTIVITIES

What became of the traditional water–development agencies during the decade of the 1970s? The Corps of Engineers, the Bureau of Reclamation, and the Soil Conservation Service were still ready, willing, and able to develop the country's remaining water resources, but the country had changed. Water was no longer looked upon as the key to regional economic development, if it ever was. Most of the best dam sites were dammed, and the remaining good sites, like the Grand Canyon, were off limits by order of the increasingly powerful environmental movement. The value of these sites in their unspoiled condition now outweighed the marginal benefits of increased water supply or flood peak reduction. The sites that could still be developed were inferior and therefore much more expensive to develop. By the 1970s large water projects had not become totally irrelevant, but they were no longer the centerpiece of regional and national economic development. They were also becoming very expensive as inflation and interest rates rose. A direct impact of the higher interest rates was that Congress asked locals to shoulder a larger percentage of the cost, which considerably dampened local appetites for capital–intensive projects.

The iron triangle started breaking down in the 1970s as the old guard congressman passed from office. But it was the election of Jimmy Carter that really brought traditional water development to a screeching halt. Call it impolitic, call it naive, but soon after taking office President Carter produced his infamous "hit list" of unsound, uneconomic, pork barrel water projects. Essentially the list included all projects that were presently under consideration for authorization or appropriation. Congress (and the water–development lobby) did not take kindly to this all–out attack by the President. The iron triangle was dying but it was not dead yet. Some political historians have speculated that the hit list was the root cause of all of President Carter's problems with Congress. Carter and the Congress were stalemated, and that meant no new construction. Carter firmly believed in water conservation and the use of nonstructural alternatives. He directed the Water Resource Council to require all project plans include at least one nonstructural alternative. By 1980 traditional federal water development had all but come to an end.

1980 TO THE PRESENT: A TIME OF CHANGING FOCUS

Ronald Reagan became President in 1980. For a moment there was a glimmer of hope that Reagan would get water development back on track, especially out West. Reagan certainly was more sympathetic to the traditional development interests, but a growing budget deficit, a deepening recession, increasing inflation, and high interest rates dictated Reagan's policy toward water development. There just was not money available for big water projects. Being a Republican, Reagan also supported reducing the size and scope of the federal government. He abolished the Water Resources Council in 1981, zero–funded federal participation in river basin commissions, and cut back appropriations to the water–resources research institutes. He also increased local cost–sharing requirements for participation in federal projects. Even though Reagan was philosophically more receptive to the traditional water–development lobby than was Carter, the result was the same—no new water projects. Indicative of the changing mood were the exploratory discussions between the federal government and the state of California for selling one of the Bureau of Reclamation's crown jewels, the Central Valley Project. It was even proposed around 1988 that Hetch–Hetchy Dam be removed.

Congress passed the Reclamation Reform Act of 1982, eliminating the 160–acre limitation and the residency requirement as conditions for receiving water from federal projects. Both of these progressive–era concepts had been roundly abused and their removal was largely a concession to large corporate farms, especially in California. But even the act's requirement of an increase in the repayment costs for irrigation facilities and the five– to sixfold increase in the charge for water in renegotiated water contracts have done little to eliminate the huge federal subsidy given to western farmers.

The concern with environmental issues remained strong throughout the decade of the 1980s. The Comprehensive Environmental Response, Compensation, and Liability Act (CERCLA), or "Superfund" as it is popularly known, became law in 1980. CERCLA was created to deal with releases of hazardous materials into the environment by accident and from old dump sites. Hazardous sites are evaluated based on their potential threat to people and the environment, and the most serious sites are placed on the National Priorities List. Superfund was funded at $1.6 billion with the money coming from appropriations and taxes on the chemical industry. While the money was intended to go toward environmental cleanup, most of it has been used to pay for litigation. CERCLA has a joint and several liability clause which means if a dump was used by multiple parties, all are liable, but if only one party can be identified, that party may be held liable for the entire cleanup cost. Obviously this is seen as unfair by industry. CERCLA was amended by the Superfund Amendment and Reauthorization Act (SARA) in 1986. SARA expanded the EPA's powers and increased the level of funding to $8.5 billion. Congress also amended and strengthened RCRA and the Safe Drinking Water Acts in 1984 and 1986, respectively. The amendments to the Safe Drinking Water Act included a program to encourage states to establish wellhead protection programs. Wellhead protection programs attempt to prevent groundwater pollution by regulating land use near water supply wells. Wellhead protection is founded on the principle that pollution prevention is the best way to ensure clean groundwater.

The Clean Water Act was reauthorized in 1987. A major focus of the reauthorization amendments was nonpoint source pollution. Experience had shown that regulation of point sources had helped clean up the Nation's waters, but further progress required addressing the nonpoint sources. Congress required states to conduct studies to identify nonpoint source problem areas and to develop abatement plans.

By the end of the 1980s people began assessing what had been achieved in terms of water quality after two decades of effort and the investment of billions of dollars. There was no question that the Nation's waters were cleaner by some criteria. For example, dissolved oxygen levels were higher in most streams and lakes than a decade earlier. But many questions remained. One was whether the benefits of improved water quality were worth the costs. Other questions concerned the health and integrity of aquatic ecosystems. The Nation's waters were in

THE CHANGING FOCUS OF THE BUREAU OF RECLAMATION

Indicative of a major change in water management is the new mission of the Bureau of Reclamation. In the late 1980s the Bureau wrote, "The arid West essentially has been reclaimed. The major rivers have been harnessed and facilities are in place or are being completed to meet the most pressing current water demands and those of the immediate future." The emphasis within the Bureau has shifted from construction to operation and management of existing facilities. The Bureau officialy redefined its mission as to

> Manage, develop, and protect water and related resources in an environmentally and economically sound manner in the interest of the American public.

(Source: Bureau of Reclamation, 1997)

some ways cleaner but aquatic ecosystems continued to suffer, with estimated extinction rates for aquatic species much higher than for terrestrial species (Chapter 9). Recent plans for water quality management have dusted off the concept of basin–wide planning. Like the phoenix from the ashes, basin–wide planning has been resurrected in the 1990s; however, this time it is driven by water quality, not water quantity.

In the early 1990s Congress opened debate on the reauthorization of the Endangered Species Act, the Safe Drinking Water Act, and the Clean Water Act. With the election of President Clinton, the Democrats recaptured the White House for the first time in 12 years. Clinton promised an environmentally responsible federal government and by and large kept that promise for two years. Everything changed, however, with the off–year elections in 1994, which saw Republicans gain a majority in both the Senate and the House, ending a half–century of Democratic control. Congress became decidedly less sympathetic to environmental issues and regulations. There has been talk of instituting a rigorous cost–benefit analysis of all environmental regulations. Presumably old regulations that generate more costs than benefits would be repealed, and new regulations would not pass. While the challenges of evaluating the costs and benefits of regulations are formidable, the 1996 amendments to the Safe Drinking Water Act do incorporate significant cost–benefit considerations (Chapter 7).

Another topic smoldering in Washington is the regulatory taking of private property. The taking issue is a classic battle between individual rights and the larger welfare of society. On one side are those who feel any governmental regulation that limits a person's use of his or her property should require the government to pay compensation, just the same as if the property had been physically taken for a public use. On the other side is the role of government to protect our health, safety, and welfare. A number of private property bills were introduced in

1995 but none passed. Should a private property bill pass it would severely restrict the government's role in environmental regulation. Regulations on the draining and filling of wetlands, the regulation of flood hazard areas, and habitat preservation for endangered species could all be curtailed. Congress has also considered restructuring the EPA, overhauling the Superfund program, and limiting the scope of the Endangered Species Act.

Other significant events in the decade of the 1990s were the major flood disasters. The so–called Great Flood along the Upper Mississippi and Missouri Rivers in the summer of 1993 inundated all or parts of nine states for most of the summer. This was followed within a few months by unprecedented winter and spring flooding throughout much of California. Floods returned to some of the same areas along the Missouri and Mississippi during the spring and early summer of 1995. Major flooding has occurred in Pennsylvania in 1996, northern California in 1996, and North Dakota and Minnesota in 1997. These flood disasters came on the heels of hurricane Andrew which ripped a $15 billion path of destruction across south Florida in 1992, and a major earthquake in southern California in 1994. And if that were not enough, there were billions of dollars in damages from flooding, landslides, and tornadoes attributed to El Niño in the 1997–1998 winter season. The price tag of these disasters has grabbed the attention of Congress. In classically reactive style Congress passed new flood–related legislation, and has taken a hard look at national flood policy. President Clinton signed the Water Resources Development Act in October 1996. The act is comprehensive and covers topics ranging from flood control and floodplain management, to aquatic ecosystem restoration, to new cost–sharing provisions and a National Dam Safety Program. Several provisions of the act modify the Corps' flood control activities. Congress has urged the Corps to intensify its nonstructural flood mitigation programs, including determining nonstructural options in the earliest stages of project formulation, and to review how existing policies and procedures impede the justification of nonstructural measures as alternatives to flood control.

Between 1991 and 1992 there were four national conferences focusing on water and renewable resources management. The first of these was held at Park City, Utah, in 1991. The recommendations that arose from the Park City conference are called the "Park City Principles" (see Box 2.3). The principles are intended to encourage the resolution of complex water problems and were endorsed, amplified, and modified by subsequent conferences. From a historical perspective some of these principles are not new but represent an evolution of ideas that have been discussed for years. The evolution of watershed–based planning into the so–called "problemshed" is an example. A problemshed is the geographical area that encompasses the water–resource problem and all the affected parties. The boundaries of the problemshed can extend beyond the physical boundary of the basin. The last principle states that freshwater sustainability should be the guiding principle for water management. Sustainable use means using freshwater resources in ways that maintain and perpetuate their quality and quantity for future generations of people and other species.

In the last two centuries our attitudes toward water and other natural resources

BOX 2.3 PRINCIPLES FOR THE RESOLUTION OF COMPLEX WATER MANAGEMENT ISSUES

Water problems should be approached holistically, on a problemshed basis, with the needs of natural systems fully integrated into the management decisions.

The framework for policy must be flexible and responsive to change, yet provide predictability and certainty to support management decisions and investment.

States play a key role in water management, and obstacles to intergovernmental partnerships must be overcome. Federal, state, and local participation should be encouraged in the development of each other's programs.

Water policy development should emphasize market–based approaches and a preference for negotiation, consultation, and cooperation between all entities with interests in policies.

The goal of freshwater sustainability should be a guiding principle for future water–resource management.

evolved from unbridled exploitation with little concern for environmental impacts to a complex, some say too complex, set of laws, regulations, and institutions designed to protect our resources from pollution and degradation. One very important lesson we should learn from history is that our future plans should be flexible—even reversible—to accommodate the needs of future generations. We live in a rapidly changing society. In the past our plans locked us into relatively inflexible resource–use patterns. When social and environmental values shifted, these patterns became archaic and were difficult to change. The result was often increased animosity and social conflict. We need to recognize from the outset that social and environmental values will be different for our children and grandchildren.

In summary, the federal government's involvement in water planning and management increased over time, relying on its various powers to control commerce and promote the general welfare. But local political action has strongly influenced federal activities, and at times even set the agenda in the traditional areas of flood control, irrigation, and navigation. The federal government's policies have been more short–term reactionary than long–term visionary, with major legislation following national crises. Interest in basin–wide planning, first offered as a model nearly a century ago, has waxed and waned, but still retains its seductive logic today, driven by concerns over water quality and ecosystem health.

3

WATER QUANTITY LAW

THE LEGAL SYSTEM

How Laws Are Made

The Purpose of Law

State Law: Surface Water

The Riparian Doctrine

The Prior Appropriation Doctrine

Regulated Riparianism

State Law: Groundwater

Absolute Ownership—the English Doctrine

Reasonable Use—the American Doctrine

Correlative Rights

Prior Appropriation

Federal Law

Federal Powers Related to Water Resources

Federal Reserved Rights

Nonreservation Indian Reserved Rights

Other Federal Reserved Rights

Other Federal Interests That Affect Water Resources

Transboundary Water Resources

Judicial Decisions

Interstate Compacts

Legislation

Water quantity laws deal with issues of ownership and the right to use water, and are some of the oldest laws of civilization. In the United States the basic laws governing water use evolved in the 1800s, a century before laws addressing water quality management. In the 1800s the focus of water planning and management was on resource development and use. Water powered the mills that ground the grain, sawed the lumber, and carded the wool. Expanding cities in both the East and the West required water for the developing industries as well as their burgeoning populations. Water–based transportation needed a minimum depth of water for navigation. And out west irrigation was "making the desert bloom." All across the country people were using water to further economic development. Water not being used in a productive activity was water wasted. The degradation in water quality that accompanied the various uses was ignored. Ignored too were the environmental values sustained by the water in the form of habitats for fish and wildlife.

With increasing water use came increased conflict between different users. If economic development was not to be stifled, it was imperative that laws be set down clarifying who could use water, and how much they were entitled to. Piece by piece the complex legal structure for water resources began to emerge. Unfortunately, many of our laws evolved ahead of our understanding of the hydrologic system, and water laws in many states today still treat surface water and groundwater as being hydrologically separate and distinct. In this chapter we first examine state law and then consider the federal legal framework.

HOW LAWS ARE MADE

There are two ways to make law. The first way is by *statute.* Statutory laws are created when a governing body, such as a state legislature, passes legislation which eventually becomes law. All western states have statutory water laws; however, many eastern states do not. This is largely an artifact of the difference in climate between the West and the East. The western states are predominantly dry and have always had to deal with the problem of limited water supplies. They were forced early on to enact laws and regulations for water use. In some cases they even established laws before they had been admitted as states into the Union. Congress creates statutory law when it passes federal legislation. Some of the more interesting federal water laws are statutes that were originally passed for one reason, and were later interpreted by the courts as being important water laws.

The second way to make law is in the courtroom. This type of law is *case law* or *common law.* When one party sues another party the dispute is settled in court. A party may be a person, a corporation, a government agency, a state government, or the federal government. The party bringing the suit is called the *plaintiff* or the *petitioner.* The party being sued is called the *defendant* or the *respondent.* If the suit is appealed to a higher court the parties may change places, with the plaintiff becoming the defendant and vice versa. If there is no statutory law pertinent to the conflict at hand, the court renders a decision based on the facts and its interpretation of fairness. This is the origin of common law. Common laws are local customs and rules that have been confirmed by the courts. If there is statutory law applicable to the case at bar, then it is the court's job to interpret the legislation. When a judicial ruling is followed by other courts in deciding similar cases, the ruling has established a *precedent.* Precedent is very important in case law. The judicial system is a fine example of incremental decision making. Courts make decisions based on the narrowest possible interpretation of the facts and rely heavily upon precedent (Trelease, 1979).

THE PURPOSE OF LAW

According to the noted water law scholar Frank Trelease there are three purposes of water law. First, the law should promote desirable behavior and discourage undesirable behavior. Second, the law should resolve conflict in an orderly fashion. Conflict is inevitable in the use of water. As a matter of fact the word "rival" has the same etymology as the word "river." Quite a few water disputes in the old West were resolved at gun point. The third function of law is to define and protect property rights with adequate security to encourage investment and development. Water rights are different from rights in real property. Water rights are *usufructuary,* which means a person has the right to *use* water, but does not own the water itself. The right holder does not own the *corpus* of the water. This means, for example, that a person can temporarily store water in a reservoir, but the water must eventually be released so that downstream users have access to it. The property right function of law is fundamental and forms the basis of our philosophical approach to natural–resource use and management. It is a philosophy that assumes individuals make decisions to improve their welfare, and the sum of all the individual decisions constitutes the total welfare of society. People will not invest time and money in an economic activity, say, irrigation agriculture, if their right to use water is not well defined and guaranteed. In fact, according to economists, it is precisely in situations where ownership of the resource is poorly defined that abuse, pollution, and degradation of the resource occur. Resources in which private rights are difficult or impossible to define are called *common property resources.* Common property resources belong to no one, and individuals cannot be excluded from their use. Individuals use the resource to the fullest extent possible since the resource is "free." Unfortunately, when every individual makes this same decision, society as a whole suffers because the resource is overused and degraded. Economists argue that the best way to manage natural resources and to achieve the level of environmental quality that people desire is through the use of private property and a free–market allocation mechanism. For various reasons society has largely chosen not to use pure market–based mechanisms to manage and allocate water resources. The role of economics in water–resources planning and management is explored in depth in Chapter 5.

STATE LAW: SURFACE WATER

The two basic state water law doctrines are the *riparian doctrine* and the *prior appropriation doctrine.* While these are the basic doctrines they should be viewed as templates from which each state has fashioned it own laws—laws that continue to change as society itself changes. Laws are after all but a reflection of the society that creates them. Surface–water law deals with water in stream channels and is also referred to as watercourse law. Surface–water law encompasses other forms of surface water including lakes and "diffuse water." Diffuse water flows over the surface, but not in a defined channel. In Chapter 1 we called this type of surface runoff overland flow. In some states the law explicitly defines what is legally considered a watercourse. Amendments to the water laws of the state of New Mexico in 1941 define a watercourse as a " . . . channel having definite bed and banks with visible evidence of occasional flow of waters." Riparian law developed in the humid eastern United States where water was abundant (Figure 3.1). The prior appropriation doctrine is found in western states and developed in response to the dry conditions. Some states have mixed doctrines. The mixture of legal doctrines in the Plains reflects the transitional climate—drier in the western part and more humid in the eastern part. The three far–western states of California, Oregon, and Washington are also mixed–doctrine states. Here, however, the riparian rights are generally older rights, and the states have followed prior appropriation for most of this century.

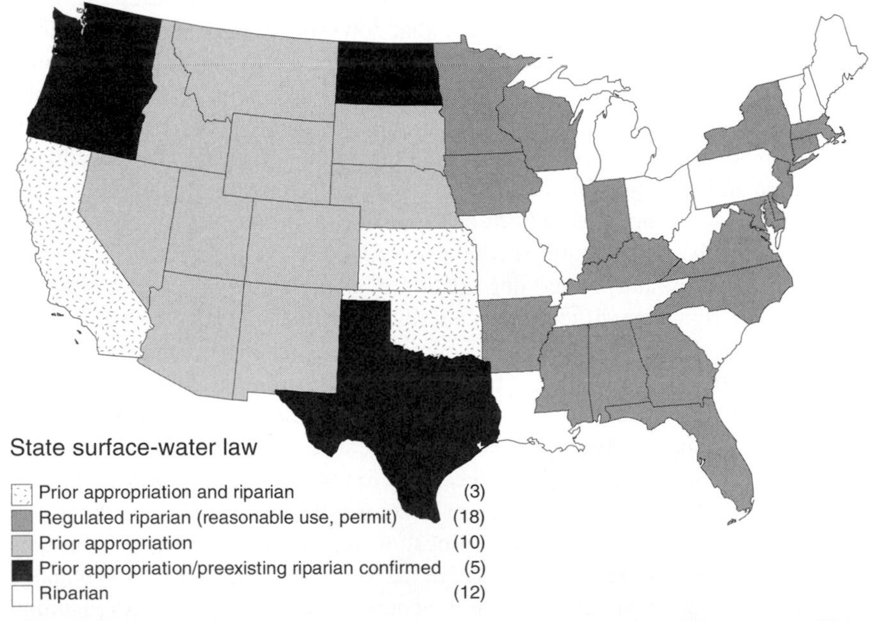

FIGURE 3.1 States categorized according to their surface-water law. (Viesmann and Welty, 1985, based on data from the John Muir Institute, 1980, and Dellapenna, 1997)

THE RIPARIAN DOCTRINE

The origin of American riparian law is generally thought to be English common law brought over with the colonists to America. Some scholars have advocated a French connection for the doctrine, though this has been vigorously disputed (see Meyers *et al.,* 1988). When the colonists began forging a new society in America they found the physical geography not too dissimilar from that of England and western Europe. The natural vegetation was mixed temperate forest and the humid climate produced ample runoff to the many streams and rivers. In a similar environment they applied the same type of law they were accustomed to back in England.

The Basic Doctrine

The riparian doctrine says if you own land that touches a stream or a lake you have a right to use water. It is not necessary that you own any portion of the bed of the stream or lake, just that your property is adjacent to it. It also does not matter how much of your land touches the water body. Since the water right is incident to the ownership of land, you can never lose the right so long as you own the land. The use of water by a nonriparian land owner is illegal.

Figure 3.2 is a hypothetical depiction of property owners along a stream in a riparian doctrine state. A general rule in riparian law is that the water must be used within the watershed of origin. This ensures that the water returns to the water body for use by other riparians. In Figure 3.2 this rule means the owner of parcel A can use water only on that part of the land contained within the watershed. But

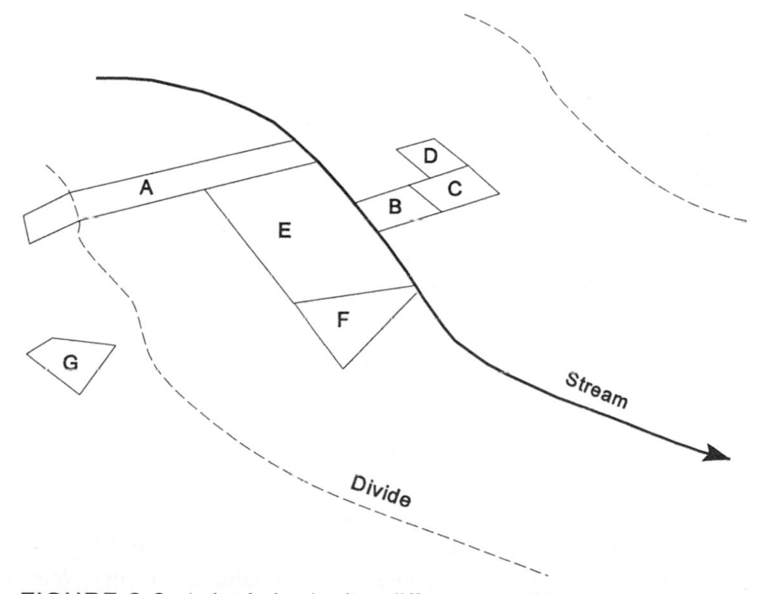

FIGURE 3.2 A sketch showing how different tracts of land may or may not be riparian depending upon location and ownership. See text for discussion and examples.

this is not a hard and fast rule; different courts in different states have reached different conclusions based on the facts of the case.

When land is bought, sold, or subdivided two possible rules apply to the riparian status of the land. These two rules are *source of title* and *unity of title.* The more restrictive, and more common, is the source of title rule. Source of title says that water may only be used on land that has been held as a single tract throughout the history of its title. For example, in Figure 3.2 assume that parcels B and C were originally a single riparian tract. Water could be used on the entire (B–C) parcel. The tract was later subdivided and parcel C sold off. According to the source of title rule, parcel C loses its riparian right. Even if the owner of B buys C back at a later date it does not regain riparian status. This is why the source of title rule is sometimes referred to as the "smallest tract" rule. The unity of title rule is more liberal and considers all land held under one owner and contiguous to a riparian tract to be riparian. Under this rule if B bought back parcel C it would again be riparian. If B bought parcel D, which was never riparian before, unity of title says it too becomes riparian, since it is contiguous and held by the same riparian owner. This is why the unity of title rule is also called the "no limit rule." The unity of title rule is applied only in Wisconsin. Parcel F, with very limited frontage on the stream, has just as much right to use water as any other parcel because the right is not conditioned by the size of the frontage. What if parcel G were owned by the same person who owned parcel A? Could water be used on parcel G? Observe that parcel G is outside the watershed and is nonriparian. This exact question was addressed by the Massachusetts Supreme Court in the case *Stratton v. Mt Hermon Boys' School.* This case is presented later after we discuss the different theories of the riparian doctrine.

Theories of the Riparian Doctrine: Natural Flow and Reasonable Use

The first version of riparian law is the *natural flow* theory. Under the natural flow theory every user is obliged to maintain the quantity and quality of the water before and after use. In other words, the use of water may not change the water in any way. The natural flow theory follows from the fact that all riparian landowners have coequal rights. The exercise of a right by one person should not infringe or diminish the value of the right of another. The natural flow theory was workable in earlier times and in sparsely settled areas where water demands were small and water pollution easily diluted or biodegraded. Population growth, urbanization, and industrial development all made the natural flow theory more and more unworkable. Most states adopted the modified riparian theory of *reasonable use.* With reasonable use each riparian landowner has a right to make reasonable use of water and can prevent unreasonable uses. Reasonable use is a relative test based on the need (mainly economic) of the user and the larger (economic) benefits to society. Reasonable use means sharing the resource, and the limit of a right is determined by the impact of the use on others. Notice the subtle change in emphasis between the natural flow theory and the reasonable use theory. With natural flow a riparian user could not hurt the stream; with reasonable use the user should not hurt other riparian users.

There are problems with having a standard based on reasonableness. It is impossible to predict for any particular time and place what uses will be allowed. Fifty years ago it was considered reasonable to dump untreated sewage into lakes and steams. This is clearly unreasonable today. What is reasonable on one stream may be judged unreasonable on another. This can be true even in the same state. A landowner might begin a new use which causes harm to his neighbor, and is therefore unreasonable. But if the new use provides large benefits to society the courts may allow it. For decades the courts have been doing what amounts to a qualitative cost–benefit analysis in deciding what is fair under the reasonable use standard.

Because all rights are coequal, the riparian doctrine requires users to continually adjust (reduce) their right as new users and uses come along. The potential for conflict is apparent because the adjustment is inevitably toward a smaller percentage of the total flow. The pure riparian doctrine does not protect established users from users that come along later. Is this fair? And what happens during a drought? Under riparian law all right holders suffer equally. Again, is this fair?

Riparian water law also raises some interesting practical problems. For example, how do nonriparian property owners within a city obtain the right to use water in and around their home? Obviously the doctrine had to be modified to serve the practical requirements of providing people in cities with water for domestic use. Or, what if a stream changes course through erosion of its banks. Do landowners who were riparian before the change, but are nonriparian after, lose their water right?

These are real problems but they pale by comparison to the problem of increasing water demand driven by population and economic growth. The riparian doctrine ostensibly creates private water rights, but it treats the stream as a common property resource. Increased use of water to the benefit of one individual cannot help but harm other riparians by reducing the available supply. Increasing demands on the abundant but finite resources have forced about half of the eastern states to institute permit systems. In some ways riparian states using permits are acting more like prior appropriation states in the West. Some water law scholars think the permit–related changes to the riparian doctrine in these states has been so fundamental a change that it justifies being called a new water law doctrine, termed *regulated riparianism* by Dellapenna (1994). We will examine regulated riparianism after we discuss the prior appropriation doctrine.

This case concerns a riparian owner making a nonriparian use of water outside the watershed. When analyzing this or any water law case consider the following questions:

1. Who is the plaintiff? Who is the defendant? Where are they located with respect to each other? Describe the facts of the case. You might draw a simple map to organize this information (see Figure 3.2).

2. Explain any previous rulings on this case. Is this an original case or an appeal from an earlier ruling?

3. What are the issues at bar? That is to say, what is the conflict about?

4. What did the court decide? Who won and what are the implications of the decision?

STRATTON v. MT HERMON BOYS' SCHOOL SUPREME JUDICIAL COURT OF MASSACHUSETTS, 1913 216 MASS 83, 103, N.E. 87

RUGG, C. J. The plaintiff, the owner of a mill upon a small stream, sues the defendant, an upper riparian proprietor upon the same stream, for wrongful diversion of water therefrom to his injury. The material facts are that the defendant owns a tract of land through which the stream flows and upon which also is a spring confluent to the stream. Upon this land it has been established pumping apparatus whereby it diverts about 60,000 gallons of water each day from the spring and stream to another estate belonging to it and not contiguous to its land adjacent to this stream, but located about a mile away in a different watershed, for the domestic and other uses of a boys' school with dormitories, gymnasium and other buildings and a farm. The number of students increased from 363 in 1908 to 525 in 1911, while the number of teachers, employed and other persons on the estate was over 100. During the latter year there were kept on the farm 103 cattle, 28 horses and 90 swine. There was a swimming pool, laundry, canning factory and electric power plant, for the needs of all of which water was supplied from this source. There was evidence tending to show that this diversion caused a substantial diminution in thc volumc of water which otherwise in the natural flow of the stream would have come to the plaintiff's land and in the power which otherwise might have been developed upon his wheel by the force of the current.

The defendant requested the court to rule in effect that diversion of water to another nonriparian estate owned by it was not conclusive evidence that the defendant was liable, but that the only question was whether it had taken an unreasonable quantity of water under all the circumstances. This request was denied and the instruction given that the defendant's right was confined to a reasonable use of the water for the benefit of its land adjoining the water course, and of persons properly using such land, and did not extend to taking it for use upon other premises, and that if there was such use the plaintiff was entitled to recover at least nominal damages even though he had sustained no actual loss. The exceptions raise the question as to the soundness of the request and of the instruction given. * * *

In the main, the use by a riparian owner by virtue of his right as such must be within the watershed of the stream, or at least that the current of the stream shall be returned to its original bed before leaving the land of the user. This is implied in the term "riparian." It arises from the natural incidents of running water. A brook or river, so far as concerns surface indication, is inseparably connected with its watershed and owes the volume of current to its area. A definite and fixed channel is a part of the conception of a water course. To divert a substantial portion of its flow is the creation of a new and different channel, which to that extent defeats the reasonable and natural expectations of the owners lower down on the old channel. Abstraction for use elsewhere not only diminishes the flow of the parent stream but also increases that which drains the watershed into which the diversion is made, and may injure thereby riparian rights upon it. Damage thus may be occasioned in a double aspect. The precise point whether riparian rights include diversion in reasonable quantities for a proper use on property outside the watershed has never been decided in this commonwealth. There are numerous decisions in other jurisdictions to the effect that the rights of a riparian proprietor do not extend to uses on land outside the watershed. These were made in cases where actual perceptible damages were wrought by the diversion. * * *

There are numerous expressions to the effect that the rights of riparian ownership extend only to use upon and in connection with an estate which adjoins the stream and cannot be stretched to include uses reasonable in themselves, but upon and in connection with nonriparian estates. * * * These principles, however, are subject to the modification that the diversion, if for a use reasonable in itself, must cause actual perceptible damage to the present or potential enjoyment of the property of the lower riparian proprietor before a cause of action arises in his favor. This was settled after an elaborate discussion by Chief Justice Shaw in *Elliot v. Fitchburg R. R.*, 10 Cush. 191, 57 Am. Dec. 85. That case has been widely cited with approval by courts of many states. The soundness of its reasoning never has been questioned. That was a case where an upper riparian proprietor granted to the defendant railroad corporation a right to erect a dam across a stream and conduct water to its depot not on a riparian estate for use in furnishing their locomotive steam engines with water. The plaintiff, a lower riparian proprietor, was denied the right to recover nominal damages in the absence of proof of actual damages, although the principle was fully recognized that an action would lie for any encroachment upon the substantial rights of the lower owner though causing no present damage. The distinction is between a diversion which causes a present or potential injury to the lower estate for a valuable use and one which cannot produce such a result. There have been numerous decisions touching the question whether appropriation of water by an institution like the defendant to the extent here shown can be regarded as within the rights of a riparian proprietor. * * * But it is unnecessary to discuss or review them or to decide whether the true rule in such cases is not the one here established.

The governing principle of law in a case like the present is this: A proprietor may make any reasonable use of the water of the stream in connection with his riparian estate and for lawful purposes within the watershed, provided he leave the current diminished by no more than is reasonable, having regard for the like right to enjoy the common property by other riparian owners. If he diverts out of the watershed or upon a disconnected estate the only question is whether there is actual injury to the lower estate for any present or future reasonable use. The diversion alone without evidence of such damage does not warrant a recovery even of nominal damages. * * * Exceptions overruled.

The lower court followed the letter of the law in saying a nonriparian use is unreasonable per se. But the Supreme Court overturned the lower court. The Supreme Court found that the plaintiff had not suffered any actual damages. Absent such explicit damage the court was not going to negate the benefits to the school from using water outside the basin. Obviously the court felt the benefits outweighed the costs. If the school changed the nature of its out–of–basin use and caused damages to the plaintiff, then presumably the court would enjoin the use.

	In-basin use	Out-of-basin use
Riparian	Reasonable uses are always allowed.	Unreasonable but might be allowed if no damage occurs to other riparians.
Nonriparian	Unreasonable and not allowed. Exceptions are granted, for example, municipalities.	The most unreasonable and not allowed. Exceptions are granted, for example, municipalities.

FIGURE 3.3 Categorizing reasonable or unreasonable uses based on combinations of land ownership and the location of water use.

This case illustrates one of four possible situations of land ownership and water use by a riparian right holder (Figure 3.3). The landowner can be either riparian or nonriparian, and the water can be used either inside or outside of the basin. In theory the riparian doctrine only allows the situation depicted in the upper left corner of Figure 3.3. The three remaining combinations are by definition unreasonable. As *Stratton* showed, this does not mean that every court will automatically disallow an unreasonable use. The court weighs the relative benefits and costs in making its decision. This is why the city of New York can transfer water from the upper Delaware River basin 150 miles south for use by city residents. The most illegal of the four combinations is the bottom right where a nonriparian makes an out–of–basin use of water. The benefits would have to far exceed the costs for the courts to ever allow this.

THE PRIOR APPROPRIATION DOCTRINE

The prior appropriation doctrine is the water law in 10 western states (Figure 3.1). Four of the Plains states and the three far–western states have mixed doctrines. The prior appropriation doctrine originated in the mining camps of the

Sierra Nevada in California. Gold miners rushing into the foothills were beyond the reach of civilized society and the law. The miners were driven by necessity to develop their own common law for allocating mining claims and resolving disputes. Their common law customs were eventually confirmed by case law and statute. Two fundamental rules developed by the miners were "first in time, first in right," and "use it or lose it." The first rule says if you are the first person to stake a claim, the claim is yours alone to work. The problem was that miners often would not stake their claim for fear of attracting attention. The rule gave preference to the first claimant, but it did not guarantee protection from claim jumpers. The second rule limited speculation. A miner had to actively work his claim. This prevented miners from staking claim to more land than they could work, thus ensuring the resource (gold) would not be monopolized. Keep in mind that the miners were staking claims on the *public domain.* The miners did not own the land, the land belonged to the state or the federal government. This fact was important in establishing the prior appropriation doctrine in California because you cannot assert riparian rights if you do not own the land.

Placer and hydraulic mining required water to wash the gold from the sediment. The locations of the mining claims were dictated by geology, not hydrology, and claims were often many miles from the nearest stream, often even beyond the watershed divide. The solution was straightforward; divert water from the stream and convey it by ditch to the mining claim. The miners thus established the principle of diverting water for use on a parcel of land without regard to the parcel's location. The miners applied their common law for mining claims to the allocation of water—first in time, first in right, and use it or lose it. These are basic precepts of western water law (see *Irwin v. Philips*). Since the gold rush had international appeal some miners undoubtedly brought with them knowledge of water law systems similar to prior appropriation used by other cultures in dry climates throughout the world.

The Basic Doctrine

The name prior appropriation signifies that water rights are allocated on a temporal basis—first in time, first in right. The first person to appropriate water from a stream has the most senior water right. The next person to make an appropriation has the next–most–senior right, and so on. All water rights subsequent in time to any particular right are junior and have lower priority. The senior rights are the best rights because they will receive water before junior appropriators. This makes senior rights the most valuable. When the supply of water is limited, appropriators are cut off from the stream in inverse order of priority. This means the most junior right is cut off first. If there still is not enough water available to fully satisfy senior rights, the next–most–junior right is curtailed. This continues until the senior rights receive all the water to which they are entitled. In the riparian system water rights are coequal; in prior appropriation they are not.

IRWIN v. PHILLIPS
SUPREME COURT OF CALIFORNIA, 1853
5 CAL. 140
HEYDENFELDT, J., DELIVERED THE OPINION OF THE COURT. MURRAY, C. J., CONCURRED.

The several assignments of error will not be separately considered, because the whole merits of the case depend really on a single question, and upon that question the case must be decided. The proposition to be settled is whether the owner of a canal in the mineral region of this State, constructed for the purpose of supplying water to miners has the right to divert the water of a stream from its natural channel as against the claims of those who, subsequent to the diversion, take up lands along the banks of the stream, for the purpose of mining. It must be premised that it is admitted on all sides that the mining claim in controversy, and the lands through which the stream runs and through which the canal passes, are a part of the public domain, to which there is no claim of private proprietorship; and that the miners have the right to dig for gold on the public lands was settled by this Court in the case of *Hicks et al. v. Bell et al.*, 3 Cal. 219. It is insisted by the appellants that in this case the common law doctrine must be invoked, which prescribes that a water course must be allowed to flow in its natural channel. But upon an examination of the authorities which support that doctrine, it will be found to rest upon the fact of the individual rights of landed proprietors upon the stream, the principle being both at the civil and common law that the owner of lands on the banks of a water course owns to the middle of the stream, and has the right in virtue of his proprietorship to the use of the water in its pure and natural condition. In this case the land are the property either of the State or of the United States, and it is not necessary to decide to which they belong for the purposes of this case. It is certain that at the common law the diversion of water courses could only be complained of by riparian owners, who were deprived of the use, or those claiming directly under them. Can the appellants assert their present claim as tenants at will? To solve this question it must be kept in mind that their tenancy is of their own creation, their tenements of their own selection, and subsequent, in point of time, to the diversion of the stream. They had the right to mine where they pleased throughout an extensive region, and they selected the bank of a stream from which the water had been already turned, for the purpose of supplying the mines at another point.

Courts are bound to take notice of the political and social condition of the country which they judicially rule. In this State the larger part of the territory consists of mineral lands, nearly the whole of which are the property of the public. No right or intent of disposition of these lands has been shown either by the United States or the State governments, and with the exception of certain State regulations, very limited in their character, a system has been permitted to grow up by the voluntary action and assent of the population, whose free and unrestrained occupation of the mineral region has been tacitly assented to by the one government, and heartily encouraged by the expressed legislative policy of the other. If there are, as must be admitted, many things connected with this system, which are crude and undigested, and subject to fluctuation and dispute, there are still some which a universal sense of necessity and propriety have so firmly fixed as that they have come to be looked upon as having the force and effect of *res judicata.* Among these the most important are the rights of miners to be protected in the possession of their selected localities, and the rights of those who, by prior appropriation, have taken the waters from their natural beds, and by costly artificial works have conducted them for miles over mountains and ravines, to supply the necessities of gold diggers, and without which the most important interests of the mineral region would remain without development. So fully recognized have become these rights, that without any specific legislation conferring or confirming them, they are alluded to and spoken of in various acts of the Legislature in the same manner as if they were rights which had been vested by the most distinct expression of the will of the law makers; as for instance, in the Revenue Act "canals and water races" are declared to be property subject to taxation, and this when there was none other in the State than such as were devoted to the use of mining. Section 2 of Article IX of the same Act, providing for the assessment of the property of companies and associations, among others mentions "dam or dams, canal or canals, or other works for mining purposes." This simply goes to prove what is the purpose of the argument, that however much the policy of the State, as indicated by her legislation, has conferred the privilege to work the mines, it has equally conferred the right to divert the streams from their natural channels, and as these two rights stand upon an equal footing, when they conflict, they must be decided by the fact of priority, upon the maxim of equity, *qui prior est in tempore, potior est in jure.* The miner who selects a piece of ground to work, must take it as he finds it, subject to prior rights, which have an equal equity, on account of an equal recognition from the sovereign power. If it is upon a stream, the waters of which have not been taken from their bed, they cannot be taken to his prejudice; but if they have been already diverted, and for as high and legitimate a purpose as the one he seeks to accomplish, he has no right to complain, no right to interfere with the prior occupation his neighbor, and must abide the disadvantages of his own selection.

It follows from this opinion that the judgment of the Court below was substantially correct, upon the merits of the case presented by the evidence, and it is therefore affirmed.

Appropriators obtain a right to a fixed quantity of water. This too is different from the riparian doctrine. Under the pure riparian doctrine the right holder is allowed to make any reasonable use of water, but there is no proscription on the amount of water that can be used. (In regulated riparian states with permits there may be restrictions on water quantity.) There are two types of appropriation rights—*flow rights* and *storage rights.* A flow right gives a right to a specified flow of water for a specified period of time. For example, a flow right might be a right to divert 2.7 cfs from March 1st to October 30th. A storage right is for a specified volume of water. An example is a right for 130 acre–feet of water per year. When the appropriator actually uses the water or how much is used at any one time is unimportant. Obviously, storage rights can only exist where there is reservoir storage. The fixed quantity in the appropriation right is the maximum amount of water the appropriator can use. Depending upon streamflow conditions and seniority the actual amount received by the right could be less. In wet years a period of free water may be declared. Appropriators can use the free water and it does not count against their allocation for that year.

To perfect a water right the appropriator would first give notice of his intent to divert water from the stream, construct the necessary diversion works, and then physically divert the water. In the earliest days giving notice meant posting a sign along the banks of the stream at the intended point of diversion and running an announcement in the local newspaper. The notification requirement today is satisfied in most states by filing for a permit with the state engineer or administrative water agency. The point is that you could not obtain a valid water right by surreptitiously sneaking water from the stream. All senior rights had to be put on notice regarding an intended appropriation so they could object if they felt the appropriation would interfere with their established rights. Originally, all states required a physical diversion of the water from the stream to perfect a valid decree. Many states have modified this requirement in recent years to accommodate instream flows for recreation and environmental uses. The state of New Mexico still requires the water to be physically diverted from the channel before a water right can be perfected.

If the appropriator proceeded with *due diligence* from the time of initial filing to the time the water was actually used, the original conditional decree ripened into final decree with a seniority date equal to the date of the initial application, not the (later) date when the water was actually used. This is the *relation back* principle; the date of the decree relates back to the date of the initial filing. In the case of a large project that could take years to complete, relation back could mean the difference between a good senior right and a less valuable junior right. The key to relation back is the due diligence criterion. The appropriator had to demonstrate continued progress toward completing the act of appropriation. If due diligence was challenged and found wanting, the appropriator could receive a later priority date. The relation back theory was an important issue in *City and County of Denver v. Northern Colorado Water Conservancy District.* Denver's water supply is a combination of water from the South Platte River, which rises on the eastern slope of the Rocky Mountains, and water diverted from the western slope through tunnels under the continental divide. In 1914 and 1921 engineers from the city undertook preliminary surveys for a transbasin diversion tunnel in the Fraser and Williams Fork River watersheds on the western slope. In 1927 Denver filed a plat showing a plan to divert water, but it was not until 1946 that the city finally began work on a diversion tunnel. The city tried to have the priority date relate back to the 1914 survey. The trial court said no. Denver appealed to the Colorado Supreme Court and lost there too. The Colorado Supreme Court said in part;

> * * * Kinney, in his great work on irrigation, says: "Probably the best definition of the word diligence was given by Lewis, C.J. in rendering the opinion in an early Nevada case, *Ophir Silver Min. Co. v. Carpenter,* 4 Nev. 534. It is there defined as 'the steady application to business of any kind, constant effort to accomplish any undertaking.' "
> * * * Denver had not even begun the actual construction of its project and had made no effort whatever as appears from the record towards financing it, but only a laudable but fruitless attempt after nine years of inaction to induce the United States Reclamation Service to finance it for the joint use of Denver and the South Platte Water Users Association. Meanwhile others have worked diligently and long to put a part of this water to actual use. The record before us does not show such conclusive evidence of "steady application" to the business of constructing the project or of such "constant effort to accomplish" it as to require us to hold that the trial court erred in refusal to date back Denver's appropriation, to the loss of such prior users. On the contrary, in order to sustain Denver's claim, we should have to establish as a law of Colorado that a great city or a great corporation, by filing of a plat of a water diversion plan and the fitful continuance of surveys and exploratory operations, could paralyze all development in a river basin for a period of nineteen years without excavating a single shovel full of dirt in actual construction and without taking any step towards bond issue or other financing plan of its own for carrying out its purpose; that for nineteen years no farmer could build a ditch to develop his farm and no other city or industry could construct a project for use of water in that area without facing loss of the water when and if the city or corporation which filed the plat should actually construct its project. This we cannot do.* * *Accordingly, the decree of the trial court herein is affirmed.
> (*City and County of Denver v. Northern Colorado Water Conservancy District,* Supreme Court of Colorado, 1954, 130 Colo. 375, 276 P.2d 992.)

The court felt that a city the size of Denver should have done something—moved dirt, developed a financing plan, whatever—to demonstrate due diligence. The court saw it as unjust to water users of more modest means, but who had continued to work on their appropriation, to be pushed down the seniority ladder by the great city of Denver, which had done nothing on its intended appropriation for nearly two decades. The relation back principle also meant, however, there was little time to consider the potential environmental impacts of a water diversion. Under prior appropriation, he who hesitated lost his place in line. In most cases negative impacts on the environment were not considered important anyway.

Water rights are perfected and owned separately from the land. This is totally opposite of the riparian system where land ownership creates the water right. In the drier climates of the western states streams are few and far between, and it is unlikely that a parcel of land will be adjacent to a stream. To make productive use of the land and water resources, the water had be brought to the land, regardless of where the two were located. This means water rights can be transferred from one parcel to another, and from one watershed to another. Transferability is an important aspect of prior appropriation. There are constraints on water rights transfers which we will discuss momentarily.

The limit and extent of a water right is based on *beneficial use*. Water can only be appropriated for a recognized beneficial use. Most states list beneficial uses in either their water law statutes or their state constitution. Irrigation, domestic use, industrial, and manufacturing uses have always been considered beneficial uses. Instream flows for recreation, aesthetic purposes, and maintenance flows for ecosystems were not considered beneficial until relatively recently. In the 1970s and 1980s states responded to growing concerns about the environment by broadening their statutory definition of beneficial to include instream uses. As mentioned before, instream flows are still not a beneficial use in New Mexico. While beneficial use theoretically limited the amount of water that could be appropriated it could encourage waste when appropriators applied for the maximum amount of water they might possibly use, not the amount they actually needed (Teclaff and Teclaff, 1973).

The prior appropriation doctrine was utilitarian. The physical diversion of water, the beneficial use requirement, the use it or lose it stipulation, and the fixed quantity limitation all promoted the use of water. Water left in the stream was considered wasted. Speculation was minimized by limiting the water right to the intended (beneficial) use and no more. Speculation was minimized too by the use it or lose it rule; if you did not use your water right for a statutory period of time the right was declared *abandoned* and the water was declared available for (re)-appropriation. The statutory period on nonuse for the presumption of abandonment is different in different states but is usually between 5 and 10 years. Municipalities are exempt from the strict abandonment rule because cities must acquire water for use in the future as a part of long–range planning. The use it or lose it rule does not promote resource conservation. Appropriators are encouraged to use the water even if they do not need it so as to avoid any presumption of abandonment.

Many streams and rivers in the West are fully appropriated and even over-appropriated, and there is no water available for new appropriations. In most states when a stream is fully appropriated the state agency granting water rights will refuse to grant new water rights. This is not the case in Colorado. Article XVI, Section 6, of the Colorado Constitution says, " . . . the right to divert the unappropriated waters of any natural stream to beneficial use shall never be denied." You can go to any stream in Colorado today and start an appropriation; however, the right may be so junior that it will never yield any water. These extremely junior water rights are sometimes called "flood rights"; the only time they yield water is during a flood. The "Colorado Doctrine" is considered the purest form of prior appropriation. Colorado even has a separate water court system for handling water–related matters.

Once an appropriator uses the water he or she must return the unused portion to a natural stream. Downstream appropriators rely on this *return flow* to satisfy their water right. The multiple reuse of water means that the total quantity of water

FIVE DIFFERENCES BETWEEN THE RIPARIAN AND PRIOR APPROPRIATION DOCTRINES

1. The right is derived from and tied to the ownership of riparian land in the riparian system; under prior appropriation the right is given to someone who diverts the water and applies it to a beneficial use.
2. Riparians never lose the water right as long as they own riparian land and can establish new uses; nonuse (abandonment) can destroy the right under prior appropriation.
3. Riparians use water on adjacent land within the watershed; appropriators may use the water anywhere but the right is appurtenant to a specific parcel.
4. Riparians share in the use of water with no fixed quantity assured them; appropriators are given the right to a specific flow or quantity based on seniority date.
5. Riparians suffer equally in times of shortage; appropriators suffer unequally according to seniority.

represented by the perfected rights on a stream can exceed the average flow of the stream (see Example 3.1). The return flow does not have to return to the stream from which the water was diverted. In fact it will not when the water is diverted for use in another watershed. Once junior appropriators perfect rights based on your return flow entering the stream at a certain point, you cannot alter the point where the return flow enters the stream to their detriment. You are, however, allowed to alter the amount of return flow. If you increase the efficiency with which you use water—not change the type of use, just the efficiency of use—you are not required to return the same amount of return flow. Any improvement in water use efficiency must occur on your property. Once the water leaves your property it has escaped your control and it must be allowed to return unimpeded to the receiving stream. If improved efficiency makes more water available for your own

EXAMPLE 3.1

This example shows the operation of a hypothetical stream under the prior appropriation system (Figure 3.4). There are three flow water rights on the stream. The long–term *average* discharge of the stream is 10 cfs. Assume the three appropriators are irrigating and their efficiency is 50 percent. In other words, half of the water they divert returns to the stream; the other half is consumed. The appropriators and their water rights, relative seniority, and locations on the stream are shown below. The average streamflow is 10 cfs, but the sum of the three water rights is 15 cfs. Let us examine the operation of the system under two different scenarios for water availability.

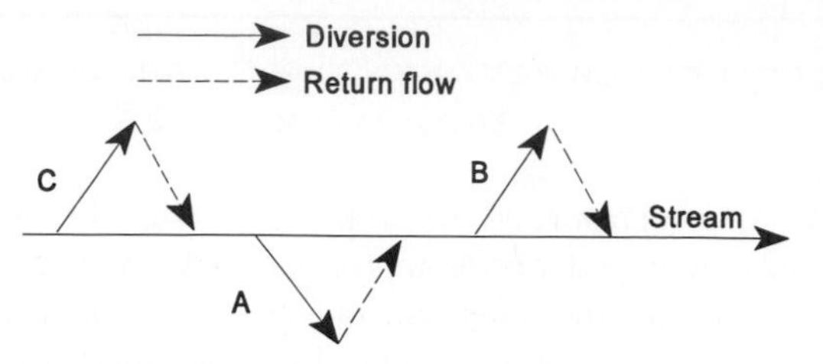

FIGURE 3.4 Hypothetical stream with three appropriators.

Appropriator	*Water right (cfs)*	*Seniority*
A	5	2
B	4	1
C	6	3

Scenario 1: Streamflow above the point of diversion for C equals the long–term average flow of 10 cfs.

In this scenario everyone gets their full entitlement (Figure 3.5). Appropriator C diverts 6 cfs, leaving 4 in the stream. The assumption of 50 percent efficiency means 3 cfs returns from C's property to the stream. These 3 cfs of return flow plus the 4 cfs left in the stream gives 7 cfs above the point of diversion for A. Appropriator A diverts 5 cfs, leaving 2 cfs in the stream. This 2 cfs, when added to the 2.5 cfs return flow from A, yields 4.5 cfs at B's point of diversion. This is enough (barely) to satisfy B, the most senior appropriator. Multiple reuse thus allows more water to be appropriated in water rights than the average flow of the stream.

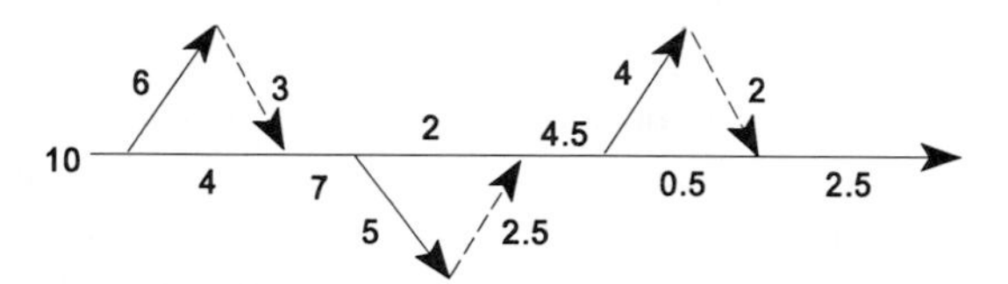

FIGURE 3.5 Diversions and return flows for scenario 1.

Scenario 2: Streamflow above the point of diversion for C is 8 cfs.

In this situation if C tried to take 6 cfs there would be just enough water for A but not for B, and both A and B are senior to C. How much water can C take? Trial and error shows that C can take about 2.5 cfs. This will satisfy A because there is 6.75 cfs in the stream above A, and 4.25 cfs available for B. If you wanted to play with smaller fractions of flow, then C could take a little more than 2.5 cfs, but such small flows cannot be measured reliably.

On a real stream with dozens or maybe hundreds of water rights it is unlikely that every appropriator is going to want water at the same time. This would allow even more water rights to be granted in excess of the average flow.

use, you can only use this water on the same parcel of land. You cannot, for example, take the water and begin irrigating a new piece of land that was not irrigated before. This is because a water right is *appurtenant* to a specific piece of land. A water right is granted for a certain use at a certain location. If through conservation you now have more water than you can use on that property, theoretically you no longer need that water and you lose your right to it, and it should be left in the stream for others to appropriate. As you can see this rule might dampen enthusiasm for water conservation.

Water brought into one basin from a hydrologically separate drainage basin is considered *foreign* or *developed* water. Since this water was never part of the natural waters of the basin, it is not subject to the return flow regulations. An example is the water that Denver brings under the continental divide from the Colorado River basin into the South Platte River basin. Appropriators cannot obtain a water right in the return flow of foreign water. Denver can do anything it pleases with this water.

Transfer and Change in Use

Changing the point of diversion, changing the type of use, or transferring the right somewhere else are all examples of changes that must be approved by the state engineer or administrative agency. The general rule is that a change or transfer may not injure other junior or senior appropriators. These other appropriators are called "third parties" because they are not directly involved in the transfer. This "no injury" rule means the amount of water that can be transferred is limited to the historical consumptive use. For example, if appropriator A in Example 3.1 sold her water right, the buyer would only be allowed to transfer the historical consumptive use of 2.5 cfs, not the 5 cfs on the original decree. The logic is that the return flow is appropriated and belongs to downstream users, and you cannot transfer what you do not own. The buyer is required to leave 2.5 cfs in the stream to keep it "whole." When a buyer goes to purchase a water right the right may say *X* cfs or *X* acre–feet on paper. How much "wet water" the right actually yields when transferred may not be known, and it can be expensive to find out.

Selling and transferring water rights is not new but the issue has taken on new importance in the last few decades. Since the earliest days of prior appropriation if a farmer needed more water but the stream was already fully appropriated he could buy additional water rights. Alternatively, he might be able to lease the rights for a season. Many states rely on the purchase of existing water rights as the mechanism for providing instream flows for recreation and environmental purposes. In this case a state agency purchases rights and leaves the water in the stream.

Nowhere has the issue of purchase and transfer been more active in recent years than between farmers and cities. Expanding cities need more water and are willing and able to pay for it. Farmers have senior water rights and may be willing to sell at the right price. *Water markets,* where willing buyers and sellers exchange water and money, have been proposed as efficient mechanisms for matching supply and demand in the West. Water markets appear to offer an efficient alternative in some situations, and water is moving from cornfields to suburban lawns in California, Arizona, Colorado, New Mexico, and Nevada. But overall, water markets have been more talk than reality because of the formidable barriers to their widespread development (McCormick, 1994b). One barrier is the uncertainty regarding how much water a right will yield upon transfer. This is where good water–use records are important, but good records are surprisingly scarce. Defining the actual quantity of water associated with a water right may take time and money, which increases the market transaction costs.

Another barrier is the latent but emerging issue of the *public trust* or *public interest* in water resources. Recent court decisions have asserted an expanded public interest in otherwise private water rights. The California Supreme Court held in *National Audubon Society v. Superior Court* (1983) that the city of Los Angeles could not divert all the water to which it was legally entitled from Mono Lake because of the potential adverse impacts on the lake's ecosystem. It was deemed in the public interest to maintain the ecosystem. This ruling limited the amount of water from an existing water right and it may have set an important precedent. The poorly defined nature and potentially large extent of the public's interest in water resources adds further uncertainty to potential water markets. Another public interest case is the *Sleeper* decision from northern New Mexico. The case involved an application to transfer water from traditional agricultural use to a proposed ski development. The Hispanic culture in the area dates back hundreds of years and water is considered the "lifeblood" of the community. The State Engineer approved the transfer but it was challenged and reversed by the District Court. The District Court said the "unique cultural heritage" was more important than economic development. The State Court of Appeals reversed the District Court, holding that the laws of New Mexico did not allow the State Engineer to consider the public interest in approving the transfer. The state's water laws were then amended to allow the State Engineer to consider the public interest in future transfer applications (McCormick, 1994b).

Why would a farmer sell his water rights? One reason is the hassle factor of farming near a city. Suburban development is bringing cities closer to the farms. Farmers have to deal with increased traffic when they move machinery between fields. Suburban commuters often seem totally devoid of patience and good manners when they come up behind a slow–moving tractor on a rural road. The farmers new neighbors complain to local officials that farms give off offensive odors and draw flies. Irrigators have to deal with trash thrown into the irrigation ditches. And, of course, property taxes go up as nearby suburban development increases the speculative value of the farmland. Let us do a totally hypothetical but realistic calculation. Take a medium–sized irrigated farm, say 320 acres. At $1000 per acre the farmland is worth $320,000. Assume the farmer irrigates 300 acres and has 900 acre–feet in water rights (3 acre–feet per acre is typical). A realistic price for a 1–acre–foot water right is around $1000. The value of the farmer's water is $900,000, which is almost three times the value of the land! The farmer is nearly a millionaire in water. Now add to this the fact that the farmer is probably approaching late–middle age and the hassle factor mentioned above, and you can see why farmers are lining up to sell their water and retire.

REGULATED RIPARIANISM

Regulated riparianism is a highly regulated system of water administration based on riparian principles (Dellapenna, 1994). Figure 3.1 shows 18 states applying some form of regulated riparianism for surface water (Dellapenna, 1997). Whether this constitutes a sufficiently different system worthy of designation as a new third doctrine in American water law is debatable. The following description of regulated riparians is taken from the draft of the American Society of Civil Engineers (ASCE) *Model Water Code* (Dellapenna, 1994). Not every state has all of the following characteristics, though some come close. The most fundamental difference between the traditional riparian doctrine and regulated riparianism is that, with few exceptions, the regulated riparian code requires a water user to obtain a permit before water is used. The basis of the state's power to require a permit is the protection of the public's health, safety, and welfare. The water right is based on the permit; the permit is conditioned on the reasonableness of the proposed use. While this all sounds similar to pure riparian law, the critical difference is that the assessment of reasonableness occurs *before* any water is used, not after. The requirement that water be used on riparian land is also discontinued, so nonriparian uses are no longer unreasonable per se. Permits are issued for a limited period of time, which allows the regulatory agency to periodically reevaluate the reasonableness of the use in light of changing socioeconomic conditions. During the permit granting and reevaluation procedure important water use data can be gathered for long–term water–resource planning. The durations for permits in regulated riparian states currently range from 3 to 20 years (Dellapenna, 1994). Other aspects addressed by the permit include enforcement provisions and a clear procedure for resolving disputes.

Out of political necessity some states have exempted certain classes of users (usually agriculture) from the permit process if they were using water before the new statutes were passed. This has the effect of vesting these water rights in preference to other rights. The *Model Water Code* takes a different approach. It would guarantee these users a permit the first time around but subject their permit to the same renewal process as all other permits (Dellapenna, 1994).

The summer of 1991 saw drought conditions across most of Pennsylvania and the Susquehanna River basin. In 1993 the Susquehanna River Basin Commission announced a plan to begin registering and charging an annual fee to all large water users. Large users were defined as withdrawing more than 100,000 gallons per day from either a surface– or groundwater source. This information might eventually be used as part of a permit system. The most vocal opposition to the plan came from farmers. One of their contentions was that if they are charged for the water they withdraw, they should be credited for the groundwater recharge occurring through their fields. Farmers assume the net water use on farms is less than the net water use on urban or industrial land occupying the same area. Whether this assumption is true could be tested by doing a careful water balance, but needless to say no research has been done, and the farmers have been given preferential treatment in the latest version of the plan.

STATE LAW: GROUNDWATER

The flow of groundwater is difficult if not impossible to observe, which makes groundwater more difficult to understand and manage than surface water. This is evident in the many state groundwater laws that completely ignore the reality of hydrologic interconnections between groundwater and surface water. While there are certainly aquifers with no discernable connection to surface water, it is more common to have water table aquifers that are in direct hydrologic communication with surface waters. Pumping from these aquifers can reduce streamflow. It may take months or even years for the impact to occur, but it will occur eventually.

States that base their groundwater laws on scientific principles of groundwater hydrology classify groundwater into two categories—*tributary* and *nontributary.* Tributary groundwater is hydrologically connected to surface water; nontributary groundwater is not. The rules and management strategies for the two categories are different. Tributary groundwater is considered part and parcel of the surface water to which it flows and is managed in conjunction with surface waters. Pumping groundwater from a tributary aquifer is considered no different than diverting water from the stream. Wells and ditches are treated as alternate points of diversion from the same hydrologic system.

Nontributary groundwater does not discharge to a surface source. Administration and management of nontributary groundwater are somewhat easier in that the aquifer can be managed as a nonrenewable stock resource. In states that ignore hydrologic principles, all groundwater is treated the same and it is considered to be hydrologically unrelated to surface water. These laws were enacted when the hydrologic characteristics of groundwater were unknown, and they have not changed. These states are now saddled with hydrologically inane laws and they just have to make do.

There are four basic legal doctrines applied to groundwater (Figure 3.6). These four doctrines are mixed in some states, and modified by the use of permits in others. The four doctrines are *absolute ownership, reasonable use, correlative rights,* and *prior appropriation.* Some states apply different doctrines depending upon the nature of the groundwater, that is, whether it is percolating groundwater or an underground stream.

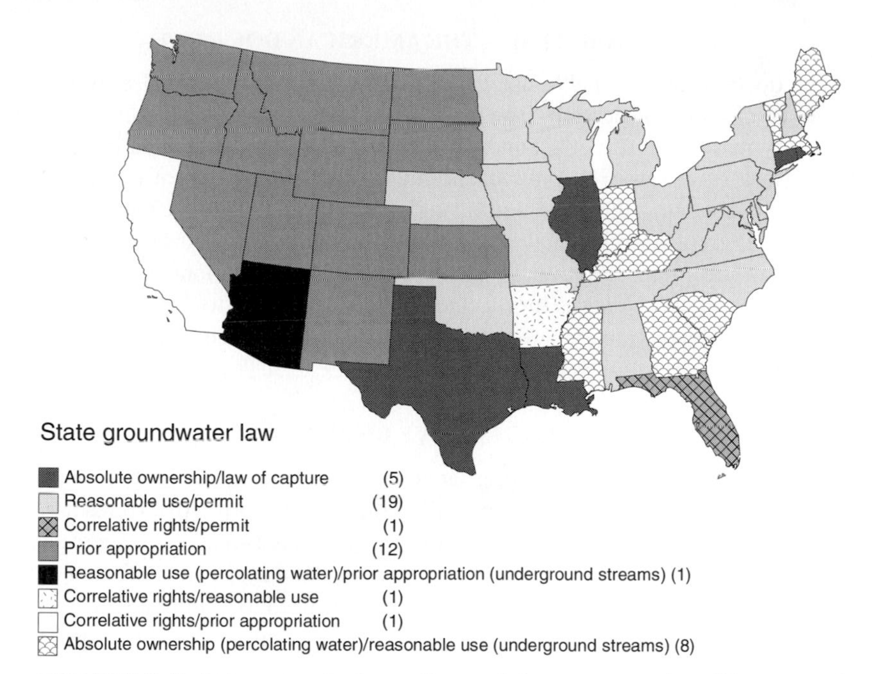

FIGURE 3.6 States categorized according to their groundwater law. (Viesmann and Welty, 1985, based on data from the John Muir Institute, 1980)

ABSOLUTE OWNERSHIP—THE ENGLISH DOCTRINE

Absolute ownership is the doctrine brought to the states from England. This doctrine is devoid of any scientific understanding of groundwater. Absolute ownership means each landowner has complete and absolute right to the water beneath his or her land. Each landowner thus has an unlimited right to pump and use groundwater. The water can be used anywhere and is not restricted to the overlying land. This doctrine means you have an unlimited right to interfere with your neighbor's use of the resource and vice versa. For example, if pumping your well lowers the water table below your neighbor's well and causes his well to go dry, he has no legal recourse to limit your use, or recover damages, so long as you are not being malicious or wasteful. Likewise there is nothing to prevent him from deepening his well and returning the favor. It is ironic that a doctrine that gives absolute ownership does not create an enforceable water right, because there is no legal remedy for injury produced by the actions of others. Absolute ownership is a simple rule of capture. Anyone who pumps the water can have it, and it encourages a race to the bottom of the aquifer.

REASONABLE USE—THE AMERICAN DOCTRINE

The doctrine of reasonable use developed in America in response to the excesses of absolute ownership. The landowner is viewed as having the right to make any reasonable use of groundwater *on the overlying land,* even if it causes injury (in a legal sense) to others. The major difference between this rule and absolute ownership is the constraint on the place of use. Otherwise the doctrine is not very different from absolute ownership. The courts have interpreted reasonable use to mean any traditional water use, and few restrictions have been imposed, even when use causes injury. This is quite different from riparian reasonable use for surface water, where the riparian rights are limited by their impact on other users.

CORRELATIVE RIGHTS

This doctrine says that overlying landowners have coequal rights to make a reasonable use of groundwater from a common source. Correlative rights are the closest thing to a groundwater analogue of the riparian reasonable use doctrine for surface water. The key factor in determining reasonableness is the effect of a use on other users. In California correlative rights were used to require a reduction in groundwater pumping by all landowners in an amount proportional to their ownership of land overlying the aquifer.

PRIOR APPROPRIATION

Prior appropriation for groundwater operates the same as prior appropriation for surface water, and as with surface water it is mainly a western doctrine. It is a seniority–based system, water must be used for a beneficial use, the right can be transferred as long as there is no injury to third parties, and the right can be lost through nonuse. With tributary groundwater, wells are administered within the same seniority system as surface–water rights. The term *conjunctive use* refers to the administration and management of groundwater and surface water together as a single integrated water–resource system.

In most western states, along most streams, surface–water rights were established long before groundwater was used. Many streams were already fully appropriated by the time people began drilling wells into the alluvium adjacent to the channels. It was first thought the wells were tapping a new, different source of water. No one realized at the time that the groundwater was flowing slowly toward and replenishing the streams. After a few decades it was obvious that the wells were "stealing" water from the river—water that was already appropriated. This is exactly what happened along many streams in Colorado and New Mexico. Surface waters were diverted and appropriated during the late 1800s and early 1900s. In the 1940s advancements in well–drilling and pumping technology seemed to have made available a whole new (ground) water resource, and over the next two decades literally thousands of wells were drilled in the alluvium adjacent to streams. By the 1960s the fact that these junior wells were diverting water that belonged to senior surface rights was well established. The only politically feasible thing to do was to grandfather many of these wells into the already established seniority system.

Prior appropriation for nontributary groundwater is easier to administer than tributary groundwater because there is no possible conflict with surface–water rights. There are two possibilities for nontributary groundwater—where recharge occurs and where there is little or no recharge. In the case of recharge the management objective should be to balance the rate of recharge and the rate of withdrawal. Ideally you try to establish dynamic equilibrium between input (recharge) and output (pumping) so that the amount of groundwater in storage stays relatively constant. This may be difficult to do in the real world, because it is not always easy to determine precisely the amount of recharge, and recharge varies with weather conditions, while pumping is more constant.

In situations where there is little or no recharge, the aquifer may be treated as a stock resource. Pumping groundwater in this situation is, by definition, groundwater mining. Assuming that the water is going to be used, the main issue is to decide how long the resource should last. Once the lifetime of the resource is set, the rate of pumping is adjusted to deplete the resource over that period of time. This requires limiting the number of wells (on a seniority basis), and it may also require establishing minimum spacing requirements between wells. Spacing requirements are used to reduce well interference. Well interference is when one or more cone(s) of depression overlap, thus reducing the groundwater flow to the wells involved.

The setting of an arbitrary life for the resource has been dealt with in a very practical manner. In a groundwater basin in eastern New Mexico the regulations allowed two–thirds of the water to be pumped in 40 years. According to the state engineer this was about two generations, and it was also the standard repayment period for a Bureau of Reclamation irrigation–water contract. In a groundwater basin in eastern Colorado, they adopted the rule that 40 percent of the water could be used in 25 years. They also implemented a 3–mile well spacing requirement.

Designating a lifetime for the resource involves a trade–off between current and future users. If the water is pumped today it will not be available for future generations. The argument for saving water for future generations is that they may have a more valuable use for the water, or they might be able to use it more efficiently since their technology will be more advanced. But a counterargument is that by using the water today we will be further along the economic development path, and future generations will benefit from the wealth we create today. Some state statutes have been interpreted as forbidding groundwater mining. The Idaho Groundwater Act, I.C. § 42–237a (g), states,

> Water in a well shall not be deemed available to fill a water right therein if withdrawal therefrom of the amount called for by such a right would affect, contrary to the declared policy of this act, the present or future use of any prior surface or groundwater right or result in the withdrawing the groundwater supply at a rate beyond the reasonably anticipated average rate of future natural recharge.

The Idaho Supreme Court in *Baker v. Ore–Ida Foods, Inc.* (1973) interpreted this statute as expressly forbidding groundwater mining. The court deemed that in doing so they were promoting the "full economic development" of the state's groundwater resources. In states that do allow groundwater mining, presumably they too feel it promotes the full economic development of the resource. In situations where there is essentially no recharge, limiting the amount of pumping to the recharge rate means no groundwater can be used.

FEDERAL LAW

The federal government derives its power and authority from the U.S. Constitution. The powers of the federal government are limited to those expressly delegated to it and those that may reasonably be implied. All other powers are reserved to the states and to the people. This is the essence of our federal system of government. It is a power–sharing arrangement between the federal, state, and local governments. One of the most difficult challenges to water planning, management, and administration is effective cooperation and coordination between the different levels of government.

FEDERAL POWERS RELATED TO WATER RESOURCES

The federal government has invoked a variety of powers in the arena of water–resource management. The most important have been the power to control interstate commerce; the proprietary power, which derives from the property clause of the Constitution and gives the government the power to control the use of resources on the public domain; and the government's power to tax and spend for the general welfare. The war power was important in justifying federal water–resource activities in the early years of the republic. Concern over national defense and our ability to move troops justified federal expenditures on navigation and harbor improvements in the early 1800s. The National Defense Act of 1916 authorized the Corps of Engineers to build and operate Wilson Dam on the Tennessee River. The dam and associated hydroelectric facilities were deemed in the interest of national defense because the electricity was to be used in a munitions plant. Often when the federal government makes decisions that affect water resources more than one federal power justifies its actions.

FEDERAL POWERS RELATED TO WATER RESOURCES

1. Commerce power
2. Proprietary power
3. General welfare power
4. Treaty-making power
5. Judicial power
6. Compact power
7. War power

Commerce Power

The Constitution gives the federal government the right to regulate commerce between the several states. The commerce power has been used since the early 1800s when the federal government first began to take an interest in and responsibility for improving internal navigation. The justification was that improved interstate commerce promoted national economic development and national defense. This power evolved into a navigation servitude where the federal government maintains supreme control over all *navigable waters* in the country. Originally, navigable waters were waters that were "navigable in fact." Navigable in fact meant they could float a boat, logs, or whatever. Later, navigable was interpreted to mean any water that could be made navigable by reasonable improvements. The government has spent billions of dollars building canals, dredging rivers, and improving harbors all justified under the commerce clause. As we saw in Chapter 2 the earliest federal water quality law was the Refuse Act of 1899, which derived its authority from the commerce clause and was intended to protect navigation. The Refuse Act restricted the discharge of any material into navigable waters that might pose a hazard to navigation. The Refuse Act was the authority for federal water quality legislation half a century later.

More recently the U.S. Supreme Court asserted a federal interest in state groundwater management via the commerce clause in *Sporhase v. Nebraska ex rel. Douglas* (1983). Mr. Sporhase owned land in both Nebraska and Colorado. He wanted to pump groundwater from his Nebraska land to irrigate crops on his Colorado land. Nebraska statutes permitted interstate transport of groundwater, but only to states that allowed reciprocal transport of groundwater to Nebraska. Colorado statutes forbade the transport of groundwater across the state line, so Nebraska officials refused to grant Sporhase a permit. The case was eventually appealed to the U.S. Supreme Court. The Supreme Court held groundwater to be an article of commerce, plain and simple, and state statutes prohibiting the interstate transport of groundwater were illegal because they violated the commerce clause. The commerce clause is intended to prevent regional protectionism and states cannot restrict interstate commerce without a compelling reason. Prior to *Sporhase* it was assumed that states had absolute ownership and control of their groundwater. In its decision the Court called state ownership of groundwater "legal fiction." As you might imagine this sent a minor shock wave resonating through the legal landscape. And it is one more example of the continuing controversy over federal versus state ownership of water resources.

Proprietary Power

About one–third of the United States, some 755 million acres of land, is public domain. When new states were created they were carved from the public domain. The federal government owns the public domain and all the resources found on it. The proprietary power gave the government authority to give away or sell land to settlers. For example, revenue from the sale of public land was originally conceived as the method for financing federal irrigation projects. The proprietary power is also the authority for federal reserved water rights. This is an extremely important topic and is discussed below.

General Welfare Power

The federal government has the power to tax and spend for the general welfare. This power supplements many of the other powers. Federal spending for irrigation projects in the West can also be justified by this power, as can spending for navigational improvements and flood control.

Treaty–Making Power

Only the federal government can make treaties with foreign powers. We have treaties with both Mexico and Canada dealing with the allocation, use, and management of international waters.

Judicial Power

The U.S. Supreme Court is the court of original jurisdiction for interstate disputes. There have been dozens of lawsuits between states over their respective rights to water resources. These interstate suits are decided by the Supreme Court.

Compact Power

The states may only enter into interstate compacts with the consent of Congress. Interstate compacts have been used in both the East and the West to resolve disputes ranging from water pollution to the division of the waters of an interstate stream. Interstate compacts are discussed later.

FEDERAL RESERVED RIGHTS

The federal government owns the public domain and all the natural resources found thereupon, including the water. Starting in the second half of the 19th century the federal government began reserving parts of the public domain. Reserved lands were no longer open to homesteading so people could not gain title to these lands. The federal government reserved land for national forests, national parks, military installations, and most importantly in terms of water resources, as homes for the Indian tribes. The federal reserved water right is based on the fact that when the federal government reserved the land it *implicitly* reserved some water along with it. How much water is the big, very big, question. The qualitative criterion is that enough water was reserved to accomplish the purpose of the reservation. The precedent–setting case for federal reserved water rights is *Winters v. United States* (1908).

WINTERS v. UNITED STATES SUPREME COURT OF THE UNITED STATES, 1908.

207 U.S. 564, 28 S.CT. 207, 52 L.ED. 340.

This suit was brought by the United States to restrain appellants and others from constructing or maintaining dams or reservoirs on the Milk River in the State of Montana, or in any manner preventing the water of the river or its tributaries from flowing to the Fort Belknap Indian Reservation.

The allegations of the bill, so far as necessary to state them, are as follows: On the first day of May, 1888, a tract of land, the property of the United States, was reserved and set apart "as an Indian reservation as and for a permanent home and abiding place of the Gros Ventre and Assiniboine bands or tribes of Indians in the State (then Territory) of Montana, designated and known as the Fort Belknap Indian Reservation." The tract has ever since been used as an Indian Reservation and as the home and abiding place of the Indians. * * *

Milk River, designated as the northern boundary of the reservation, is a non-navigable stream. Large portions of the lands embraced within the reservation are well fitted and adapted for pasturage and the feeding and grazing of stock, and since the establishment of the reservation the United States and the Indians have had and have large herds of cattle and large numbers of horses grazing upon the land within the reservation, "being and situate along and bordering upon said Milk River." Other portions of the reservation are "adapted for and susceptible of farming and cultivation and the pursuit of agriculture, and productive in the raising thereon of grass, grain and vegetables," but such portions are of dry and arid character, and in order to make them productive require large quantities of water for the purpose of irrigating them. * * * It is alleged with detail that all of the waters of the river are necessary for all those purposes and the purposes for which the reservation was created, and that in furthering and advancing the civilization and improvement of the Indians, and to encourage habits of industry and thrift among them, it is essential and necessary that all of the waters of the river flow down the channel uninterruptedly and undiminished in quantity and undeteriorated in quality.

It is alleged that "notwithstanding the riparian and other rights" of the United States and the Indians to the uninterrupted flow of the waters of the river the defendants, in the year 1900, wrongfully entered upon the river and its tributaries above the points of the diversion of the waters of the river by the United States and the Indians, built large and substantial dams and reservoirs, and by means of canals and ditches and waterways have diverted the waters of the river from its channel, and have deprived the United States and the Indians of the use thereof. And this diversion of the water, it is alleged, has continued until the present time, to the irreparable injury of the United States, for which there is no adequate remedy at law. The allegations of the answer, so far as material to the present controversy, are as follows: * * *

That the individual defendants and the stockholders of the Matheson Ditch Company and Cook's Irrigation Company were qualified to become settlers upon the public land and to acquire title thereto under the homestead and desert land laws of the United States. And that said corporations were recognized and exist under the laws of Montana for the purpose of supplying to their said stockholders the water of Milk River and its tributaries, to be used by them in the irrigation of their lands.

That for the purpose of reclaiming the lands, and acting under the laws of the United States and the laws of Montana, the defendants, respectively, posted upon the river and its tributaries, at the points of intended diversion, notices of appropriation, stating the means of diversion and place of use, and thereafter filed in the office of the clerk and recorder of the county wherein the lands were situated a copy of the notices, duly verified, and within forty days thereafter commenced the construction of ditches and other instrumentalities, and completed them with diligence and diverted, appropriated, and applied to a beneficial use more than 5,000 miners' inches of the waters of the river and its tributaries, or 120 cubic feet per second, irrigating their lands and producing hay, grain and other crops thereon. * * *

MR. JUSTICE McKENNA, after making the foregoing statement, delivered the opinion of the court.

The case, as we view it, turns on the agreement of May, 1888, resulting in the creation of Fort Belknap Reservation. In the construction of this agreement there are certain elements to be considered that are prominent and significant. The reservation was a part of a very much larger tract which the Indians had the right to occupy and use, and which was adequate for the habits and wants of a nomadic and uncivilized people. It was the policy of the Government, it was the desire of the Indians, to change those habits and to become a pastoral and civilized people. If they should become such the original tract was too extensive, but a smaller tract would be inadequate without a change of conditions. The lands were arid and, without irrigation, were practically valueless. And yet, it is contended, the means of irrigation were deliberately given up by the Indians and deliberately accepted by the Government. The lands ceded were, it is true, also arid; and some argument may be urged, and is urged, that with their cession there was the cession of the waters, without which they would be valueless, and "civilized communities could not be established thereon." And this, it is further contended, the Indians knew, and yet made no reservation of the waters. We realize that there is a conflict of implications, but that which makes for the retention of the waters is of greater force than that which makes for their cession. The Indians had command of the lands and the waters—command of all their beneficial use, whether kept for hunting, "and grazing roving herds of stock," or turned to agriculture and the arts of civilization. Did they give up all this? Did they reduce the area of their occupation and give up the waters which made it valuable or adequate? And, even regarding the allegation of the answer as true, that there are springs and streams on the reservation flowing about 2,900 inches of water, the inquiries are pertinent. If it were possible to believe affirmative answers, we might also believe that the Indians were awed by the power of the Government or deceived by its negotiators. Neither view is possible. The Government is asserting the right of the Indians. But extremes need not be taken into account. By a rule of interpretation of agreements and treaties with the Indians, ambiguities occurring will be resolved from the standpoint of the Indians. And the rule should certainly be applied to determine between two inferences, one of which would support the purpose of the agreement and the other impair or defeat it. On account of their relations to the Government, it cannot be supposed that the Indians were alert to exclude by formal words every inference which might militate against or defeat the declared purpose of themselves and the Government, even if it could be supposed that they had the intelligence to foresee the "double sense" which might some time be urged against them.

Another contention of appellants is that if it be conceded that there was a reservation of the waters of Milk River by the agreement of 1888, yet the reservation was repealed by the admission of Montana into the Union, February 22, 1889, c. 180, 25 Stat. 676, "upon an equal footing with the original States." The language of counsel is that "any reservation in the agreement with the Indians, expressed or implied, whereby the waters of Milk River were not to be subject of appropriation by the citizens and inhabitants of said State, was repealed by the act of admission." But to establish the repeal counsel rely substantially upon the same argument that they advance against the intention of the agreement to reserve the waters. The power of the Government to reserve the waters and exempt them from appropriation under the state laws is not denied, and could not be. The *United States v. The Rio Grande Ditch & Irrigation Co.*, 174 U.S. 690, 702 (1899); *United States v. Winans,* 198 U.S. 371 [1905]. That the Government did reserve them we have decided, and for a use which would be necessarily continued through years. This was done May 1, 1888, and it would be extreme to believe that within a year Congress destroyed the reservation and took from the Indians the consideration of their grant, leaving them a barren waste—took from them the means of continuing their old habits, yet did not leave them the power to change to new ones.

Appellants' argument upon the incidental repeal of the agreement by the admission of Montana into the Union and the power over the waters of Milk River which the State thereby acquired to dispose of them under its laws, is elaborate and able, but our construction of the agreement and its effect make it unnecessary to answer the argument in detail. For the same reason we have not discussed the doctrine of riparian rights urged by the Government.

Decree affirmed.

When the federal government created the Fort Belknap Reservation it implicitly reserved enough water to accomplish the purpose of the reservation. In this case the purpose was to convert the nomadic "uncivilized" Indians into sedentary farmers. The only way to succeed at farming in this part of the country is with irrigation. The court determined the reserved right was implicit, since there was no mention of reserving water when the Fort Belknap Reservation was established. Montana is a prior appropriation state and the federal government recognized it as such. The court held that the priority date of the reserved right was the date the reservation was established, not the date the water was used. This is a federal version of the relation back principle. Another difference between reserve rights and state–granted rights is that reserve rights cannot be abandoned or lost by nonuse.

When *Winters* was decided at the turn of the century no one paid much attention. By the 1960s everyone was paying attention. The *Winters* decision threatened to drastically reorder the existing priority system on many western streams. Indian reservations, most of which were established in the late 1800s, had effectively moved to the front of the seniority line, with rights to potentially large amounts of water. The decision that made established water users stand up and take notice was *Arizona v. California* (1963). The states of Arizona and California have been fighting over the waters of the Colorado River for most of this century. There have been four separate lawsuits, and the 1963 Supreme Court decision was the third in the series. The 1963 decision dealt with many different aspects of the Colorado River, one of which was the reserved rights of the Indians in the basin. Citing the purpose of the reservations to as homes for the Indians the Court recommended that the standard for quantifying Indian reserved rights be the *practicably irrigable acreage* (PIA). The court allocated 6.63 acre–feet per acre to 136,636 acres on five Indian reservations (Chemchuevi, Cocopah, Yuma, Colorado River, and Fort Mojave). This totals 905,897 acre–feet of water per year (Black, 1987). The annual average flow of the Colorado River is around 13.5 million acre–feet per year. The Navajo Reservation was not included in the decision and its potential reserved water right is huge by comparison. Some estimate that the Navajo's water allocation using the PIA standard could be anywhere from 2 to 5 million acre–feet of water per year. The combined Indian claims could potentially equal one–third of the Colorado River's annual average discharge. When you consider that the Colorado River is already overappropriated by users in Los Angeles, San Diego, and Denver, and farmers both in and out of the basin, the potential conflict is enormous. Table 3.1 shows the proportion of Indian lands in 19 western states.

In the latest suit between Arizona and California (*Arizona v. California,* 1983) the Supreme Court qualified the PIA standard. Irrigable land now must be economically irrigable, not just practicably irrigable. This means a cost–benefit analysis must be done to determine whether the benefits from irrigating a piece of land exceed the costs of bringing the water to the land. Undoubtedly for many thousands of acres the costs will outweigh the benefits.

Is the PIA a good criterion for quantifying Indian reserve rights? The answer is probably yes if the reservation has a lot of irrigable land. In *Wyoming v. United States* (1989) the Shosone and Arapaho tribes on the Wind River Reservation received a right to more than 477,292 acre–feet with a high priority. In the same year the Mescalero Apaches in New Mexico (*State of New Mexico ex rel. Reynolds v. Lewis,* 1989) received virtually no water. The court rejected their PIA claim because most of their land was not economically irrigable. Should there be a dif-

TABLE 3.1 Indian Lands in the Western States, 1996[a]

State	Land area[b]	Indian lands[b]	Percent of state in Indian land
Alaska	365,039	1,140	0.3
Arizona	72,730	20,718	28.5
California	99,823	592	0.6
Colorado	66,386	800	1.2
Hawaii	4,111	0	0.0
Idaho	52,961	754	1.4
Kansas	52,367	34	0.1
Montana	93,156	5,479	5.9
Nebraska	49,202	66	0.1
Nevada	70,276	1,232	1.8
New Mexico	77,673	8,349	10.8
North Dakota	44,156	867	2.0
Oklahoma	43,955	1,062	2.4
Oregon	61,441	797	1.3
South Dakota	48,573	5,002	10.3
Texas	167,625	525	0.3
Utah	52,587	2,331	4.4
Washington	42,612	2,602	6.1
Wyoming	62,147	1,889	3.0
Subtotal	1,526,820	54,239	3.55
Rest of the U.S.	736,398	3,046	0.41
Total	2,263,218	54,239	2.40

[a] U.S. Bureau of the Census, 1990; U.S. Bureau of Indian Affairs, 1996)
[b] In thousands of acres.

ferent standard, say, one based on population size or the land's potential for industrial or retail development? Not all Indian reservations are susceptible to irrigation and not all Indians want to be farmers. Generally, non–Indians dislike the PIA standard because it potentially represents such a large amount of water.

NONRESERVATION INDIAN RESERVED RIGHTS

Some Indians do not currently live, and never have lived, on reservations. Technically, therefore, they do not have a federal reserved right. The Pueblo Indians along the Rio Grande in New Mexico are an example. The pueblos (Indian villages) are quite old, dating back at least 700 years—a wee bit longer than the federal government in Washington. The Pueblo Indians have always used water for domestic uses and irrigation, and they reject any need to establish their rights based on federal reserve rights. The Pueblos claim aboriginal rights to water from a time immemorial. By in large their rights are now recognized as *prior and paramount* to all other rights, and they can never be lost by nonuse. But the question is again, how much water do the Indians own with these prior and paramount rights? In the *Aamodt* decision (*State of New Mexico ex rel. Reynolds v. Aamodt,* 1985) the court based the aboriginal rights on the amount of water used since 1924, the date of the Pueblo Lands Act. That act compensated the Pueblos for land conveyed to non–Indians, and according to the federal judge, "fixed the measure of Pueblo water rights to acreage irrigated as of that date." This decision will likely be appealed and should not be regarded as a final determination of the Pueblo water right.

Defining and quantifying the water rights of all appropriators in a single drainage basin is called a *general adjudication.* The time and effort involved in a general adjudication can be enormous. As an example, the general adjudication of the relatively small Nambe (*nam–bay*) and Pojoaque (*po–wa–kee*) rivers in Northern New Mexico (which are about 40 miles in length, in a wet year) was begun in 1966 and was still in progress in the mid 1990s (*State of New Mexico ex rel. Reynolds v. Aamodt,* 1985).

OTHER FEDERAL RESERVED RIGHTS

The federal government has reserved land for national forests, national parks, and other purposes. What is the extent of the federal reserved rights for these reservations? The answer came in *United States v. New Mexico* (1978). This is referred to as the "Mimbres" case since it dealt with waters of the Rio Mimbres in the Gila National Forest in southwest New Mexico. The United States claimed a reserved right in the Gila National Forest for various conservation and environmental purposes, including aesthetic values, recreational uses, and instream flows for fish. The Supreme Court disagreed. In a narrow reading of the Organic Act (1897), which authorizes the reservation of land as national forests, the Court held that Congress intended to reserve national forests for two and only two purposes: "to conserve the water flows and to furnish a continuous supply of timber for the people." Those are the only purposes for which these reservations are made, period. National forests are not reserved for aesthetic, environmental, recreational, or wildlife preservation, and never mind that they are used for those purposes today. The Court denied the existence of reserved water rights for these expanded purposes. The decision was hailed by states–rights advocates and other resource–development interests because it restricted the federal right to only the purposes for which the land was originally reserved.

OTHER FEDERAL INTERESTS THAT AFFECT WATER RESOURCES

Many other federal statutes and activities either directly or indirectly affect water resources. The Wild and Scenic Rivers Act (1968) is intended to preserve streams in their pristine condition. The Endangered Species Act (1973) is potentially very important in affecting water resources. The federal government has also passed important environmental legislation dealing with water quality, drinking water standards, and the handling of toxic and hazardous substances. We will explore these issues in later chapters.

TRANSBOUNDARY WATER RESOURCES

The geographically arbitrary nature of political boundaries means most major rivers and many lakes and aquifers are *transboundary resources* contained in more than one political jurisdiction. In terms of water management in the United States the *interstate* character is probably the most significant, though on certain rivers the *international* aspect is clearly significant. The Rio Grande and the Colorado Rivers are both interstate and international rivers since they are shared between Mexico and the United States. These are the only two rivers in the world that are shared between a more developed country, a G7 nation no less, and a less–developed country. Conflicts over water quantity and quality on these rivers have been settled by treaty between the two countries and by interstate compacts between the several states.

The interstate nature of most rivers in the United States has both pros and cons. On the one hand it causes conflict because each state attempts to use the resource to promote its own best interest, but on the other hand it prevents one state from monopolizing the resource (Black, 1987). On interstate waters the federal government is, by definition, involved in resolving disputes between the states. The three ways the federal government resolves interstate disputes is through Supreme Court decisions, allowing states to negotiate interstate compacts, and by legislation.

JUDICIAL DECISIONS

The Supreme Court is the court of original jurisdiction in disputes between the states. In hearing a case the Supreme Court does not take evidence. The Court appoints a Special Master who investigates the claims of the states, collects data, and makes recommendations. The Court may accept, reject, or modify the Special Master's recommendations. In deciding disputes over water quantity the Court uses the principle of *equitable apportionment.* Equitable apportionment does not mean the waters are divided equally. The Court may consider the prevailing state water laws (riparian or prior appropriation), but other factors may be considered as well. In a dispute between Colorado and New Mexico, both prior–appropriation states, over the Vermejo River the Court did not follow strict priority in equitably apportioning the water (*Colorado v. New Mexico,* 1982 and 1984). The Court considered the relative benefits and costs involved, the pattern of existing uses, the physical nature of the river itself, and the potential for water storage (Matthews, 1994). In a case involving two riparian states (*Connecticut v. Massachusetts,* 1931) the Court stated "the laws in respect of riparian rights that happen to be effective for the time being in both states do not necessarily constitute a dependable guide or just basis for the decision of controversies such as are here presented" (Sherk, 1994.) The Court prefers that states resolve their disputes by interstate compact.

INTERSTATE COMPACTS

Interstate compacts have been created for four purposes: to allocate the waters of interstate streams, pollution control, flood planning and control, and for comprehensive regulatory and development purposes (Black, 1987). Table 3.2 lists the water–related interstate compacts in the United States. In addition to managing the water, the compacts create various administrative bodies and may include procedures for decision making, enforcement, and resolving disputes.

Water Allocation Compacts

Compacts dividing water between states are the most common and are found mainly in the West. Currently there are 22 water allocation compacts on western streams (Table 3.2). Some compacts allocate the waters of small streams that are of consequence only to local water users, while others divide the waters of major rivers affecting millions of people living hundreds of miles from the river. Water allocation compacts divide the waters using either storage allocation or flow allocation procedures (McCormick, 1994a). This is similar to the two basic types of water rights in western states. Storage allocation allows the upper basin state(s) to store a certain amount of water for later use. This type of allocation is easy to monitor and enforce because reservoir storage is open and visible. Flow allocation is more complex. States have used hydrologic models, percentage of total flow, and guaranteed quantities of flow as methods for dividing the waters. Each method has limitations in terms of enforcement, and they differ in how the risk of shortage is shared between the states. In a careful analysis of western water allocation compacts McCormick (1994a) concluded that using a percentage of flow was probably the best overall allocation method. In the two cases where hydrologic models are used they have proven to be inaccurate over time. The method that guarantees an absolute quantity of water to downstream states can be extremely burdensome on upstream states during times of drought. Interstate compacts are intended to resolve disputes; however, the compacts themselves may engender conflict. Common arguments are that one state is not living up to the terms of the compact, or the language of the compact is interpreted differently by each state with the passage of time.

Pollution Control Compacts

Before passage of federal water quality legislation, interstate water quality disputes were resolved in court through nuisance lawsuits or by interstate compact. There are five interstate pollution control compacts between eastern states (Table 3.2). While the purposes of pollution control compacts have largely been superseded by the federal Clean Water Act (1972), they still provide useful examples of how to coordinate planning and management between states on a basin-wide scale.

Planning/Flood Control Compacts

There are planning/flood control compacts on five eastern rivers—the Wabash River, Wheeling Creek, Merrimack River, Thames River, and Connecticut River (Sherk, 1994). All but the Wheeling Creek compact give the compact commissions authority to undertake studies and make plans for flood protection works.

Comprehensive Regulatory and Development Compacts

Two compacts fall into this category, the Delaware River Basin Compact and the Susquehanna River Basin Compact. One of the earliest interstate conflicts on the Delaware River involved water supply and the proposed interbasin diversion of water from the Catskill Mountains to New York City. First New Jersey initiated an equitable apportionment suit against New York to prevent the diversion. New Jersey lost as the Court reaffirmed the Special Master's determination that the diversion would not materially injure New Jersey (*New Jersey v. New York,* 1931). Ongoing concern about water supply, pollution, flooding, and overlapping administrative authority in the basin finally prompted the states of New Jersey, New York, Pennsylvania, and Delaware to enter into and ratify the Delaware River Basin Compact in 1961. By making the United States signatory to the compact, and by giving the compact commission authority to allocate water and approve any project having "substantial effect" on the basin, the compact is a "prime example of interstate cooperation and commitment to dynamic regional water resource management" (Weston, 1984).

The concerns that prompted the Susquehanna River Basin Compact included water quality due primarily to acid mine drainage, increasing interest in recreational and environmental uses of the river, flood control, and the multiplicity of governmental entities exercising authority in the basin (Sherk, 1994). The compact was modeled on the Delaware River Basin Compact and ratified by the three states of New York, Pennsylvania, and Maryland and the federal government in 1970. The authority of the Susquehanna River Basin Commission (SRBC) on water matters within the basin exceeds the authority of the individual states. As mentioned earlier in the section on riparian water rights, the SRBC recently initiated registration of large water users. The SRBC also requires large water users to supply compensatory water to the river during droughts to make up for their consumptive use.

Interstate compacts and agreements have been created to manage water resources other than interstate rivers. The Great Lakes Basin Compact was created primarily to address issues of comprehensive planning and flood control. The compact is not very effective since it gives the compact commission only an advisory function. The Chesapeake Bay Agreement was first developed and signed between Maryland and Virginia in 1983. The agreement was prompted by increasing pollution and eutrophication of the bay. Pennsylvania signed the agreement in 1985, which was significant because Pennsylvania has no shoreline on the bay.

TABLE 3.2 Interstate Compacts in the United States[a]

WATER ALLOCATION COMPACTS

River	Year	States	Method of allocation	Dispute resolution
Animas-La Plata	1968	CO, NM	None	None
Arkansas	1949	CO, KS	Limit on maximum draws from John Martin Reservoir	Arbitration, if both states consent
Arkansas	1965	KS, OK	Storage limitations	None
Arkansas	1970	AR, OK	Percentage of flow	Arbitration (not compulsory)
Belle Fourche	1943	SD, WY	Percentage of flow for postcompact uses	None
Bear	1955, 1978	ID, UT, WY	Percentage of flow; some storage limits	6 of 9 votes (each state has 3 votes)
Big Blue	1971	KS, NE	Postcompact use restricted in NE to provide minimum flow to KS	None
Canadian	1950	OK, NM, TX	Storage limitations	None
Colorado	1922	AZ, CA, CO, NM, NV, UT, WY	Fixed quantity from Upper Basin to Lower Basin	Ad hoc committee may be appointed by states; actions subject to ratification by all states
Costilla Creek	1944, 1963	CO, NM	Detailed priority of individual water rights	None
Klamath	1956	CA, OR	Priority of type of use for postcompact uses	Mandatory arbitration
La Plata	1922	CO, NM	Equal division of flow if state line discharge < 100 cfs	None
Pecos	1949	NM, TX	Division of flow based on hydrologic model	None
Red	1978	AR, LA, OK, TX	Divided into 5 reaches; percentage allocation within each reach	6 of 8 votes on some matters (each state has 2 votes)
Republican	1943	CO, KS, NE	Specific quantities within subbasins assigned to each state	None

Rio Grande	1938	CO, NM, TX	Divides flow based on model	None
Sabine	1953, 1962, 1977, 1992	LA, TX	Flow in interstate reach divided	3 of 4 votes (each state has 2)
Snake	1949	ID, WY	Percentage of flow for postcompact uses	Casting vote (federal representative)
South Platte	1923	CO, NE	Fixed quantity for NE, but limits on actions CO must take	None
Upper Colorado	1948	CO, NM, UT, WY	Percentage of flow; AZ gets 50,000 acre-feet per year	4 of 5 votes (federal representative has 1 vote)
Upper Niobrara	1962	NE, WY	Storage limitations; interstate priorities	None
Yellowstone	1950	MT, ND, WY	Percentage of flow for postcompact uses	Casting vote (for disputes between WY and MT)

POLLUTION CONTROL COMPACTS

Water body or region	Year	States	Compact name and purpose
New York Harbor	1935	NJ, NY	The New York Harbor Interstate Sanitation Compact. Established the Interstate Sanitation Commission which classifies and designates different waters for specific purposes. States agree to restrict the release of contaminants into different waters.
Ohio	1940	NY, IL, KY, IN, OH, WV	The Ohio River Valley Water Sanitation Compact. Created the Ohio River Valley Water Sanitation District and Commission. The commission makes reports and recommendations on pollution control for different classifications of water. Commission may order pollution abatement under certain circumstances.
Potomac	1970	D.C., MD, PA, VA, WV	The Potomac River Basin Compact. Created the Interstate Commission on the Potomac River. Commission is authorized to promote uniform laws and regulations and to recommend minimum water treatment standards.

continued

TABLE 3.2 *Continued*

POLLUTION CONTROL COMPACTS

Water body or region	Year	States	Compact name and purpose
New England	1947	CT, MA, RI	The New England Interstate Water Pollution Control Compact. Established the New England Interstate Water Pollution Control Commission. The commission was authorized to establish standards and classify waters for pollution control.
Tennessee	1958	No states have ratified the compact	The Tennessee River Basin Water Pollution Control Compact. Established the Tennessee River Basin Water Pollution Control Commission. Commission authorized to draft and recommend pollution control legislation. Commission may order pollution abatement under certain circumstances.

PLANNING/FLOOD CONTROL

River	Year	States	Compact name and purpose
Wabash	1959	IN, IL	The Wabash Valley Compact. Created the Wabash Valley Interstate Commission. Commission recommends plans and studies, supports research, and publishes reports on development and flood control in the Wabash Valley.
Wheeling	1967	PA, WV	The Wheeling Creek Watershed Protection and Food Prevention District Compact. Created the Wheeling Creek Watershed Protection and Food Prevention District. District can contract with the federal government for flood control and acquire property.
Merrimack	1957	MA, NH	The Merrimack River Flood Control Compact. Created the Merrimack River Valley Flood Control Commission. Commission authorized to conduct flood control studies and coordinate activities with the U.S. Army Corps of Engineers. Compensation between states for interstate benefits and costs associated with flood control.
Thames	1958	CT, MA	The Thames River Flood Control Compact. Created the Thames River Valley Flood Control Commission. Commission authorized to conduct flood control studies and coordinate activities with the U.S. Army Corps of Engineers. Compensation between states for interstate benefits and costs associated with flood control.

Connecticut	1953	CT, MA, NH, VT	The Connecticut River Flood Control Compact. Created the Connecticut River Valley Flood Control Commission. Commission authorized to conduct flood control studies and coordinate activities with the U.S. Army Corps of Engineers. Compensation between states for interstate benefits and costs associated with flood control.

COMPREHENSIVE REGULATORY AND DEVELOPMENT COMPACTS

River	Year	States	Compact name and purpose
Delaware	1961	DE, NJ, NY, PA	The Delaware River Basin Compact. Created the Delaware River Basin Commission. Commission is given broad authority to approve plans and to manage and allocate waters of the basin.
Susquehanna	1970	MD, PA, NY	The Susquehanna River Basin Compact. Created the Susquehanna River Basin Commission. Powers and duties similar to the Delaware River Basin Commission but with more attention to floodplain management.

OTHER INTERSTATE AGREEMENTS

Water body	Year	States	Name and purpose
Chesapeake Bay	1980	MD, VA, (PA)	Created the Chesapeake Bay Commission. PA became a member in 1985.
	1983	D.C., MD, PA	Chesapeake Bay Agreement
	1987	D.C., MD, PA	Chesapeake Bay Agreement (expanded and refined). Under the agreement the parties agree to manage the bay as an integrated ecosystem and are committed to a comprehensive program to improve water quality and protect wildlife habitat. Parties agree to reduce point and nonpoint source pollution, specifically a 40 percent reduction in nutrient loading by the year 2000.
Great Lakes	1968	IL, IN, MI, MN, NY, OH, PA, WI	The Great Lakes Basin Compact. Established the Great Lakes Commission. Commission authorized to collect and report information, and to make recommendations for basin improvement, policy, and uniform laws and policies. Commission is only advisory.

[a] Sources: McCormick, 1994a by permission of the American Water Resources Association; Sherk, 1994; and Black, 1987.

Land use in Pennsylvania directly affects the bay through point and nonpoint source pollution of the Susquehanna River, the largest freshwater tributary to the bay. The agreement was expanded in 1987 and the District of Columbia and the U.S. Environmental Protection Agency became signatories. One of the goals of the agreement is to reduce nutrient loading to the bay by 40 percent by the year 2000. For its part Pennsylvania has passed nutrient management regulations to control nutrient pollution from farms. New modeling studies have found that there is more nutrient pollution to the bay than originally thought, and the goal of a 40 percent reduction may be unattainable.

LEGISLATION

The third method the federal government has used to apportion interstate rivers is legislation. The first time this happened Congress probably did not even realize it had done so. The Boulder Canyon Project Act (1928) gave the Secretary of the Interior the authority to make water contracts with the states in the Colorado River basin. In the act Congress included what it considered to be a fair division of the Colorado River among the states of California, Arizona, and Nevada. Arizona sued California in 1952 over rights to the river. In announcing it decision 11 years latter in *Arizona v. California* (1963) the Supreme Court held that the Boulder Canyon Project Act had allocated the waters of the Colorado River back in 1928. Congress had divided the waters of an interstate stream but did not know it until the Supreme Court interpreted the legislation 35 years later. Significantly, the Court rejected equal apportionment in favor of the legislative division. More recently Congress allocated the waters of the Carson and Truckee Rivers between California and Nevada based on existing decrees (Sherk, 1994).

The tangle of state and federal cases, legislation, and regulations that constitute water quantity law provide a fascinating glimpse into how we allocate water and resolve conflicts, and how our priorities for water and the environment have changed over time. No river in the United States demonstrates conflict over water better than the Colorado River. The short case study that follows touches on some of the more significant events in the evolution of management of the Colorado River.

CASE STUDY: THE COLORADO RIVER BASIN

The Colorado River Basin encompasses 244,000 square miles in parts of seven states (Colorado, Wyoming, Utah, New Mexico, Nevada, Arizona, and California) (Figure 3.7). From its headwaters in the Colorado Rockies to its mouth at the Gulf of California the river flows 1440 miles. The Colorado no longer reaches the Gulf as it is completely consumed by upstream uses. The climate of the basin ranges from permanent snow and ice at elevations above 14,000 feet in the Rocky Mountains to the driest desert found in North America. Most of the Colorado Basin has a semiarid to arid climate.

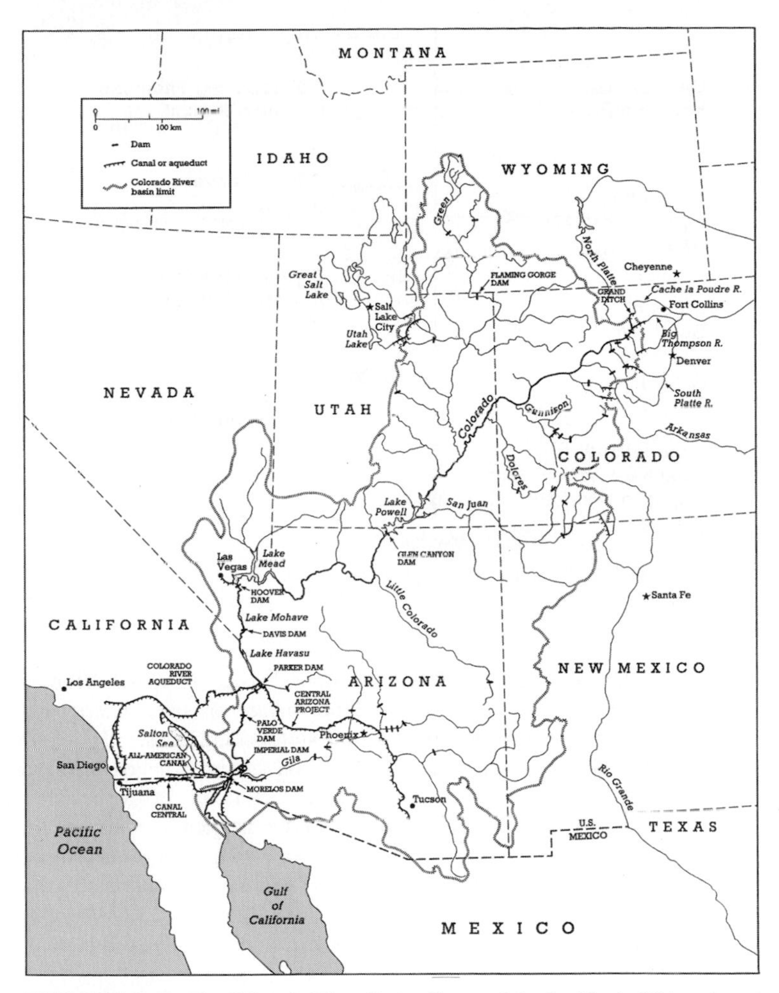

FIGURE 3.7 The Colorado River Basin. (Source: Map by Martin Walz, adapted from the NATIONAL GEOGRAPHIC magazine for use by the Geographic Education Program, National Geographic Society, 1993)

Managing the waters of the Colorado affects the management of other natural resources both inside and outside the basin (Black, 1987). The river is used to generate 12 billion kilowatts of electricity, provides water to farmers and cities, and is an important source of water–based recreation. The electrical energy generated at the dams would otherwise have to be supplied by either fossil fuels or nuclear power. The diversion of water out of the basin reduces demands on surface– and groundwater resources in the basins where the water is imported. California diverts approximately one–third of the average flow of the river to the coastal megalopolis of Los Angeles and San Diego and for irrigation of the Imperial and Coachella Valleys just north of Mexico. Much of the land in the Colorado Basin is federally owned and

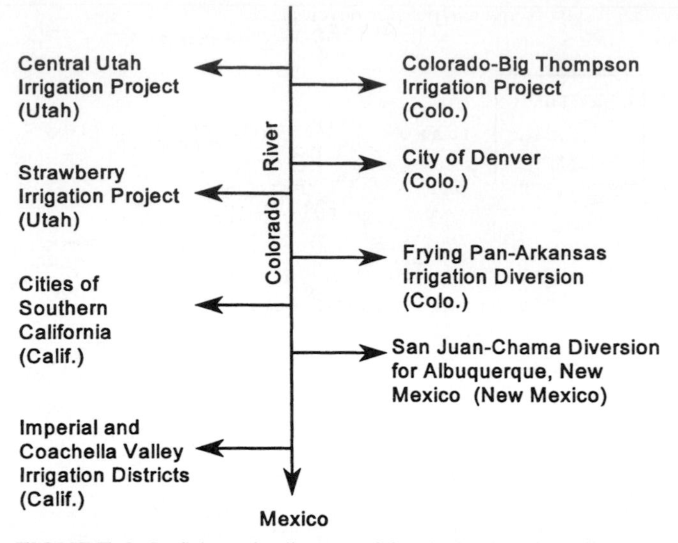

FIGURE 3.8 Schematic diagram of interbasin diversions from the Colorado River.

managed by the National Forest Service, National Park Service, or the Bureau of Land Management. The actual and potential impacts on water quantity and quality constrain management options for the surrounding lands.

Approximately 17 million people and 1 million acres of irrigated farmland receive Colorado River water. Figure 3.8 shows major out–of–basin diversions from the Colorado. Some diversions are for irrigation, some are for municipal water supply, and some, like the Colorado–Big Thompson project, started out as a supplemental irrigation project and have become more of a municipal diversion over time. The diversions for Denver and Los Angeles were built by local agencies, while the other diversions are federal Bureau of Reclamation projects.

As early as 1901, 100,000 acres in the Imperial Valley were irrigated using water from the Colorado. The water was conveyed through a patchwork of canals and abandoned stream channels and followed a meandering path in and out of old Mexico. Problems with the Mexican government led to lobbying of Congress to help build an "All American Canal" to convey water to the Imperial Irrigation District. Major floods during the period 1905–1907 destroyed the irrigation diversion structures on the river and for two years the river's entire flow went into the Salton depression, creating the Salton Sea. The Salton Sea is maintained today by irrigation return flows. A major dam on the main river was advocated as the only way to control flooding and provide the storage needed for a stable water supply. Hoover Dam was built in 1928. Grand Canyon National Park was created in 1919, preserving a 132–mile stretch of the river. Imperial Dam and the All American Canal were constructed in the 1930s. Parker Dam and the 242–mile Colorado River Aqueduct were completed in the 1940s to supply water to the city of Los

Angeles. Most of the early water development occurred in the lower basin. This prompted great concern in the upper basin that the lower basin would gain control of the river through prior appropriation. To prevent this the basin states divided the waters of the Colorado by interstate compact.

The Colorado River Compact (1922) was the first interstate compact. The compact divides the basin two different ways. One division is for water use; the other is based on water supply. Based on water use the basin is divided into the Upper Division (Wyoming, Utah, Colorado, and New Mexico) and the Lower Division (California, Arizona, and Nevada). This is a politically unbalanced division so the basin is divided differently for water supply. For consideration of water supply the basin is divided into an Upper Basin and a Lower Basin. The Upper Basin states are all of those parts of Colorado and Wyoming, and parts of Arizona, New Mexico, and Utah. The Lower Basin is all of those parts of California and Nevada, and parts of Arizona, New Mexico, and Utah. For water supply each subbasin has two "whole" states and parts of three others, providing a balance in political power. The point of division between the Upper and the Lower Basins is Lee's Ferry, Arizona, located just south of the Utah–Arizona state line. In the 1800s fugitive John D. Lee operated a ferry on the river for 14 years until he was discovered and executed.

The river was measured downstream at Yuma, Arizona, for about 20 years before the compact was negotiated, providing the only discharge data on which to base an allocation. The annual average discharge upriver at Lee's Ferry was estimated as 16.5 million acre–feet (MAF) per year. The compact divided 15 MAF per year equally between the Upper and the Lower Divisions. In theory this provided 7.5 MAF to each division and left 1.5 MAF of surplus water that might be needed to satisfy Mexico and the Indians. The compact makes no provision for dividing the water between the states within the divisions. The compact states that the

> States of the Upper Division will not cause the flow of the river at Lee's Ferry to be depleted below an aggregate of 75,000,000 acre–feet for any period of ten consecutive years.

This method of allocation was, in retrospect, naive on the part of the Upper Division. By guaranteeing an absolute quantity of 7.5 MAF per year, on average, to the Lower Division, the Upper Division assumed all the risk in times of water shortage. Meeting this obligation to the Lower Division meant building storage reservoirs in the Upper Basin. Now for the bad news. Based on the analysis of proxy data, i.e., tree ring–based streamflow reconstructions, the 20–year period during which discharge was measured was wetter than any time in the last 400 years. Instead of 16.5 MAF the long–term annual average discharge is more like 13.5 MAF per year. (The average discharge for the 80–year historical streamflow record is 15.2 MAF per year.) As it turns out there is no surplus water and the Upper Division is obligated to deliver 7.5 MAF per year to the Lower Division. Instead of 7.5 MAF the Upper Divisions' share is now more like 6 MAF per year. The United States signed a treaty with Mexico in 1944 guaranteeing 1.5 MAF at the border. Meeting the treaty obligation to Mexico is divided equally between the two divisions.

The compact suspended the law of prior appropriation between the Upper and the Lower Divisions, but it did not protect Arizona from California. Arizona refused to ratify the compact in 1922. Arizona claimed California was exceeding its share, and Arizona wanted the Gila River excluded from the compact. This was the beginning of many lawsuits between Arizona and California. At one point Arizona even sent its state militia to the banks of the Colorado to prevent California from constructing diversion works. Arizona did not ratify the compact until 1944. Even without Arizona's ratification the compact was adopted because Congress passed the Boulder Canyon Project Act (1928). The act provided for construction of Boulder (renamed Hoover) Dam and the All American Canal if the following three conditions were met:

—California passed a limitation act restricting its diversions to a maximum of 4.4 MAF per year
—six of the seven basin states ratified the compact
—the President signed the bill

All conditions were met and the compact shows a perfection data of 1922. Hoover Dam was the dam everyone had been waiting for. Hoover Dam is the world's first *mulitpurpose* dam. The dam and reservoir provide flood control, water supply, hydroelectricity, and recreation. Never before had multiple purposes been combined into a single structure. The multipurpose dam became the holy grail of water development for the next 40 years.

In 1948 the Upper Basin states divided their allocation in the Upper Colorado River Basin Compact (1948). Learning from their past mistake, the waters were divided on a percentage basis. The division is as follows:

Colorado	51.75 percent
Utah	23.00 percent
Wyoming	14.00 percent
New Mexico	11.25 percent
Arizona	50,000 acre–feet per year

We saw earlier that *Arizona v. California* (1963) was a landmark case for the quantification of Indian reserved water rights. Among other issues the case dealt with the allocation of water between the states of the Lower Division. California had been using more water than it was entitled to because Arizona was not physically able to use its allocation. California tried to claim a right to the water according to prior appropriation but the Court said no. The Court found that the waters of the Lower Division had been allocated, to everyone's surprise, by the Boulder Canyon Project Act back in 1928. The Court held that the Colorado River had been allocated with 4.4 MAF to California, 2.8 MAF to Arizona, and the remaining 0.3 MAF to Nevada.

In 1968 the Colorado River Basin Project Act was passed by Congress. The act authorized the Central Arizona Project (CAP). Without the project Arizona was unable to use its share of the river. However, the act stated that CAP would not be built unless Arizona passed groundwater management regulations. Arizona passed a Groundwater Management Act in 1980. The CAP was constructed and now brings 1.2 MAF of water per year to farmers and the cities of Phoenix and Tucson. (It is interesting to note that because of the decline in irrigated agriculture, the CAP is delivering more water than Arizona can currently use.) There are many other statutes, court cases, and regulations that combine to form the "law of the river" but they are too numerous to cover here.

Another issue plaguing the Colorado Basin is salinity. There are a number of natural sources adding salt to the river but the main culprit is irrigation agriculture. With every use of water for irrigation, water evaporates and salt concentrates in the return flow. The 1944 treaty guaranteed 1.5 MAF per year to Mexico but did not mention water quality. When the Welton–Mohawk Irrigation District in Arizona began discharging highly saline water into the river in 1961 it caused salinity levels at the Mexican point of diversion to nearly double. The Mexican government was furious and threatened to take the United States to the International Court of Justice (Hundley, 1986). In typical fashion the proposed solution was technological, including tile drains in the Mexican fields, a diversion canal to take the saline water around the Mexican's point of diversion, and a multimillion dollar desalinization plant. The desalinization plant was built but has not been needed. A simpler and more cost effective option would have been to simply take some land out of irrigation.

4

Water Use in the United States

Water Availability: An International View
A Comment on Water–Use Estimates
Water Supply and Demand: National View
Water Supply and Demand: Regional View
Water Supply and Demand: State View
Total Withdrawals (Surface and Ground)
Surface Water
Groundwater
Geographic Information Systems (GIS)
Types of GIS

This chapter considers two separate but closely related topics. The first is the characteristics of water use in the United States. For this topic we examine how much water is used for different purposes, how uses have changed over time, and the spatial patterns of uses. Later chapters look at specific water uses in more detail. In describing the spatial patterns we wade through a fair amount of tabular data. The copious amounts of data provide the rational for the second topic in this chapter—the use of Geographic Information Systems (GIS) for the storage and management of water data. GIS can do much more than just store data, as we shall shortly see.

The terms *instream use* and *offstream use* distinguish two basic categories of water use. Instream uses include transportation (navigation), hydroelectric power generation, recreation, and instream flows for environmental purposes. Offstream uses include urban and domestic uses, irrigation, and water used for industry and manufacturing. Offstream use requires a water *withdrawal,* which is the removal of water from a surface or groundwater source. The *consumptive use* is that part of the water withdrawn that is evaporated, transpired, or incorporated into manufactured products or plant tissue. The water consumed is unavailable for use by others. Some people have expanded the concept of consumptive use to include water pollution and interbasin transfers, reasoning that they too make water unavailable for downstream users. Such an expanded definition is not conceptually incorrect. Polluted water can be cleaned up, and water transferred out of one basin is available to users in the basin where it is imported.

In transporting water from a source to a point of use there can be *conveyance losses.* In irrigation agriculture conveyance losses include evaporation and seepage from ditches and canals. In urban water systems conveyance losses occur through leaky pipes. The term *yield* refers to the amount of water that can be withdrawn from a surface or groundwater source. The term *safe yield* is not as easily defined. When applied to a renewable source such as a stream, lake, or water table aquifer, safe yield usually means the yield equal to the long–term rate of recharge. Using the water balance concept, safe yield implies a water–use system in dynamic equilibrium. The safe yield (output from the system) equals the rate of renewal (input to the system) and the amount of water in storage remains the same over time. In situations where groundwater is mined, safe yield may have a very different meaning. It might mean the amount of water that can be withdrawn so that the supply is exhausted in a prescribed period of time. The yield from reservoirs is discussed in Chapter 6.

WATER AVAILABILITY: AN INTERNATIONAL VIEW

Our focus is water use in the United Sates, but it is instructive to compare our situation to other regions around the world. Figure 4.1 shows trends in per capita water availability for continental regions around the world with projections to the year 2000. The relative positions of the various regions have not changed in the last 40 years. Oceania and South America have the greatest water availability per capita, while Europe and Asia have the least. What has changed over time is the absolute availability per capita. All regions have seen availability decrease with increasing population. South America and Oceania have experienced the largest decreases. In South America the projected decrease from 1950 to the year 2000 is 73 percent. On a per capita basis the United States ranks fairly low compared to other world regions, though we still have a generous supply. Asia is projected to replace Europe as the region with the lowest per capita water availability by the year 2000. *Water–stressed* countries are defined as those with water availability between 1000 (35,000 gallons) and 1600 cubic meters (56,800 gallons) per person per year and have severe shortages during droughts. *Water–scarce* regions have less than 1000 cubic meters per person per year. Recent studies predict the number

PER CAPITA WATER AVAILABILITY
(1000 cu. m/person/yr)

	YEAR					
REGION	1950	1960	1970	1980	1990	2000
Oceania	112	91.2	74.6	64		50
S. Am.	105	80.2	61.7	48.8		28.3
N. & C. Am.	37.2	30.2	25.2	21.3		17.5
Africa	20.6	16.5	12.7	9.4		5.1
U.S.	10.6	8.8	7.6	6.8		5.6
Asia	9.6	7.9	6.1	5.1		3.3
Europe	5.9	5.4	4.9	4.6		4.1

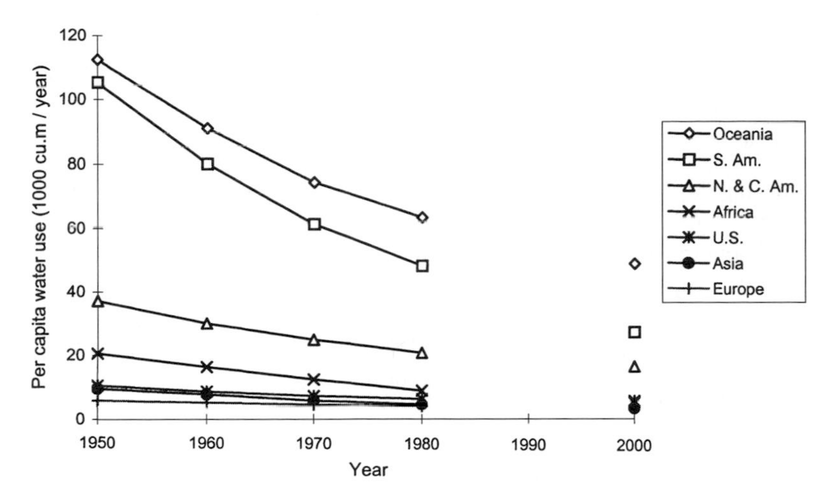

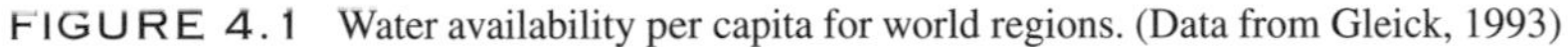

FIGURE 4.1 Water availability per capita for world regions. (Data from Gleick, 1993)

of water–stressed countries in the world doubling over the next 30 years. The single most important reason for water scarcity worldwide is wasteful use promoted by water prices that are too low.

A COMMENT ON WATER–USE ESTIMATES

The water–use estimates reported for 1990 in the following sections come from the U.S. Geological Survey (Solley *et al.,* 1993). The estimates were compiled at the county level and aggregated to the larger state and regional scales. A variety of national databases were used, including the EPA's Industrial Facilities Discharge files, the EPA's Public Drinking Water Supply files, the Census Bureau's population files, and the Department of Energy's Energy Information Administration reports. It was up to the individual district office to determine the most reliable set of data for that state.

Numerical estimates of water use are average daily values. For some sectors, like irrigation, water is not applied year round, so the actual application rates are much higher than the average daily rates given in the tables. In the U.S. Geological Survey's report, numerical data are generally rounded to three significant figures for values greater than 100, and two significant figures for values less than 100. All numbers were rounded independently so the sum of the rounded numbers may not equal the totals (Solley *et al.,* 1993, p. 4).

In the book *Water in Crisis* (1993), Gleick makes the following observation:

> Good water data are hard to come by. Data are often not collected regularly or systematically. . . . Standards and techniques of measurement differ from region to region, or worse year to year. Units of measure vary around the world—different quantitative measures may have the same name, or a single measure may be called different things in different places. Data are collected by individuals with differing skills, goals, and intents. Some data are collected objectively; other data are collected to support particular ideological or political biases. The more we know about the data set and how it was collected, the better equipped we will be to evaluate it and use it (p. 117)

In limiting our analysis to the United States some of the potential problems identified by Gleick are minimized, but none are eliminated. In Chapter 1 we discussed the problem of English versus metric unit systems, which hinders the direct comparison of data. Chapter 9 addresses some of the problems created by nonstandardized methods for water quality data collection and monitoring. Skill levels will always vary between individuals, and while it may not be as prevalent today, ideological and political motives can still color the data.

WATER SUPPLY AND DEMAND: NATIONAL VIEW

Table 4.1 shows estimated data on the total water withdrawals in the United States in billions of gallons per day (bgd) for the year 1990. Of the total 408 bgd withdrawn, 339 bgd (83 percent) was freshwater, and the remaining 69.4 bgd (17 percent) was saline. Surface sources made up 80 percent of the total withdrawals while groundwater composed 20 percent. In terms of freshwater alone, surface sources made up 76 percent of the total withdrawals and groundwater 24 percent.

Table 4.2 gives estimated total water withdrawals in the United States from

TABLE 4.1 Total Offstream Withdrawals (bgd) by Source, 1990[a]

	Surface	Ground	Total
Freshwater	259.0 (76%)	79.4 (24%)	339.0 (83%)
Saline water	68.2 (98%)	1.2 (2%)	69.4 (17%)
Total	327.0 (80%)	80.6 (20%)	408.0 (100%)

[a] Source: Solley *et al.,* 1993.

TABLE 4.2 Trends in Water Withdrawal in the United States, 1900–1990[a]

	Year													
	1900	1910	1920	1930	1940	1950[b]	1955[b]	1960[c]	1965[c]	1970[d]	1975[e]	1980[e]	1985[e]	1990[e]
Withdrawal bgd	40	66	92	111	136	180	240	270	310	370	420	440	399	408
cu. km/yr	56	92	127	153	189	250	333	375	431	514	583	611	554	567
Population ($\times 10^6$)			105	122	132	151	164	179	194	206	216	230	242	252
gpcd			871	906	1033	1192	1463	1508	1598	1796	1944	1913	1649	1619

[a] Source: Solley *et al.*, 1993.
[b] 48 States and the District of Columbia.
[c] 50 States and the District of Columbia.
[d] 50 States, the District of Columbia, and Puerto Rico.
[e] 50 States, the District of Columbia, Puerto Rico, and Virgin Islands.

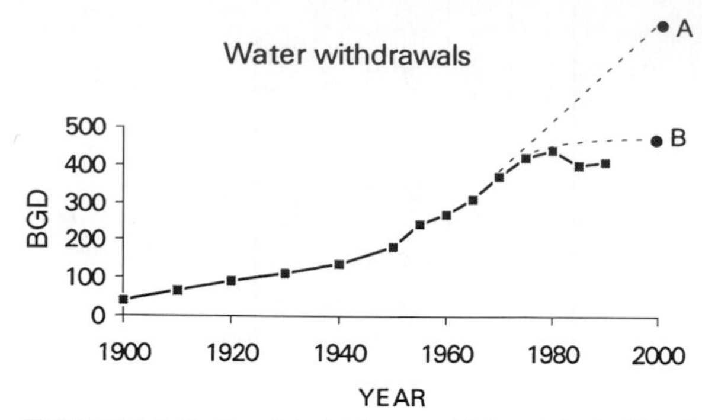

FIGURE 4.2 Trends in total water withdrawals in the United States in billions of gallons per day (bgd).

1900 to 1990 and average withdrawals per person in gallons per capita per day (gpcd). The data in Table 4.2 are shown graphically in Figures 4.2 and 4.3.

In 1900 total withdrawals were 40 bgd. By 1990 total withdrawals exceeded 10 times this amount (408 bgd), an increase of 920 percent. Withdrawals increased slowly for the first 50 years. The 30–year period from 1950 to 1980 saw withdrawals jump dramatically from 180 bgd to a peak of 440 bgd. The two reasons for the rapid increase at this time were the expansion of irrigation agriculture in the West and the massive increase in the use of water for cooling thermal electric power plants. The two points A and B at year 2000 in Figure 4.2 represent future withdrawal projections made by the Water Resources Council in its first (point A) and second (point B) national water assessments. In its first assessment the WRC (1968) used a simple linear projection to forecast future withdrawals. The result was an estimated total withdrawal of 806 bgd by the year 2000, and a whopping 1368 bgd by the year 2020. The latter figure nearly equals the annual average runoff for the 48 conterminous states! Linear extrapolation is not a very reliable method of forecasting, especially over long time periods. In its second national

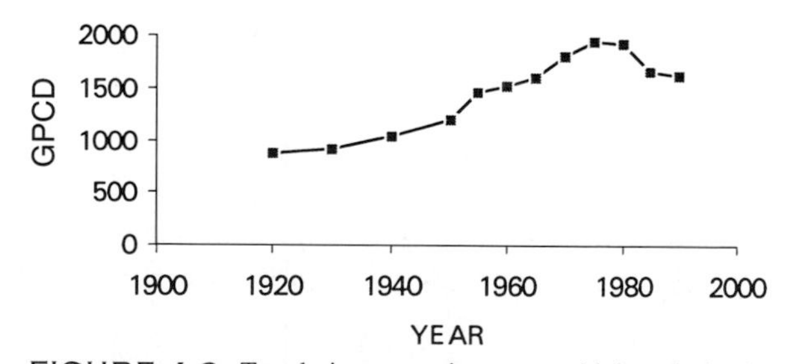

FIGURE 4.3 Trends in per capita water withdrawals in the United States in gallons per capita per day (gpcd).

TABLE 4.3 Estimated Total Water Withdrawal (bgd) by Sector, 1950 to 1990[a]

	Year								
	1950	1955	1960	1965	1970	1975	1980	1985	1990
Public	14.0	17.0	21.0	24.0	27.0	29.0	34.0	36.5	38.5
Rural	3.6	3.6	3.6	4.0	4.5	4.9	5.6	7.8	7.9
Irrigation	89.0	110.0	110.0	120.0	130.0	140.0	150.0	137.0	137.0
Thermal Elect.	40.0	72.0	100.0	130.0	170.0	200.0	210.0	187.0	195.0
Other Indust.	37.0	39.0	38.0	46.0	47.0	45.0	45.0	30.5	29.9
Total	183.6	241.6	272.6	324.0	378.5	418.9	444.6	398.8	408.3

[a] Source: Solley *et al.*, 1993.

assessment 10 years later the WRC (1978) used a more sophisticated forecasting method and projected total withdrawals to be 425 bgd by the year 2000. The 1978 projection has held up pretty well when compared to the estimates of actual withdrawals.

Per capita water use nearly doubled from 1920 to 1990 (Figure 4.3). However, the peak per capita water use came in the mid–1970s when it reached nearly 2000 gpcd.

Total withdrawals in Table 4.2 from 1950 to 1990 are decomposed into withdrawals by sector in Table 4.3. The totals in Table 4.3 do not quite match the totals in Table 4.2 because of rounding and revisions to some estimates. The data in Table 4.3 are displayed graphically in Figures 4.4 and 4.5 and show that from 1950

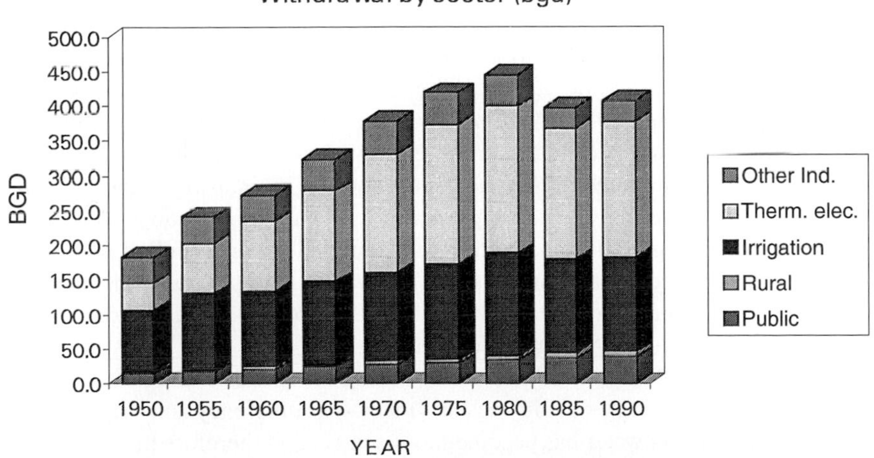

FIGURE 4.4 Total water withdrawals by water–use sector (bgd). (Solley *et al.*, 1993)

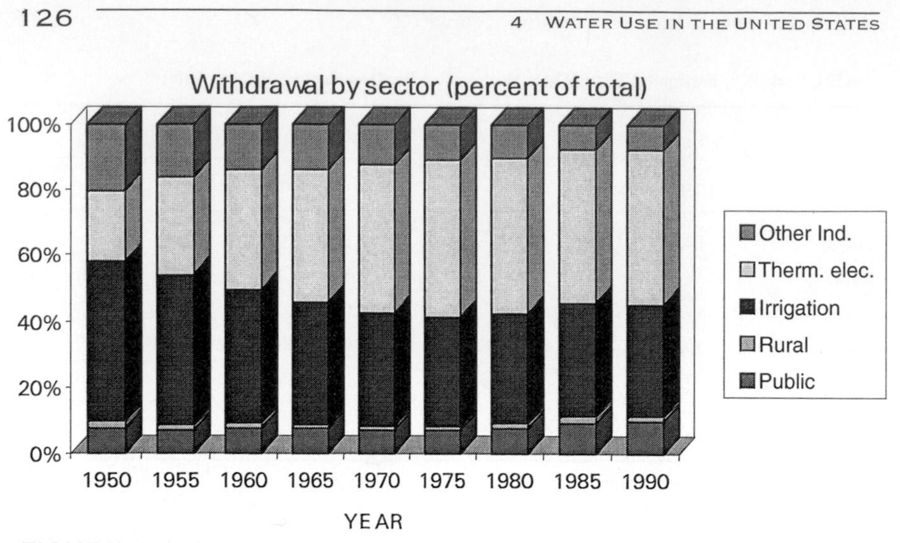

FIGURE 4.5 Water withdrawals by water–use sector as a percent of total. (Solley *et al.*, 1993)

to 1980 withdrawals increased for every sector. Withdrawals for cooling thermal electric power plants increased the most, from 40 bgd in 1950 to 210 bgd in 1980. In 1950 withdrawals for thermal electric power represented only 21 percent of the total; in 1980 they represented fully 47 percent of the total (Figure 4.5). Of the 195 bgd withdrawn for thermal electric power in 1990, 131 bgd (67.2 percent) was freshwater, while the remaining (32.8 percent) was saline. Electricity composes about one–third of all the energy used in the United States. Our voracious demand for electricity creates a concomitant demand for water to cool the power plants. Most of the saline water used to cool thermal power plants was withdrawn in California, Florida, and the mid–Atlantic states.

The second largest water withdrawal is for irrigation agriculture. Irrigation was the largest withdrawal in 1950 at 89 bgd (48 percent of the total). By 1980 irrigation withdrawals had increased to 150 bdg yet the sector had fallen to second place behind thermal electric power. The decrease in total water withdrawals between 1980 (444.6 bgd) and 1985 (398.8 bgd) was caused by reductions in all three of the largest water–use sectors—thermal electric power, irrigation, and other industrial uses. Of the three the 32–percent reduction by industry is the most impressive. This reduction was accomplished by the widespread adoption of water recycling technologies. The motivation for industrial water recycling was not an altruistic concern for water conservation, but water quality regulations that constrained or prohibited the discharge of pollutants. Table 4.4 shows the remarkable progress in recycling for various industries.

The reduction in irrigation withdrawals during the decade of the 1980s resulted from two trends. First was the increase in water–use efficiency and irrigation water conservation. As water has become more scarce, and therefore more valuable, farmers have invested in ways to improve efficiency and reduce waste. This can

TABLE 4.4 Average Number of Times Water is Recycled[a]

Industry	1954	2000 (projected)
Paper	2.4	11.8
Chemicals	1.6	28.0
Petroleum	3.3	32.7
Primary metals	1.3	12.3
Manufacturing	1.8	17.1

[a] Source: Gleick, 1993.

be seen in the reduction in the amount of irrigation water applied to farm fields. The average amount of water applied in the late 1970s was 2.9 acre–feet per acre; in the late 1980s it dropped to 2.7 acre–feet per acre. The second trend has been the reduction in acreage in some western states. Irrigation water use is covered in more detail in Chapter 7.

The two largest water withdrawal sectors (thermal electric power and irrigation) are very different in their consumptive use. In 1990 irrigation withdrew 137 bgd and consumed 76.2 bgd (55.6 percent). By contrast, thermal electric power withdrew 131 bgd of freshwater but consumed only 3.9 bgd (2.9 percent). Even though consumption is low, an important pollutant from power plants is the waste heat. Ninety percent of all the water withdrawn for irrigation was withdrawn in the West. By contrast 85 percent of all the freshwater withdrawn for thermal electric power was withdrawn in the East. The third largest sector in water withdrawals is public supplies. As with thermal electric power, most of this occurs in the East. For the Nation as a whole, 54 percent of the total withdrawals occurred in the East, and 46 percent in the West.

WATER SUPPLY AND DEMAND: REGIONAL VIEW

Figure 4.6 shows water–resource regions identified by the Water Resources Council. Table 4.5 provides data on annual renewable supplies, freshwater withdrawals, and consumption by region. Most regions in the East, with the exception of the Great Lakes, have withdrawals of less than 26 percent of the annual renewable supply. Consumption is likewise low, and generally runs 15 percent or less of the water withdrawn, though three regions in the East have consumptions greater than 30 percent of withdrawal. In the West withdrawals are a higher percentage of available supply. In the Rio Grande and Lower Colorado regions withdrawals *exceed* the renewable supply. Withdrawals in excess of supply indicate groundwater mining and interbasin transfers of water. Consumption is also high in western regions because much of the water is used for irrigation. High consumption means

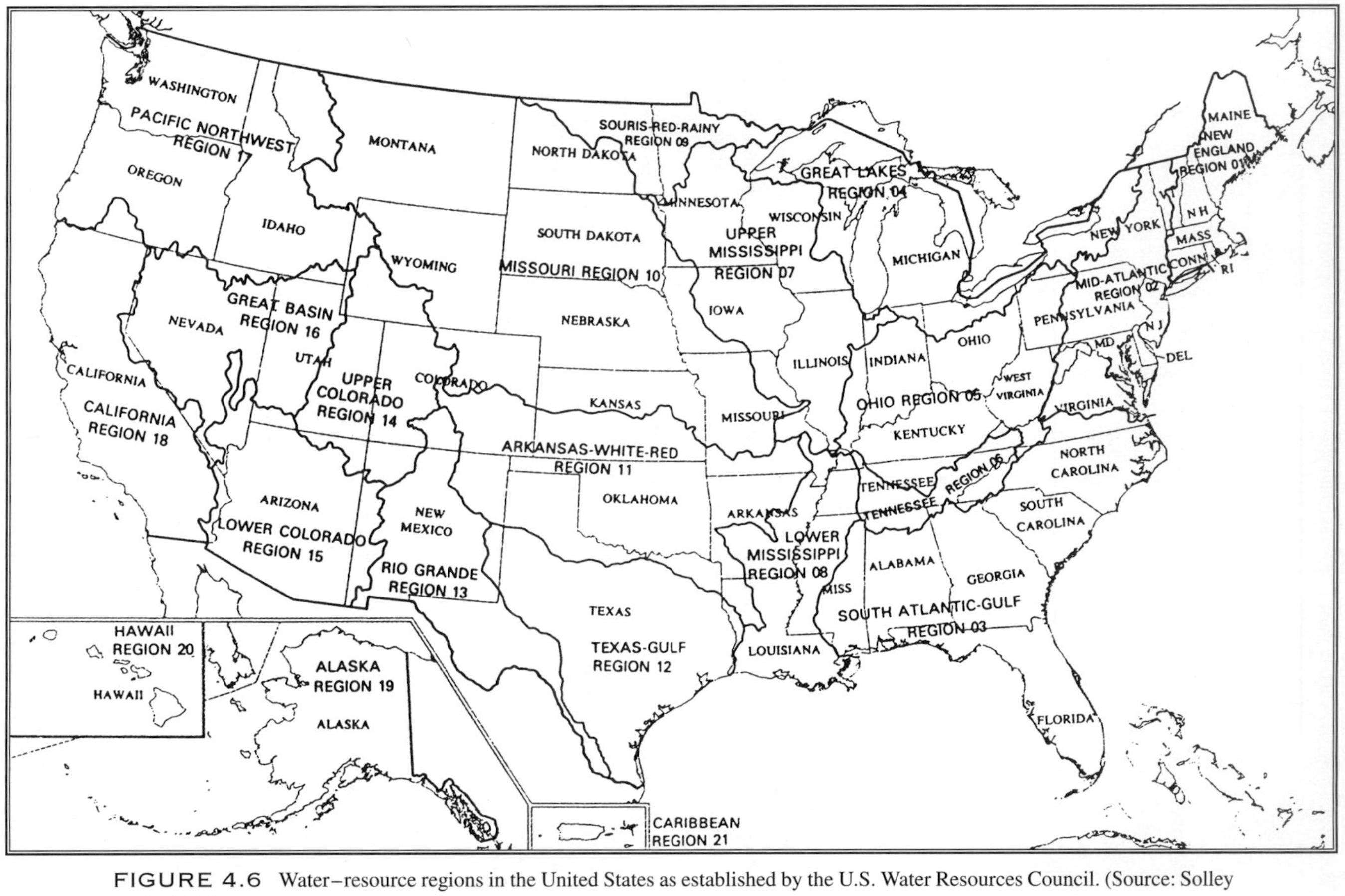

FIGURE 4.6 Water–resource regions in the United States as established by the U.S. Water Resources Council. (Source: Solley *et al.*, 1993)

TABLE 4.5 Supply and Demand (bgd) Characteristics by Water–Resource Region[a]

	Supply	Demand, 1990			
Region	Annual renewable supply	Total freshwater withdrawals	Total freshwater consumption	Withdrawal as a percent of supply	Consumption as a percent of withdrawal
New England	77.3	4.7	0.4	6.0	9
Middle Atlantic	96.4	21.1	1.3	21.8	6
South Atlantic–Gulf	213.4	33.4	5.1	15.6	15
Great Lakes	76.8	32.4	1.6	42.1	5
Ohio	140.2	30.4	2.1	21.6	7
Tennessee	43.3	9.2	0.32	21.2	4
Upper Mississippi	79.7	20.8	2.0	26.1	10
Lower Mississippi	68.0	18.0	6.9	26.5	38
Souris–Red–Rainy	7.6	0.3	0.14	3.9	47
Missouri	67.3	37.5	12.1	55.7	32
Arkansas–White–Red	63.7	15.4	7.9	24.1	51
Texas–Gulf	35.9	13.5	5.9	37.6	44
Rio Grande	5.0	6.0	3.5	120.0	58
Upper Colorado	12.3	7.1	2.5	57.7	35
Lower Colorado	7.2	7.7	5.0	106.9	65
Great Basin	8.3	7.2	3.4	86.7	47
Pacific Northwest	291.1	36.3	12.1	12.5	33
California	86.9	35.4	20.8	40.7	59
United States (conterminous)	1384	337	93	24.4	28
Alaska	905.1	0.28	0.03	0.03	11
Hawaii	7.4	1.19	0.63	16	53
Caribbean	5.2	0.59	0.20	11	34
Grand total	2302	339	94	15	28

[a] Source: WRC, 1978; Foxworthy and Moody, 1986; Solley *et al.*, 1993.

there is less water available downstream, and in some areas the increasing concentration of salt in the irrigation return flows cause serious water quality problems.

Table 4.6 gives total withdrawals by water–use sector and by region. Withdrawals are skewed toward the coastal regions. This follows the pattern of population distribution throughout the country. Nearly one–half of all withdrawals in 1990 occurred in the five coastal water–resource regions—the Pacific Northwest and California regions on the west coast, and the New England, Mid–Atlantic, and South Atlantic–Gulf regions in the East (Table 4.6). Western regions have the largest withdrawals for irrigation, led in order by the Pacific Northwest, California, and the Missouri Basin. The largest freshwater withdrawals for thermal electric power were in the three eastern regions of Ohio, the Great Lakes, and the

TABLE 4.6 Total Water Withdrawals[a] by Water–Use Category and Water–Resources Region, 1990

Region	Public supply (Fresh)	Domestic (Fresh)	Commercial (Fresh)	Irrigation (Fresh)	Livestock (Fresh)	Industrial		Mining		Thermoelectric		Total	
						Fresh	Saline	Fresh	Saline	Fresh	Saline	Fresh	Saline
New England	1400	169	133	120	8.1	479	68	20	0.0	2400	9090	4730	9160
Mid–Atlantic	5980	396	133	197	107	1730	1470	387	30	12,200	25,000	21,100	26,500
S. Atlantic–Gulf	4850	659	134	4450	350	2810	94	437	9.1	19,700	10,700	33,400	10,800
Great Lakes	4340	283	108	290	92	4190	3.7	249	7.7	22,800	0	32,400	11
Ohio	2530	360	89	68	132	2370	0	1000	22	23,900	0	30,400	22
Tennessee	511	56	56	27	201	1190	0	92	0	7070	0	9200	0
Upper Mississippi	1890	371	260	392	269	967	0	154	4.2	16,500	0	20,800	4.2
Lower Mississippi	1040	90	92	7380	1070	2620	67	40	0	5640	1060	18,000	1120
Souris–Red–Rainy	72	22	0.3	98	21	49	0	8.2	0	26	0	295	0
Missouri Basin	1620	139	40	24,800	415	171	0	279	37	10,000	0	37,500	37
Arkansas–White–Red	1400	118	165	8390	359	368	0	74	291	4530	0	15,400	291
Texas–Gulf	2520	79	57	5100	156	741	1460	130	399	4710	3150	13,500	5010
Rio Grande	533	23	20	5290	33	12	0	66	39	18	0	6000	39
Upper Colorado	118	10	6.3	6590	117	5.4	0	51	28	177	0	7080	28
Lower Colorado	1070	39	29	6060	98	174	0	170	0.7	109	0.4	7750	1.2
Great Basin	610	15	16	6300	37	106	2.3	83	110	31	0	7200	112
Pacific Northwest	1580	220	718	31,800	620	1030	36	15	0	355	0	36,300	36
California	5750	313	271	28,300	405	130	25	22	310	246	11,400	35,400	11,700
Alaska	92	6.9	18	0.6	0.6	111	0	25	357	31	0	284	357
Hawaii	238	9.9	40	755	7.2	43	0.6	1.4	0	95	1550	1190	1550
Caribbean	411	8.2	0.8	140	8.9	11	51	2.6	0	2.6	2570	585	2620
Total	38,500	3390	2390	137,000	4500	19,300	3270	3310	1650	131,000	64,500	339,000	69,400

[a] Figures may not add to totals because of independent rounding. All values in million gallons per day. (Source: Solley et al., 1993)

South Atlantic–Gulf. Saline water withdrawals for thermal electric power were also greatest in the East, with the exception of California.

Withdrawals can also be analyzed by source—surface or groundwater. Table 4.7 shows withdrawals of surface water by region and water–use sector. Virtually all saline surface water (95 percent) was withdrawn for thermal electric power generation and most of this was in the California, New England, the Mid–Atlantic, and the South Atlantic–Gulf regions. Most of the fresh surface water was withdrawn in the East for thermal electric power, and in the West for irrigation. Industry and public water supply were the next largest users of fresh surface water in the East. Of all the regions, the Great Lakes had the single largest withdrawal of fresh surface water in 1990, and most (73 percent) was for thermal electric power, with industry (12.7 percent) and public supply (12.4 percent) the next two largest uses. The Missouri region had the second largest withdrawals of fresh surface water, of which 61 percent was for irrigation and 34 percent was for thermal electric. Overall, of the *fresh* surface water withdrawn in 1990, thermal electric power claimed 50 percent, irrigation 33 percent, public supply 9 percent, and industry 6 percent.

Table 4.8 shows withdrawals of groundwater by region and sector. Irrigation agriculture is the largest user of fresh groundwater, withdrawing 64 percent of all the fresh groundwater in 1990. California led all regions in groundwater withdrawals for irrigation at 10.6 bgd, followed by the Pacific Northwest, the Missouri Basin, the Arkansas–White–Red, and the Lower Mississippi. The second largest use of fresh groundwater (19 percent) was for public water supply. California and the South Atlantic–Gulf were ranked first and second, respectively, in fresh groundwater withdrawals for public supplies.

WATER SUPPLY AND DEMAND: STATE VIEW

TOTAL WITHDRAWALS (SURFACE AND GROUND)

Figure 4.7 shows total water withdrawal by state in 1990. Table 4.9 gives withdrawals for the 10 states with the largest total withdrawals. California led all states in water withdrawals with 46.8 bgd. Of that total 43.8 percent was fresh surface water, 24.4 percent was saline surface water, and 31.2 percent was fresh groundwater. California's leading position is not surprising given that California has both the largest population (29.7 million in 1990) and the most irrigated land. California alone withdrew 11.5 percent of *all* the water (fresh and saline) in the United States in 1990. California's withdrawals were more than the combined withdrawals of Texas and Idaho, the next two largest states. Some states use mostly freshwater, e.g., Idaho and Illinois, while others use mostly saline water, e.g., Florida and New Jersey. Western states use the most fresh groundwater. While most of the states in Table 4.9 have large populations, Idaho is unique by being third on the list in total withdrawals yet its population was only 1 million

TABLE 4.7 Withdrawals of Surface Water (bgd)[a] by Water–Use Category and Water–Resources Region, 1990

Region	Public supply (Fresh)	Domestic (Fresh)	Commercial (Fresh)	Irrigation (Fresh)	Livestock (Fresh)	Industrial		Mining		Thermoelectric		Total	
						Fresh	Saline	Fresh	Saline	Fresh	Saline	Fresh	Saline
New England	1070	0.0	51	111	2.8	382	68	19	0.0	2400	9090	4040	9160
Mid–Atlantic	4580	0.1	39	95	36	1370	1470	176	29	12,200	25,000	18,500	26,500
S. Atlantic–Gulf	2340	0	14	2610	152	1920	94	53	0	19,700	10,700	26,300	10,800
Great Lakes	3880	10	81	158	41	3950	0	227	6.5	22,800	0	31,200	6.5
Ohio	1750	8.1	31	40	79	1840	0	218	0.6	23,800	0	27,800	0.6
Tennessee	402	0	0.2	23	169	1170	0	67	0	7070	0	8900	0
Upper Mississippi	724	0	126	38	52	618	0	143	0	16,500	0	18,200	0
Lower Mississippi	334	0	72	1150	358	2120	67	32	0	5570	1060	9630	1120
Souris–Red–Rainy	37	0	0.2	42	4.9	47	0	8.0	0	26	0	166	0
Missouri Basin	1010	10	6.4	17,600	172	57	0	182	0	9980	0	29,000	0
Arkansas–White–Red	1040	0	138	1790	200	301	0	25	0	4500	0	7990	0
Texas–Gulf	1470	0	11	1130	102	600	1460	47	0	4660	3150	8020	4610
Rio Grande	163	0	20	3670	13	13	0	1.5	0	1.8	0	3850	0
Upper Colorado	86	0.3	0.7	6560	112	112	0	12	0	177	0	6950	0
Lower Colorado	552	2.1	6.3	3820	68	68	0	30	0.2	62	0.4	4670	0.6
Great Basin	256	7.7	9.3	4890	9.9	29	0	5.7	93	24	0	5230	93
Pacific Northwest	850	103	669	23,900	29	691	36	10	0	345	0	26,500	36
California	2530	0.7	213	17,700	204	4.8	25	70	0.4	241	11,400	21,000	11,400
Alaska	58	6.9	90	0.5	0.5	106	0	20	308	26	0	221	308
Hawaii	17	9.9	0.6	555	3.8	23	0	0	0	0	1550	600	1550
Caribbean	330	8.2	0.3	87	3.7	0	50	0.7	0	0	2570	426	2620
Total	23,500	132	1480	85,500	1810	15,400	3260	1280	438	130,000	64,500	259,000	68,200

[a] Figures may not add to totals because of independent rounding. All values in million gallons per day. (Source: Solley *et al.*, 1993)

TABLE 4.8 Withdrawals of Groundwater (bgd)[a] by Water–Use Category and Water–Resources Region, 1990

Region	Public supply (Fresh)	Domes- tic (Fresh)	Com- mercial (Fresh)	Irriga- tion (Fresh)	Live- stock (Fresh)	Industrial		Mining		Thermo- electric (Fresh)	Total	
						Fresh	Saline	Fresh	Saline		Fresh	Saline
New England	328	169	82	8.8	5.4	96	0.0	1.3	0.0	2.9	694	0.0
Mid–Atlantic	1400	396	94	102	71	360	0.2	210	10	4.6	2640	1.2
S. Atlantic–Gulf	2510	659	120	2300	198	896	0	384	9.1	41	7110	9.1
Great Lakes	460	282	27	132	51	235	3.7	22	1.2	3.2	1210	4.9
Ohio	774	352	58	28	54	532	0	784	22	65	2650	22
Tennessee	109	56	56	3.8	32	23	0	25	0	0	305	0
Upper Mississippi	1160	371	134	354	217	349	0	11	4.2	21	2620	4.2
Lower Mississippi	701	90	20	6230	710	501	0.6	7.9	0	75	8340	0
Souris–Red–Rainy	34	22	0.1	56	16	1.3	0	0.2	0	0	130	0
Missouri Basin	612	139	33	7200	242	114	0	96	37	50	8490	37
Arkansas–White–Red	364	118	27	6600	158	67	0	49	291	31	7420	291
Texas–Gulf	1050	79	47	3970	54	141	1.1	84	399	46	5480	400
Rio Grande	370	23	19	1620	20	11	0	65	39	16	2140	39
Upper Colorado	32	9.9	5.6	32	5.3	2.9	0	39	28	0	127	28
Lower Colorado	513	37	23	2240	30	49	0	139	0.6	47	3030	0.6
Great Basin	345	13	70	1410	27	77	2.3	77	17	7.2	1970	19
Pacific Northwest	727	212	49	7850	591	336	0	50	0	10	9730	0
California	3210	210	58	10,600	201	126	0	15	310	4.6	14,400	310
Alaska	34	6.2	8.7	0.1	0.1	5.2	0	4.3	48	4.7	64	48
Hawaii	221	8.5	39	200	3.4	20	0.6	1.4	0	95	589	0.6
Caribbean	81	3.7	0.4	54	5.2	11	1.2	1.9	0	2.6	159	1.2
Total	15,100	3260	908	51,000	2690	3950	9.7	2020	1210	525	79,400	1220

[a] Figures may not add to totals because of independent rounding. All values in million gallons per day. (Source: Solley *et al.*, 1993)

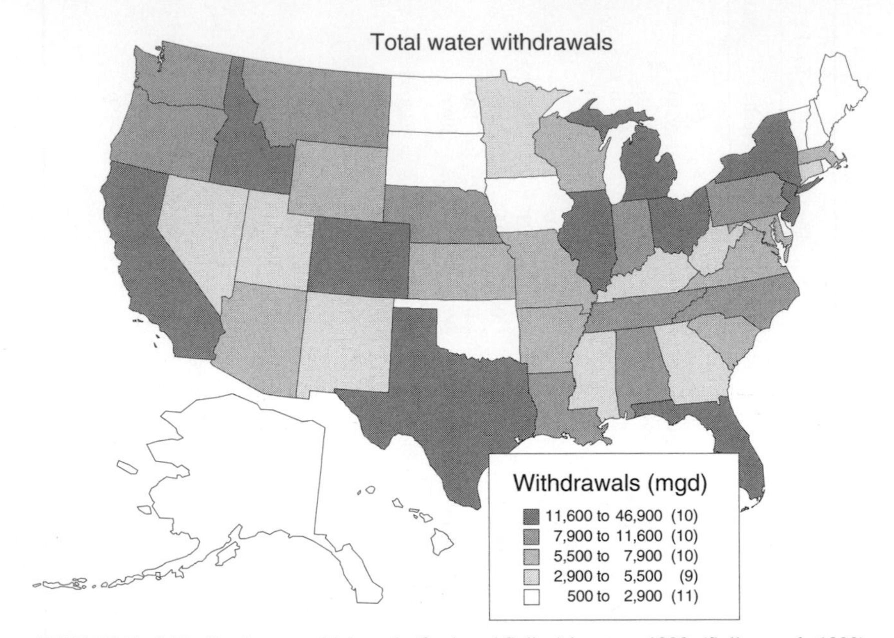

FIGURE 4.7 Total water withdrawals (fresh and Saline) by state, 1990. (Solley *et al.*, 1993)

TABLE 4.9 The 10 States with the Largest Withdrawals (bgd) of Water by Source, 1990[a]

State	Population (millions)	Groundwater		Surface water		Total
		Fresh	Saline	Fresh	Saline	
California	29.76	14.60	0.31	20.50	11.40	46.8
Texas	16.97	7.38	0.49	12.70	4.61	25.2
Idaho	1.00	7.59	0.00	12.10	0.00	19.7
New York	17.99	0.84	0.02	9.65	8.49	19.0
Illinois	11.43	0.92	0.03	17.10	0.00	18.0
Florida	12.94	4.66	0.00	2.87	10.40	17.9
New Jersey	7.73	0.57	0.00	1.65	10.60	12.8
Colorado	3.29	2.77	0.30	9.91	0.00	12.7
Ohio	10.84	0.90	0.00	10.80	0.00	11.7
Michigan	9.29	0.70	0.05	10.90	0.00	11.6

[a] Source: Solley *et al.*, 1993.

people in 1990. The explanation is the large amounts of water withdrawn for irrigation agriculture in the southern part of the state.

SURFACE WATER

Figure 4.8 shows the pattern of surface water withdrawals in 1990. The surface water pattern follows the withdrawal patterns for the three largest water–use sectors (thermal electric power, irrigation, and public supply). The patterns for thermal electric power and public supply are strongly correlated with population distribution along the East and Gulf Coasts. The large surface water withdrawals from Colorado up to Montana and Idaho, and even over to Washington and Oregon, are caused mainly by irrigation withdrawals. The large surface withdrawals in California are driven by all three uses.

GROUNDWATER

The spatial pattern of groundwater use results mainly from the withdrawals for irrigation. Irrigation agriculture withdrew nearly two–thirds of all groundwater in

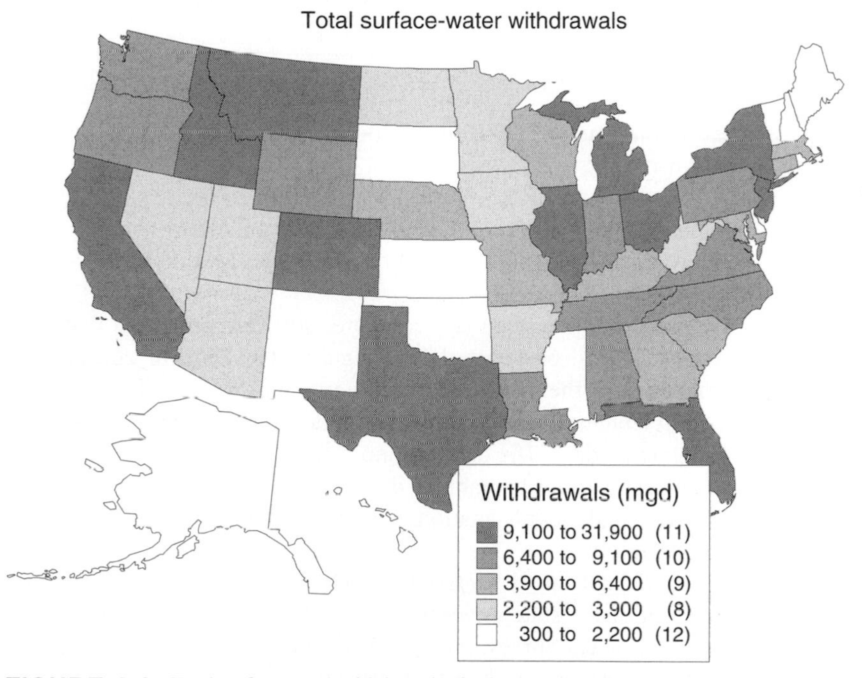

FIGURE 4.8 Total surface water withdrawals (fresh and saline) by state, 1990. (Solley *et al.*, 1993)

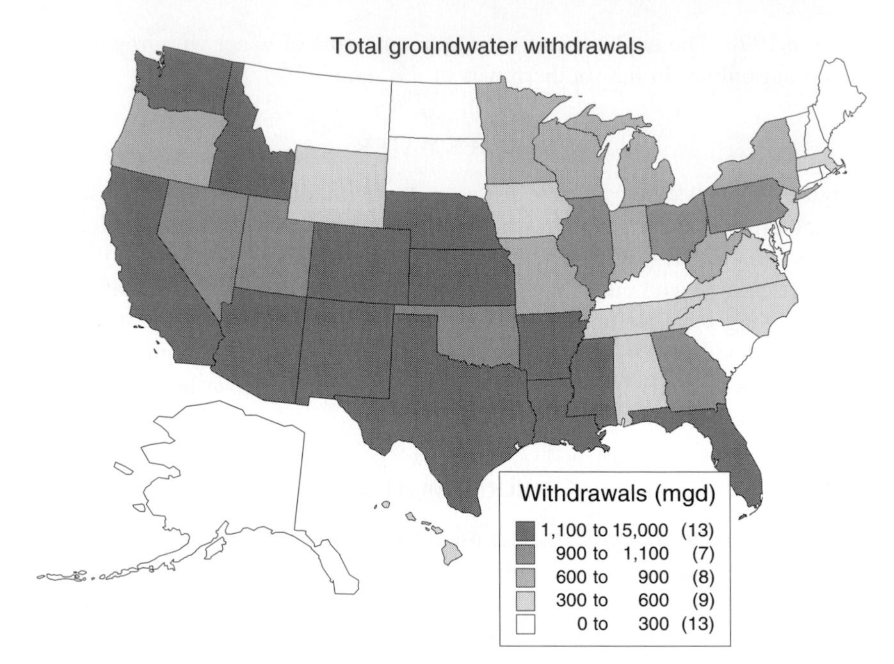

FIGURE 4.9 Total groundwater withdrawals (fresh and saline) by state, 1990. (Solley *et al.*, 1993)

1990. As a result, the pattern in Figure 4.9 shows the largest groundwater withdrawals in the West.

GEOGRAPHIC INFORMATION SYSTEMS (GIS)

Geographic information systems are computer software programs that store, manipulate, retrieve, and display spatial data (Figure 4.10). Data for a GIS include spatial data from existing thematic maps and remote sensing (aerial photography and satellites), and tabular data such as census data, weather records, and hydrologic and hydrographic data. The data are input in digital form using digitizers and scanners, or keyboarding for tabular data. The GIS software analyzes and manipulates the data, and the information is output as maps, charts, or tables to a monitor or a printer/plotter.

In a GIS the spatial data are *georeferenced,* which means their geographic position is known. All of the state–level water–use data are now maintained in a GIS by the U.S. Geological Survey. Water–use data can be downloaded from the World Wide Web (see Appendix 2). A wide variety of water– and land–resource data are now being archived in GIS format, including land–use/land cover data,

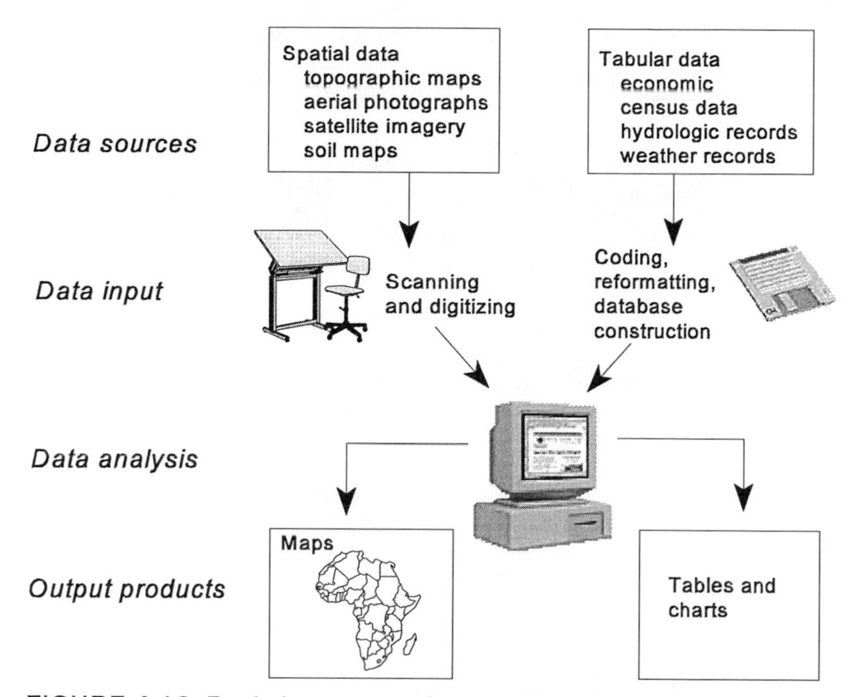

FIGURE 4.10 Four basic components of a geographical information system (GIS). GIS use existing spatial and tabular data sources that are input in digital form, analyzed by the computer (software), and turned into information outputs (after Marsh and Grossa, 1996).

water quality data, stream channel locations, and even floodplain maps for the National Flood Insurance Program. It is safe to say that eventually all spatial data will be stored in GIS formats.

TYPES OF GIS

There are two types of GIS—vector systems and raster systems. Each type has certain advantages and disadvantages. For both types space is defined using some geographic coordinate system. The coordinate system may be an arbitrary (*X, Y*) grid, but preferably a real–world coordinate system like latitude/longitude, State Plane, or UTM coordinates is used. The difference between vector and raster systems is in how they represent and store spatial data.

Vector GIS

A vector GIS defines and describes spatial data as either a *point,* a *line,* or a *polygon* (area). A point feature is located and defined by a single set of spatial coordinates (Figure 4.11). A water supply well, a precipitation gauge, and an

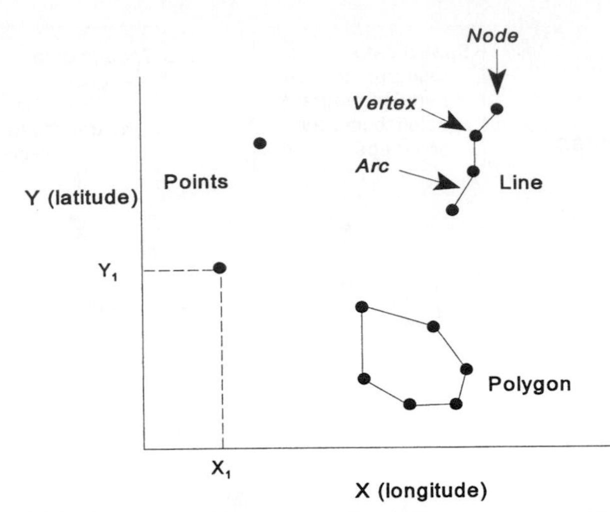

FIGURE 4.11 Representation of spatial features in a vector GIS. Spatial data are represented as either a point, a line, or a polygon. (Source: Thompson, 1998)

underground storage tank are examples of features that are represented as points and are georeferenced with a single pair of coordinates. A line feature is composed of two or more points. In the language of GIS the two end points on a line are the *nodes.* A line must have a starting node and an ending node. Each node is assigned a separate coordinate. Between the nodes there may be any number of *vertices.* The line segment between two points on a line is an *arc.* A river, a road, or a sewer line would be represented as line features. A polygon feature is an enclosed area. A polygon is made up of lines (arcs) with the characteristic that the starting node is the same as the ending node. A drainage basin, a soil polygon, or a distinct type of land use would be represented as polygons. The GIS knows where things are (absolute location) based on the feature's coordinate system. The GIS also knows the topology (relative location); topology is the relative spatial relationship between features. The GIS knows, for example, that one polygon is north of another polygon, or that one arc is west of another arc.

Once features are defined in a vector GIS, any amount of related information can be assigned to that feature. This related information is stored in a separate database. Take as an example a point representing a water supply well. Related information that could be stored in a database about the well might include the owner, when it was drilled, the depth, the diameter of the casing, and the aquifer in which it was completed. For a polygon feature representing a distinct type of soil, related information might include soil texture, horizon thickness, pH, parent material, or infiltration characteristics. This is one of the advantages of a vector GIS. The geographic features—points, lines, and polygons—are defined once, and then any amount of related information can be assigned to them. This is why the databases associated with vector GIS are referred to as relational databases. The combination of a georeferenced feature and the related database about that feature constitutes a *coverage*. So, for example, there could be a "well coverage," a "soil coverage," or a "stream coverage." The water–use data presented earlier in this chapter are stored in a vector GIS. The states or water–resource regions are polygons and the water–use data are assigned to the polygon, e.g., irrigation withdrawals, groundwater withdrawals, or saline water withdrawals.

Raster GIS

A raster GIS divides space into a uniform grid and each cell of the grid is georeferenced to the coordinate system (Figure 4.12). Spatial data are stored in the cells by assigning the cell a numerical value. In a raster system georeferencing and data storage are combined in the cells. Data are stored in what are called *layers* rather than coverages, and a layer in a raster GIS contains only one type of information. As an example, suppose you wanted to store information about soil depth, soil pH, and soil texture in a raster system. For the "soil depth layer" each cell in the grid would have a quantitative value equal to the thickness of the soil at the cell's location. Since a cell can contain only one value, you need separate layers for pH and texture. In this example three separate layers are needed to store the three different types of soil information. By contrast, a vector system would have one "soil coverage" composed of the soil polygons and a related database containing all the information (thickness, pH, texture) for each polygon. This is a drawback of the raster GIS; it requires a separate layer for each type of data and thus requires large amounts of storage space on the computer. Another disadvantage is that the smallest mappable unit is the individual cell. If the cell size is, say,

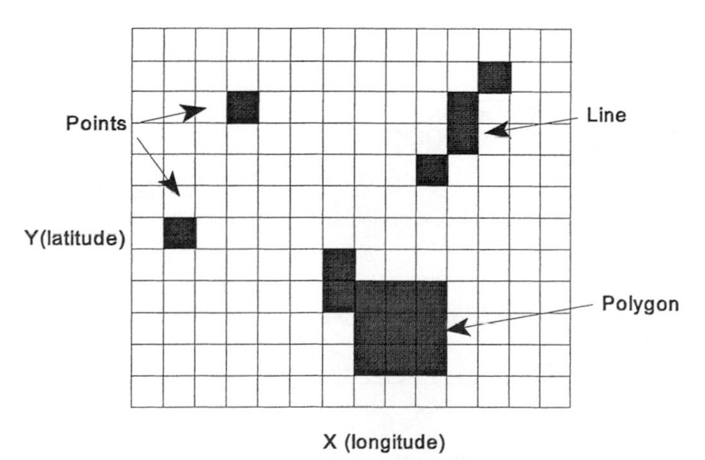

FIGURE 4.12 Representation of spatial features in a raster GIS. The cells produce a blocky representation of point, line, and polygon features compared to the vector system in Figure 4.11. (Source: Thompson, 1998)

100 by 100 meters, and you are mapping well locations, an individual well occupies one entire 100–meter square cell. This is why raster systems produce a "blocky" representation of spatial features. However, there are advantages to the raster system. The uniform grid structure makes mathematical manipulation of data much more efficient. Another advantage is that digital remotely sensed data are often stored in a raster–like format. This makes it easy to bring remotely sensed data directly into a raster GIS, and remote sensing has become an important means of gathering spatial environmental data.

While GIS serve an important role in the storage and display of data, it is their analytical capabilities which are their most powerful feature. A GIS can perform simple arithmetic operations such as calculating areas or the proportion of some area underlain by different types of soil. But one of the most powerful applications of GIS is the creation of new information. For example, a GIS could combine separate data (either coverages or layers) on soil type, land slope, and land use, and create new classifications of areas based on their soil–erosion potential (Figure 4.13). Resource analysts traditionally did this type of derivative analysis using a base map and transparent overlays of the different variables. Now it can be done more accurately and much faster using GIS. Geographic information systems con-

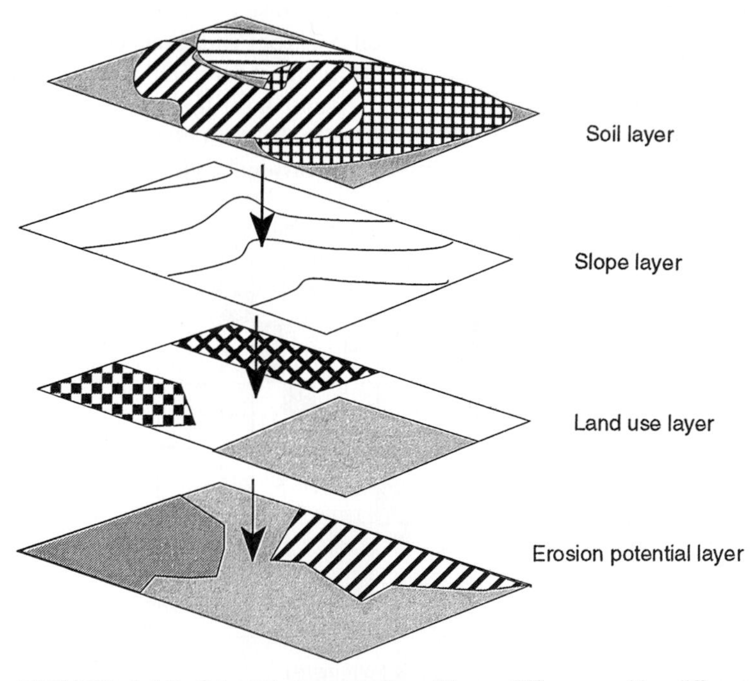

FIGURE 4.13 Schematic representation of how a GIS can combine different types of environmental data to derive new information. Here data on soil, slope, and land use are combined to create a map of erosion potential. (Source: Thompson, 1998)

tain a wide variety of analytical functions for manipulating spatial data. They can perform arithmetic operations (addition, subtraction, multiplication, and division) on data layers and statistical analyses, including spatial correlation and regression; can calculate distance, area, and drainage patterns; and can even do mathematical interpolation.

5

ECONOMICS AND WATER RESOURCES

Price Theory: Supply and Demand

Water Markets

Market versus Government Allocation

Welfare Economics

The Federal Subsidy to Water Users

Cost–Benefit Analysis

Structuring the Cost–Benefit Analysis

Cash Flow Diagram

Discount Factors

Hypothetical Cost–Benefit Analysis

Optimality Criteria

Identification of Benefits and Costs

Problems with Cost–Benefit Analysis

Random Nature of Future Payments and "Expected Value"

Evolution of Water Resource Planning Objectives and Procedures

Financial Analysis

Resource and Environmental Economics

The discipline of economics studies how individuals, firms, and governments make choices that determine the way resources are used. Resources are scarce, which means the demand exceeds the supply, and this is precisely why choice

TABLE 5.1 Inputs of Water for the Production and Manufacture of Various Products[a]

Product	Water use (gallons)
1 automobile	100,000
1 pound of cotton	2,000
1 pound of aluminum	1,000
1 pound of corn	170
1 pound of steel	25

[a] Miller, 1993, based on data from the U.S. Geological Survey.

matters. Water, like aluminum, wood, or oil, is an economic good because it has utility (value) to people. Water is often considered special because it is an economic good we cannot live without. In every society people need and use water for the domestic purposes of drinking, bathing, and cooking, though the amounts used vary dramatically between different cultures. Water is a production input to irrigation agriculture, just like fertilizer and seed. Water is also an important input to manufacturing and industrial processes. Making 1 pound of aluminum takes about 1000 gallons of water, and a typical car can require 100,000 gallons of water to make. Table 5.1 shows some commodities and the associated inputs of water required to produce them.

In some cases people pay very little for the water they use. The cost of water to farmers under Bureau of Reclamation contracts negotiated in the 1940s and 1950s was as little as $5 per acre-foot (325,828 gallons). In recently renegotiated contracts the price has risen to around $30 per acre-foot. But even at these higher prices irrigators in the West are still receiving a substantial subsidy. Owners of water rights do not actually pay for the water they use, but they do incur costs associated with conveying the water to the point of use, and treating and storing it if necessary. A private well owner does not pay for the groundwater itself, but does pays for the pump and the energy used to pump the water. Until the Clean Water Act in 1972 most cities and industries freely used the Nation's lakes and rivers to dispose of their wastes. There were costs associated with using streams for waste disposal, but they were borne by people downstream, and by the environment, but often not by the people who were polluting the water. This situation developed as cities and towns located along rivers. There was a rough balancing of the cost between communities though. While a community might not pay the costs associated with its own pollution, it would pay the cost from polluters upstream. Environmental damage to ecosystems was largely ignored.

The discussion in this chapter is limited to four general topics within economics that are relevant to water resources. First, *price theory* describes supply and demand relationships between producers and consumers. Clean water is produced by water utilities and people buy different quantities depending upon its price. Water markets are used in some western states to reallocate water between farmers, who have the water supply, and cities that have an increasing demand. A second area is *welfare economics*. Welfare economics evolved from the realization that private markets did not fully recognize the larger social costs and benefits associated with the production and consumption of resources. The goal of welfare economics is to maximize net social welfare. Profit maximization and economic efficiency are important to welfare, but so are issues of equity (fairness) in the use of resources. Welfare economics is primarily the domain of governments, because it is through the political system that decisions are made to address larger social goals. Since government has invested heavily in water resources, welfare theory guides much of the economic analysis of water projects and programs. A third area of economics relevant to water resources is *investment theory*. Investment theory provides the concepts and tools to analyze the value of money over time. Price theory, investment theory, and welfare economics are all important components of cost–benefit analysis, one of the most widely used techniques for evaluating investment alternatives. Finally, we use concepts from the field of *resource economics* and *environmental economics* to analyze externalities and economically efficient spending to achieve different levels of environmental quality.

PRICE THEORY: SUPPLY AND DEMAND

One of the most fundamental concepts in economics is the law of downward-sloping demand. Figure 5.1 is a graph showing supply and demand for an economic good. The ordinate (*Y* axis) shows the price (*P*) of the good, and the abscissa (*X* axis) shows the quantity (*Q*) of the good supplied by producers, or

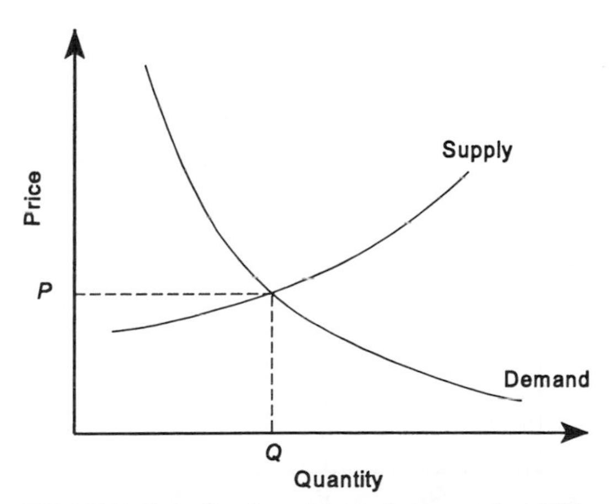

FIGURE 5.1 Supply and demand curves at the equilibrium price.

demanded by consumers, at that price. At a higher price producers would be willing to supply more but consumer demand would decrease; at a lower price consumers would demand more but producers would cut back on supply. Hence, Figure 5.1 shows the theoretical equilibrium condition between the price and the quantity supplied and demanded.

Water is an economic good and the law of downward–sloping demand applies; however, the special nature of water modifies the basic demand curve. Figure 5.2 shows a hypothetical demand curve for a residential water consumer. When the price of water increases demand decreases as expected. People lower their demand by reducing or eliminating low–priority or wasteful uses like washing a car or using water to clean sidewalks. As price continues to rise people may actually take shorter showers, or turn the water off while they brush their teeth. But with increasing price the amount of reduction in water use decreases because further reductions may require changes in behavior that are inconvenient or contrary to personal or social norms. And at even higher prices there will be no reduction at all if it means cutting into essential uses like cooking and waste disposal. In Figure 5.2 this relationship is shown by an increasingly steep demand curve (on the left) as price rises. At low prices people will buy and use more water, but there is a limit on how much water anyone can use, even if it is free. So as price falls, demand eventually drops off as well.

The rate at which demand changes as price changes is called the *price elasticity of demand.* (Similarly, there is an *price elasticity of supply.*) Conceptually, when demand changes a great deal for a given change in price demand is said to be *elastic.* When demand does not change much compared to the change in price demand is said to be *inelastic.* Economists define the elasticity of demand e as

$$e = \frac{dQ/Q}{dP/P}. \tag{5.1}$$

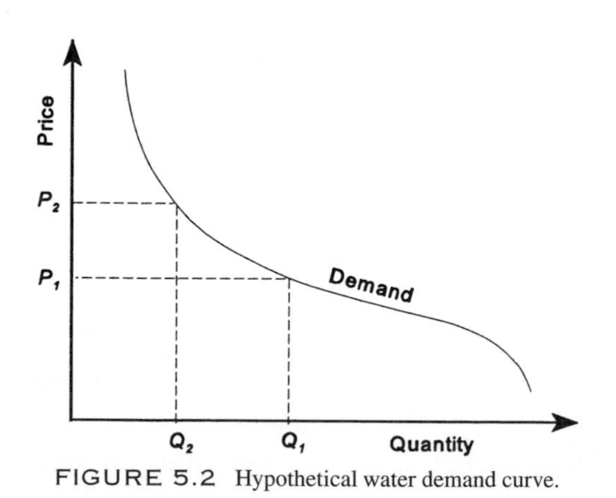

FIGURE 5.2 Hypothetical water demand curve.

Equation (5.1) is a differential equation from calculus and describes how demand changes *continuously* with price. Dividing the change in the quantity demanded, dQ, by Q, and the change in price, dP, by the price P, gives elasticity as the *percentage change in the quantity demanded divided by the percentage change in price.* In practice elasticity is calculated not by Equation 5.1 but by a simple algebraic difference equation. Equation (5.2) is the algebraic version of the price elasticity of demand:

$$e = \frac{\Delta Q/\overline{Q}}{\Delta P/\overline{P}} = \frac{(Q_1 - Q_2)/(Q_1 + Q_2)/2}{(P_1 - P_2)/(P_1 + P_2)/2}. \quad (5.2)$$

In Equation (5.2) ΔQ is the change in quantity demanded, and ΔP is the change in price. The subscripts 1 and 2 refer to two points on the demand curve (Figure 5.2). The overbar quantities are the average quantity and average price between points 1 and 2. The law of downward–sloping demand means elasticity is always a negative number. Technically, demand is considered elastic if $e < -1.0$ and is considered inelastic if $e > -1.0$. Economic theory says demand is more inelastic for necessities, and more elastic for luxury items. Figure 5.3 shows how the hypothetical residential water demand curve becomes inelastic at high and low prices and is more elastic in between. In a classic study of residential water use in the United States, Howe and Linaweaver (1967) found an average annual weighted price elasticity of demand of -0.40. In other words, if the price were doubled (raised 100 percent) demand would decrease 40 percent. Elasticity varied between the East and the West, and between indoor and outdoor water uses. The average elasticity for indoor use was -0.225 and for outdoor water use in the West it was -0.703. In a more recent study Jordan (1994) found a value of -0.33 for residential water use in Georgia.

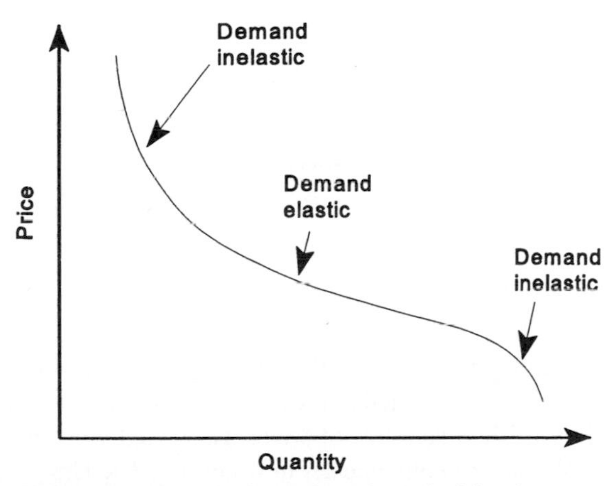

FIGURE 5.3 This figure shows how elasticity changes at different points along the water demand curve.

Elasticity has been used to predict the change in water demand with changing price. The assumption is that demand and price are related as

$$Q = aP^e, \tag{5.3}$$

where Q and P are quantity and price, respectively, a is a constant, and e is the price elasticity. Equation (5.3) describes a curve with a shape similar to the left two–thirds of the curve in Figure 5.3. Example 5.1 demonstrates a calculation of elasticity and its use in predicting water demand. The principles of supply and demand for water apply to industry and farmers just as they do to homeowners.

EXAMPLE 5.1

This example demonstrates how to calculate the price elasticity of demand and use the result to estimate water demand. Given below are hypothetical water use and price data.

Price of water (per 100 gallons)	*Quantity demanded (gallons per day)*
P_1 = \$0.146	$Q_1 = 220$
P_2 = \$0.180	$Q_2 = 190$

The price elasticity is

$$e = \frac{(220 - 190)/(220 + 190)/2}{(0.146 - 0.18)/(0.146 + 0.18)/2} = -0.701.$$

Assuming a price–demand relationship given by Equation (5.3), calculate the demand in gallons per day if the price were increased to \$0.19 per 100 gallons. Use a value of $a = 57.1$ for the constant in Equation (5.3).

$$Q = aP^e$$
$$Q = 57.1(0.19)^{-0.701}$$
$$Q = 183 \text{ gallons per day.}$$

WATER MARKETS

Supply and demand are the underlying forces driving the creation of water markets in western states. Water markets were discussed at the end of the section on prior appropriation in Chapter 3 in the context of transferring water rights. A number of obstacles were identified to the development of water markets, including the uncertainty of the quantities of water associated with the rights and the latent but potentially large public interest in water use. The U.S. Supreme Court's decision in the *Sporhase* case opened the door for development of interstate water markets.

The most well–developed water market in the West is found in northeast Colorado. The Northern Colorado Water Conservancy District (NCWCD) is a special water district formed to contract with the Bureau of Reclamation for construction and repayment of the Colorado–Big Thompson (C–BT) irrigation project. The C–BT project diverts water from the western slope under the continental divide into the Big Thompson River basin and storage reservoirs on the eastern slope. The service area for the NCWCD is wedge–shaped region extending from just north of the city of Fort Collins, south along the Front Range to Boulder, and tapering toward the northeast, following the course of South Platte River to the Colorado–Nebraska state line. The maximum amount of water the C–BT project can divert in any one season is 310,000 acre–feet. Thus the NCWCD created 310,000 "shares," with each share nominally worth 1 acre–foot of water. Each April the NCWCD declares the "quota" of water that will be delivered to a share. The long–run yield of C–BT shares has been 0.73 acre–feet per share (Michelsen, 1994). Shares can be freely traded within the C–BT project area. Shares are exchanged between farmers and between farmers, cities, and industry. Information on buyers and sellers and the price per share is readily available. Urbanization has changed the NCWCD. In 1962 agriculture held 86 percent of the shares and municipalities held 18 percent. By 1992 agriculture held 55 percent, municipalities 41 percent, and industry 4 percent (Michelsen, 1994).

The NCWCD is a unique water market because of the well–developed information network that matches buyers and sellers, the relative certainty of the quantities of water involved, and the ability to easily move water via project canals to any part of the service area. Throughout the 1970s and 1980s the price of a C–BT share fluctuated between $1200 to $2000. Table 5.2 gives some prices for water rights sold in western states in 1987.

The issue of western water markets becomes more interesting when Indian

TABLE 5.2 Examples of Water Rights Transactions in 1987 in the Western United States[a]

Location	Amount[b] (acre–feet)	Price (per acre–foot)	Type[c]	Purchaser
Denver	735	$2300	Shares	City
Northeast Colorado	615	$1050	Share	City
New Mexico (Rio Grande)	30	$1250	Ground	Developer
New Mexico (Pecos River)	3,000	$1100	Surface	Irrigator
Phoenix	700	$650	Ground	Developer
Reno	16	$2500	Surface	Developer
Salt Lake City	60,000	$164	Surface	Water District

[a] Shupe and Folk–Williams, 1988.
[b] Amount in acre–feet of permanent annual withdrawal right.
[c] Shares=delivery under shares held by an irrigation district.

reserved water rights are considered. There is no clear legal precedent either for or against a tribe's ability to market its water. Generally speaking, the sale of Indian water rights is prohibited by the Indian Non–Intercourse Act (1983), which prohibits the alienation of Indian resources without the consent of Congress. While Congress has been reluctant to endorse off–reservation marketing of rights, it has allowed tribes to negotiate some limited agreements. The agreements negotiated to date are unique unto themselves and reflective of local circumstances (McNally, 1994). But one constant remains in all of the agreements, they all prohibit the alienation of the water rights. Indian water marketing is still in its infancy, but it seems that in some circumstances marketing could allow tribes to make the most productive use of their water resource.

MARKET VERSUS GOVERNMENT ALLOCATION

How does a water market compare as a resource allocation method to government–controlled allocation? We can compare markets versus governmental allocation in terms of five characteristics—*security of ownership, flexibility, fairness, effectiveness in capturing all costs,* and *social responsibility.* Both markets and government control provide adequate security of ownership. Under existing laws and administrative systems water right holders either own water rights as private property or have indefinite or long–term renewable permits. It is possible that the government may change the conditions of a permit at the time of renewal, which is more uncertain than outright ownership of the water rights. But even ownership of water rights is not an absolute guarantee of security. There has been at least one case where a municipality condemned the water rights of an irrigation company. Under a market system, ownership is secure because owners do not have to sell if they do not want to. Neither allocation system would appear to hold an advantage here.

Ideally, resource allocation mechanisms should be flexible so they can adjust to changing conditions. Flexibility has both short–term and long–term dimensions. Short–term flexibility refers to the ability to reallocate the resource on short notice within an existing water–use pattern. A good example is temporary arrangements to move water between users during drought. Long–term flexibility refers to the ability to respond to changes in the water use patterns over time, for example, as irrigated farms are converted to suburban lawns. Government mechanisms do an adequate job at long–term reallocation, but are not well suited to short–term reallocation. Water markets, at least potentially, are flexible enough to respond to both short– and long–term changes. The advantage here would seem to be with the market.

A third characteristic that resource allocation mechanisms should possess is that water users should pay the full cost of the allocation. The full costs includes the social costs and the opportunity costs. The social costs are the internal costs plus the external costs. The opportunity costs are the cost associated with using water for certain purposes and not for others. With government allocation users frequently do not pay the full cost. Farmers receiving water from federal irrigation projects certainly have not paid the full cost because revenues from hydropower generation have been used to subsidize irrigation water development. Industries still use water to carry waste products, and they do not pay all of the external costs from their pollution. Market mechanisms can theoretically incorporate all costs, but markets never operate at their theoretical ideal. Another problem for both markets and governments is that they may stop at political boundaries, while externalities can flow beyond the boundary.

Resource allocation mechanisms should be fair. Overall, government probably does a better job at achieving fairness between users than do markets. Market systems are fair if all parties are adequately compensated in a transaction. Buyers and sellers are guaranteed to be treated fairly since they are willing participants. Third parties are not guaranteed fair treatment and might be injured by market transactions.

The last characteristic is that resource allocation mechanisms should be capable of reflecting larger social values. These larger values include clean water and a healthy environment. Governments do this by changing policy, laws, and regulations to reflect changing values. Changing western water law to make instream flows a beneficial use is an example of how some state governments have attempted to address changing social concerns about the environment. Markets are generally poor at incorporating public values. Markets are designed to promote efficient individual decision making, not to pursue broad social goals. The answer as to which allocation mechanism is better at reflecting social values is not clear, but the advantage may be with government. Markets can do some things better than government, and vice versa. But economists would certainly prefer water be allocated by market mechanisms.

WELFARE ECONOMICS

Welfare economics attempts to determine the allocation of resources that best promotes social welfare. Welfare economics is largely the province of government. Private markets work well for individual economic decision making, but rarely consider the larger goals of society. The allocation of resources according to welfare economics is *normative* in the sense that it attempts to find the allocation that best achieves a consensus on social values (James and Lee, 1971). People need material resources to promote their economic well–being, a healthy environment to promote their physical well–being, and psychologically satisfying experiences to promote their mental well–being. Water is one of many resources that can be used to meet these needs. A clean water supply and wastewater removal maintain our health, water can satisfy our needs for energy through hydropower, and water–based recreation is an important outlet for relaxation and psychological rejuvenation. While water can provide for these needs, so can other resources, and in some cases at a lower cost.

A variety of social goals have guided government planning and development of water resources, including *maximizing national income, maintaining or improving environmental quality, maintaining or improving public health, redistribution of income, regional development, national security,* and *institutional stability.* These goals are not mutually exclusive. Increasing national income, regional development, income redistribution, and a cleaner environment could all improve public health. As we saw in Chapter 2, some of these goals, like increasing national income, regional development, and environmental quality, have been pursued explicitly through water development. Other goals, such as redistribution of income and institutional stability, have been pursued more implicitly. The one goal of water–resource development that has been pursued with unwavering constancy is national economic development.

THE FEDERAL SUBSIDY TO WATER USERS

The federal government has invested billions of dollars in water–resource development. Much of this money was well spent, and the goals of the various projects and programs were largely achieved. But clearly millions of dollars were wasted and represented nothing more than pork barrel politics.

The federal government continues to subsidize water–related uses and users. Irrigators in the West still get water at less–than–market prices. People living on floodplains and along coastlines receive subsidies in the form of flood insurance and Corps of Engineer beach nourishment programs, as well as emergency relief, low–interest loans, and tax breaks when a flood occurs. Barges traveling federally maintained waterways do not pay user charges equal to the true marginal costs of construction, operation, and maintenance of the facilities. Water–resource projects are certainly not alone when it comes to federal subsidies to select groups of resource users. For years the federal government paid farmers not to grow certain types of crops. Ranchers in the western states do not pay market rates for grazing their cattle on public lands. The enormous subsidy to logging companies in the form of taxpayer–built logging roads and below–cost timber sales is legion. And the subsidy to the mining industry in the form of the ridiculously low prices for land within a mining claim is mandated by mining legislation that is more than 130 years old.

The federal government's policy of subsidizing the use of natural resources may have made sense at one time given the existing conditions, and the government's goals. Some subsidies never made sense, like when the government declared in 1928 that it would pay all the costs of flood control. As we approach the 21st century, maintaining subsidies to a select group of users is generally bad policy and bad economics. Subsidies distort the operation of the market and result in inefficient resource use. If we raise the price of water from federal irrigation projects, farmers will use less water on economically marginal crops. If we make people pay the full cost of living in floodplains, eventually fewer people will live there.

COST–BENEFIT ANALYSIS

One of the most significant contributions economists made to water resources was the development and refinement of cost–benefit analysis. The origin of cost–benefit analysis was the 1936 Flood Control Act where Congress stated that the government should pay for flood control as long as the benefits to whomsoever they accrued exceeded the costs. Congress had declared a very broad principle, and it took years to develop a formal cost–benefit analysis procedure. Cost–benefit analysis has become, for better or for worse, one of the most important components in decision making. In this section we develop the basic procedures of a cost–benefit analysis, discuss what are considered benefits and costs as related to water resources, analyze certain optimality criteria, and examine some of the method's problems.

STRUCTURING THE COST–BENEFIT ANALYSIS

Water–resource planning and management requires making decisions and then spending money to implement those decisions. For example, a water utility might need to increase its water supply to meet the increasing demand of a growing population. There are different ways (alternatives) to achieve the objective of increased water supply. The utility might drill a new well. It might be able to purchase water rights from a farmer and transfer them to the city, or it might even purchase water from another water utility. The basic question is, "which of these alternatives is the most *economically efficient*"? In other words, which alternative will accomplish the objective at the least cost? Cost–benefit analysis allows different alternatives to be compared to determine which is the most economically efficient. This is *all* cost–benefit analysis can do. If there are noneconomic objectives to be considered, cost–benefit analysis is of no help. Cost–benefit analysis can only analyze alternatives that are specified in monetary terms. The two obstacles that must be addressed so that different alternatives may be directly compared are *differences in kind* and *differences in time.*

Differences in Kind

Cost–benefit analysis requires that all benefits and costs associated with a given alternative be specified in monetary terms. For example, suppose two irrigation projects (alternatives) could be built for the same cost. Farmers in one project would use the water to produce cotton, while farmers in the other would produce corn. Which project should be built? Or more precisely, which project is more economically efficient? The answer depends on which is worth more—X bales of cotton or Y bushels of corn? The only way to directly compare the two projects is to convert bales of cotton and bushels of corn into dollars. Our earlier example of a water utility presents the same problem. The only way to directly compare the alternative of drilling a well to, say, the alternative of purchasing water rights is to compare the cost and benefits of each alternative expressed in dollars.

The first step in a cost–benefit analysis is to make all costs and benefits *equivalent in kind* by converting them into dollars. This requirement is also a major limitation of the cost–benefit method. Not all benefits and costs can be measured in dollars. What is the dollar value of an unobstructed scenic vista, or what is the value of preventing a species of plant from going extinct? Things that cannot be measured in monetary terms are called *intangibles* and cannot be included in a cost–benefit analysis. Intangibles need to be considered but they must be evaluated in a different decision–making framework, not in the cost–benefit analysis. Back in the early 1960s Senate Document 97 formally adopted multiobjective criteria for federal water–resources development. The "national and regional economic development" objective could be evaluated using cost–benefit analysis; the "social well–being" objective had to be evaluated some other way. Throughout the decade of the 1970s and into the early 1980s the Water Resources Council modified and updated the procedures used for evaluating the multiple objectives of federal water–resource investment. We will discuss this later on in the chapter.

Difference in Time

A water project provides a time series of benefits for many years into the future. Similarly, the project incurs a time series of costs associated with project construction and its operation and maintenance. Is one–dollar–worth of cotton today equal to one–dollar–worth of cotton at some time in the future? No, according to investment theory. Why? Because one dollar today, in the present (P), can be invested today, and it will be worth *more* than one dollar in the future (F). Exactly how much more depends upon the interest rate (i) and the length of time the dollar is invested (n). To a much smaller degree the future value also depends on the frequency of compounding, i.e., whether interest on the investment is compounded daily, monthly, or annually. Therefore a dollar today cannot be directly compared to a dollar in the future. Just as we must have equivalence in kind (all values in dollars), cost–benefit analysis requires equivalence in time. All payments must be compared over the same time period. Investment theory provides both the principles and the tools for comparing monetary values that occur at different points in time.

Equivalence in time is achieved using *discount factors.* The appropriate *discount rate* is, in theory, that rate where the demand for investment capital just equals the supply of capital from savings. For our purposes we will use the discount rate in the same way we use the interest rate (i). The interest rate causes an investment to increase in value over time; the discount rate is used to *discount* (reduce) a future amount of money to a smaller *present value.* Generally speaking the discount rate used for government spending is lower than the prevailing interest rate, while the discount rate for private investment is usually closer to the prevailing interest rate. The choice of the discount rate is very important because a small change in the discount rate for a capital–intensive project could change the project from being economical to uneconomical and vice versa.

CASH FLOW DIAGRAM

A cash flow diagram is a convenient way to visualize monetary values occurring at different points in time. In Figure 5.4 benefits are drawn as upward–directed arrows and costs as downward–directed arrows. The length of an arrow is proportional to the size of the payment. We will assume that all payments are made annually, and that the payment occurs as a lump sum at the end of the year. The analysis can be done for shorter (monthly or weekly) periods as well. In Figure 5.4 the first and largest cost occurs at the end of year 1. This might represent a large construction cost. There is another large construction cost in year 2, though not as large as in year 1. From year 3 to year 20 there are 18 uniform operation and maintenance costs. In years 8 and 17 there are additional costs. For example, these may be replacement costs for certain types of equipment. Benefits begin in year 3 and increase uniformly up to year 6, and remain constant thereafter to year 20, the final year of the analysis. For planning purposes costs and benefits may need to be estimated as average or "expected values." Expected values are considered in a later section.

DISCOUNT FACTORS

Discount (investment) factors are used to decrease (increase) the value of money over time. To find the value of a payment at any time, multiply the amount of money by the appropriate factor. Four factors are discussed below. The first is

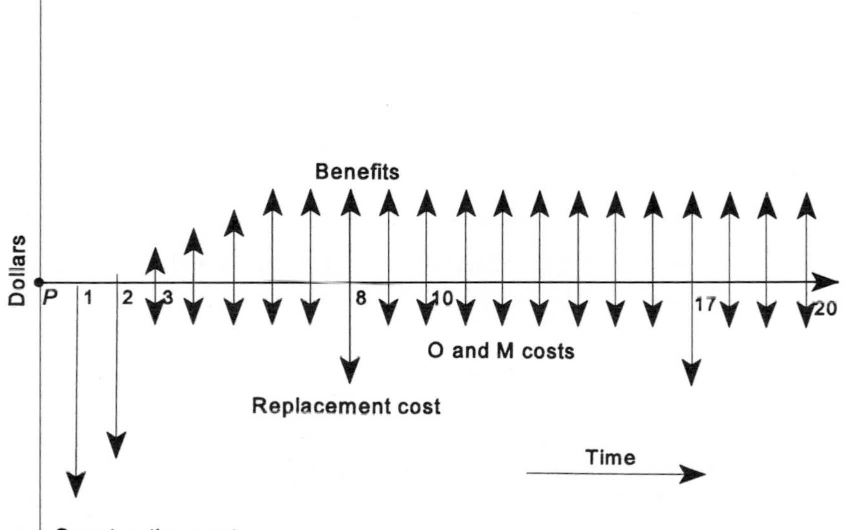

FIGURE 5.4 Cash flow diagram showing costs as down arrows and benefits as up arrows. The length of each arrow is proportional to the size of the payment.

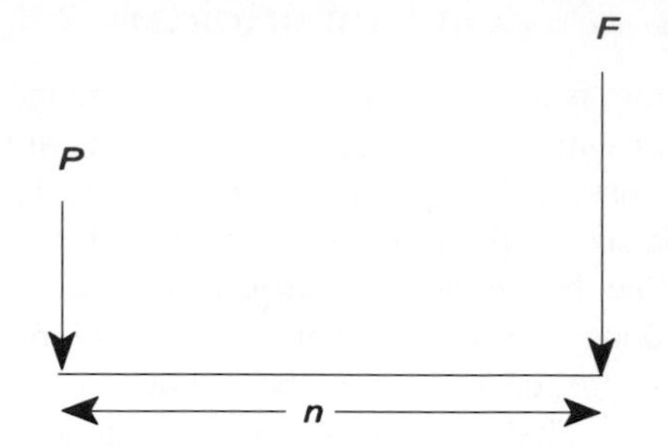

FIGURE 5.5 Schematic representation of how a single payment grows by compounding into the future or is reduced by discounting to the present.

the basic compound interest factor; the other three are discount factors. The factors are considered in two groups—single payment factors and multiple payment factors. Since all the factors are developed from the basic compound interest factor, we start with that factor.

Single Payment Compound Amount (*F/P, i, n*)

The single payment compound amount factor gives the amount of money that will accumulate after n years for each dollar invested at i percent. This factor describes how a single payment invested in the present grows to a larger value in the future through the compounding of interest. This is the factor used to describe the growth of money deposited in a fixed–interest savings account. Figure 5.5 schematically shows the effect of this factor. An amount of money invested in the present (P) grows to a larger future amount (F) in n years. A shorthand notation for this factor is $(F/P, i, n)$, which reads "the *F*uture value of a *P*resent amount invested at i percent for n years." The single payment compound amount factor is calculated as

$$(F/P,\ i,\ n) = (1 + i)^n. \tag{5.4}$$

To determine a future value, calculate the value of the factor and multiply it times the amount of money (see Example 5.2).

EXAMPLE 5.2

How much money would you have if you invested \$100.00 today at 5 percent interest for 15 years? Assume interest is compounded on an annual basis at the end of every year. Since you are putting \$100.00 in the bank one time and leaving it there this is a "single payment."

present amount P = \$100.00
interest rate i = 0.05
investment period n = 15 years

Using the shorthand notation the factor is

$$(F/P,\ 5,\ 15) = (1.05)^{15} = 2.0789.$$

Multiplying this factor by the $100.00 investment gives

$$\$100.00(2.0789) = \$207.89.$$

Notice that you use the interest rate expressed as a fraction, not as a percent, in Equation (5.4).

How much money would you have if you invested $100 at 7 percent for 15 years?

Single Payment Present Worth (*P/F, i, n*)

This factor discounts a single monetary value occurring n years in the future to an equivalent present worth. It operates as the inverse of the single payment compound amount factor. Looking once again at Figure 5.5, a future value (F) is discounted back over n years to a smaller present value (P). The shorthand notation is for the factor is $(P/F,\ i,\ n)$, which reads "the *P*resent value of a *F*uture amount discounted at i percent for n years." The factor is calculated as

$$(P/F,\ i,\ n) = \frac{1}{(1 + i)^n}. \tag{5.5}$$

Mathematically the single payment present worth factor is the inverse of the single payment compound amount factor (see Example 5.3).

EXAMPLE 5.3

Suppose someone offered to pay you $207.89, but not until 15 years from now? How much is $207.89 in 15 years worth to you today? In other words, what is the present value of $207.89 in 15 years? The answer depends upon the discount rate. Let us use the same 5 percent rate as in Example 5.2.

future amount F = $207.89
discount rate i = 0.05
discount period n = 15 years.

Using the shorthand notation the factor is

$$(P/F,\ 5,\ 15) = 1/(1.05)^{15} = 0.48102.$$

Multiplying this factor by the future amount gives the present value

$$\$207.89(0.48102) = \$100.00.$$

Not too surprising the answer is $100.00, because we simply reversed Example 5.2.

What is the present value of $207.89 discounted at 7 percent? What does a higher discount rate do to the present value of a future amount?

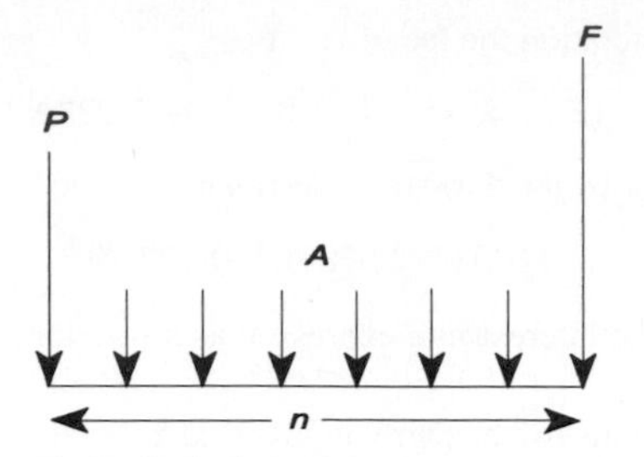

FIGURE 5.6 Schematic representation of how a series of uniform annual payments are discounted to a lump sum present value.

These are the two single payment factors. They are used to reduce a single future payment to a smaller present worth, or to grow a single payment in the present to a larger future amount. The next two factors are used for multiple payments. The first factor is from a group of factors called "uniform annual series factors."

Series Present Worth (*P*/*A*, *i*, *n*)

The series present worth factor calculates a lump sum present value from a series of uniform payments occurring annually into the future (Figure 5.6). Read the shorthand notation $(P/A, i, n)$ as "the *P*resent worth of an *A*nnual series discounted at i percent for n years." The factor is calculated as

$$(P/A,\ i,\ n) = \frac{(1 + i)^n - 1}{i(1 + i)^n}. \tag{5.6}$$

See Example 5.4.

EXAMPLE 5.4

Suppose you won a small lottery and the prize was \$100.00 per year for 15 years. How much would these 15 uniform annual payments be worth expressed as a single lump sum present value today discounted at 5 percent? From Example 5.3 you saw that a future payment is worth less when discounted to the present. Following this logic each successive \$100.00 payment is worth less than the previous payment because it occurs farther into the future; hence, it is discounted more to the present. You could solve this problem by using Equation (5.5) fifteen separate times, one time for each \$100.00 payment, and using the appropriate value for n. That solution is tedious in the extreme. Equation (5.6) does the same calculation in one easy step.

annual amount A = \$100.00
discount rate i = 0.05
discount period n = 15 years.

Using the shorthand notation the factor is

$$(P/A, 5, 15) = [(1.05)^{15} - 1]/[0.05(1.05)^{15}] = [2.0789 - 1]/[0.05(2.0789)] = 10.3797.$$

Multiplying this by the annual payments gives

$$\$100.00(10.3797) = \$1037.97.$$

The present value calculation demonstrates that $100.00 per year for 15 years is not worth $1500.00; rather it is only worth $1037.97 today. At a higher discount rate this uniform series of payments would be worth even less.

Uniform Gradient Series (*P/G, i, n*)

This factor is used to convert a series of annual payments that uniformly increase over time. In Figure 5.4 the annual benefits increase uniformly over the 4–year period from year 3 to year 6. The uniform gradient series factor converts this type of payment series into an equivalent lump sum value occurring at the beginning of the gradient, i.e., at the beginning of year 3. The factor is calculated as

$$(P/G, i, n) = \frac{(1 + i)^{n+1} - (1 + ni + i)}{i^2(1 + i)^n}. \tag{5.7}$$

There are factors to calculate the present or future value of nonuniform gradient series, but they are beyond the scope of this discussion. Table 5.3 lists these four

TABLE 5.3 Investment and Discount Factors

Factor	Notation	Formula
SINGLE PAYMENT FACTORS		
Single payment compound amount	(*F/P, i, n*)	$(1 + i)^n$
Single payment present worth	(*P/F, i, n*)	$\frac{1}{(1 + i)^n}$
UNIFORM SERIES FACTORS		
Sinking fund	(*A/F, i, n*)	$\frac{i}{(1 + i)^n - 1}$
Series compound amount	(*F/A, i, n*)	$\frac{(1 + i)^n - 1}{i}$
Capital recovery	(*A/P, i, n*)	$\frac{i(1 + i)^n}{(1 + i)^n - 1}$
Series present worth	(*P/A, i, n*)	$\frac{(1 + i)^n - 1}{i(1 + i)^n}$
GRADIENT SERIES FACTORS		
Uniform gradient series	(*P/G, i, n*)	$\frac{(1 + i)^{n+1} - (1 + ni + i)}{i^2 (1 + i)^n}$

factors along with three others that are commonly used in economic analyses. For example, the "capital recovery factor" is used to calculate the repayment schedule of a loan. The capital recovery factor gives the amount that can be withdrawn in equal amounts at the end of each of n years if one dollar is initially deposited at i percent. In this case the bank "deposits" the money with you (the loan) and recovers the capital as loan payments. Appendix 3 contains tables with values for the factors given different combinations of discount (interest) rate i and time period n.

HYPOTHETICAL COST–BENEFIT ANALYSIS

In doing cost–benefit analyses, all costs and benefits are converted into present values. There are two rules to observe in doing this. First, figure all benefits and costs using the same discount rate (i), and second, compare all alternatives over the same time period (n).

Table 5.4 gives hypothetical data for a simple water supply project and is correlated to the cash flow diagram in Figure 5.4. Table 5.5 demonstrates the cost–benefit procedure. The hypothetical example contains single payments, uniform annual payments, and a uniform gradient series.

Costs

The two construction costs are single payments and are converted to present values using the single payment present worth factor (P/F, i, n). The first construction payment is discounted one year; the second payment is discounted two years. The operation and maintenance (O and M) costs are a uniform series at $10,000 per year. These payments are handled in two steps. First, the present value of the

TABLE 5.4 Data for a Hypothetical Water Supply Cost–Benefit Analysis

discount rate: $i = 5$ percent	
period of analysis: $n = 20$ years	
COSTS	
Construction costs	
Year 1	$1,500,000
Year 2	$500,000
Operation and maintenance	
Years 3–20	$10,000 per year
Replacement costs	
Year 8	$100,000
Year 17	$100,000
BENEFITS	
Water supply benefits	
Years 3–6	Uniform increase at $40,000 per year
Years 7–20	$160,000 per year

TABLE 5.5 Hypothetical Cost–Benefit Analysis Using Data from Table 5.4

COSTS

Construction costs

$1,500,000(*P/F*, 5, 1) = $1,500,000(0.9524) = $1,428,600

$500,000(*P/F*, 5, 2) = $500,000(0.9070) = $453,500

Operation and maintenance

$10,000(*P/A*, 5, 18)(*P/F*, 5, 2) = $10,000(11.6896)(0.9070) = $106,025

Replacement costs

$100,000(*P/F*, 5, 8) = $100,000(0.6768) = $67,680

$100,000(*P/F*, 5, 17) = $100,000(0.4363) = $43,630

Total present value of all costs = $2,099,435

BENEFITS

$40,000(*P/G*, 5, 4)(*P/F*, 5, 2) = $40,000(8.6488)(0.9070) = $313,778

$160,000(*P/A*, 5, 14)(*P/F*, 5, 6) = $160,000(9.8986)(0.7462) = $1,181,814

Total present value of all benefits = $1,495,592

B/C = ($1,495,592/$2,099,435) = 0.71

series is calculated using the series present worth (*P*/*A*, *i*, *n*), where $n = 18$. This converts the series to a "present" value occurring at the beginning of year 3. The second step is to bring this single value the rest of the way back to the present (beginning of year 1) by multiplying it by the appropriate single payment present worth factor. The final two replacement costs are single payments and are converted to present values using the a single payment present worth factor. The total present value of all costs is $2,099,435.

Benefits

The water supply benefits uniformly increase from $40,000 in year 3 to $160,000 in year 6. These are converted to present value using a similar two–step procedure. The four gradient series payments are converted to a present value at the beginning of year 3 by means of the uniform gradient factor, and then brought to the beginning of year 1 as a single payment. Observe that the value which is multiplied by the uniform gradient factor is the annual *increase* in the gradient payment ($40,000). From year 7 to year 20 the benefits are a uniform series. These are brought to the present using the same two–step procedure used for the O and M costs. The total present value of all benefits is $1,495,592. The final ratio of benefits to costs is B/C = 0.71. The project thus fails to achieve a benefit–cost ratio of at least 1.0, and it should not be built. There are other techniques besides cost–benefit analysis which are used to evaluate economic efficiency and include the *internal rate of return* method and the *annual cost* method. These are covered in standard texts on engineering and resource economics.

Cost–benefit analysis is used in one of three ways. The first application is to assess the economic characteristics of a particular project. This is what we did in the hypothetical example above and it is probably the most common application. Alternatively, cost–benefit analysis can be used to determine which of a number of different alternatives designed to serve a given purpose results in the largest ratio of benefits to costs. Lastly, cost–benefit analysis can be used to determine which of a number of projects designed to serve different purposes confers the largest net benefit to society (Sewell *et al.*, 1965).

OPTIMALITY CRITERIA

Cost–benefit analysis evaluates the economic efficiency of a particular alternative. If we develop multiple alternatives and perform a cost–benefit analysis on each alternative, we can see the change in economic efficiency as the size of the project changes. This is shown graphically as Figure 5.7. The *X* axis in Figure 5.7 is total project cost while the *Y* axis is total project benefit. The 45–degree line from the origin represents all projects that have total benefits equaling total costs. The solid curve rising above the 45–degree line is the hypothetical plot of the costs and benefits for different project scales. This curve is called the "project curve." Three points along the project curve are significant. Point *X* is the project configuration that achieves the highest benefit–cost ratio. Everywhere along the project curve from the origin to point *X* the marginal benefits (MBs) are increasing faster than the marginal costs (MCs). But *X* is not the optimal project scale because between *X* and *Y* the MBs are still greater than the MCs; however, beyond *X* the MCs are increasing faster than the MBs. At point *Y* the MB = MC, which

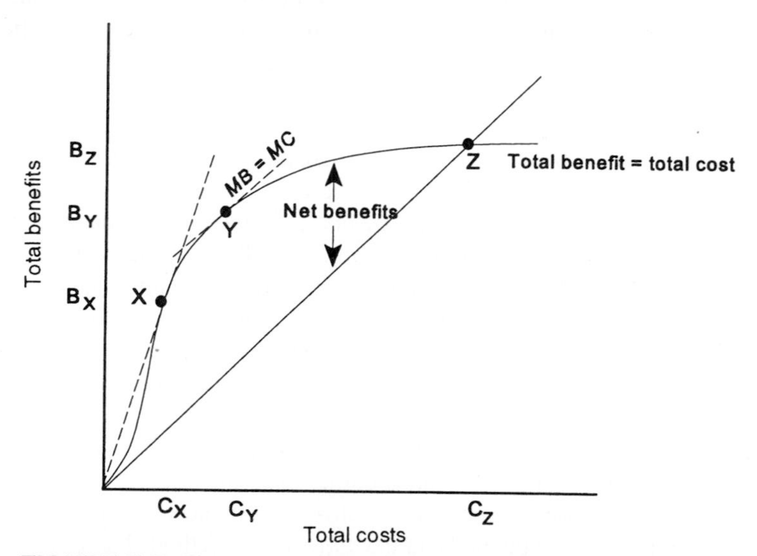

FIGURE 5.7 The project curve describes different scale projects and their associated benefits and costs. The optimal project is *Y* where marginal benefits (MB) equal the marginal costs (MC) and net benefits are maximized. The benefits and costs for project *X* are indicated on the axes as B_x and C_x, respectively, and similarly for projects *Y* and *Z* (after Sewell *et al.*, 1965).

is the necessary condition for optimality. Project *Y* is also the project size that generates the greatest *net* benefits. Both of these criteria indicate that *Y* is the optimal project scale. Everywhere along the project curve from the origin up to point *Y* the MB > MC. In other words, one dollar in cost generates more than one dollar in benefit. At point *Y* the slope of the project curve equals the slope of the 45–degree line. Beyond *Y* the MB < MC, which is shown by the slope of the project curve being less than 45 degrees. At *Z* total benefits equal total costs and the benefit–cost ratio is 1.0. Everywhere from the origin to point *Z* the ratio is greater than 1.0.

IDENTIFICATION OF BENEFITS AND COSTS

An important part of the evolution of cost–benefit analysis has been the correct identification and specification of benefits and costs. *Direct benefits* are gains to people who make direct use of the goods and services provided by the project. For an irrigation project these are gains in income to farmers using irrigation water. In some cases direct benefits include the change in the value of the land, since land with irrigation water is more valuable than the same land without water. For a flood control project direct benefits are calculated as the *damages avoided.* For a navigation project the direct benefits are the direct gains to the people using the waterway. *Indirect benefits* are benefits induced by the project. For example, as a result of a new irrigation project, related businesses may open, e.g., a farm equipment dealership, gas stations, trucking companies, and feed stores. These indirect benefits must be carefully considered. If the new activity is only a transfer of an existing activity from some location to the new location, it does not represent a new benefit and should not be counted from a *national viewpoint.* From a *local* or *regional viewpoint* it may represent an economic gain; however, some other region suffered an economic loss as a result of the transfer. *Intangible benefits* are not quantifiable and are not included in the cost–benefit analysis. They must be considered some other way.

Direct costs include the costs in goods and services used to construct, operate, and maintain the project. Labor and materials, rights–of–way purchases, and engineering and administrative costs are all included as direct costs. *Associated costs* are costs incurred by the direct beneficiaries in order to utilize the project output. A farmer may have to build on–farm facilities (water control structures, ditches, etc.) in order to use the irrigation water. *Indirect costs* are the costs associated with the production of the indirect benefits. *Intangible costs,* like intangible benefits, are not priced in the market and are not included in the analysis, but are evaluated along with the intangible benefits some other way.

PROBLEMS WITH COST–BENEFIT ANALYSIS

As much has been written about the limitations and problems of cost–benefit analysis as with how to properly conduct one. The problems can be grouped into two categories—analysis–level problems and theoretical problems.

Analysis–Level Problems

There are at least four basic problems with cost–benefit analysis. The first is the problem of correctly identifying and specifying *all* benefits and costs. Some of the more obvious benefits and costs were discussed above but there are others as well, especially external costs and benefits that may be created by the project. Building a new irrigation project increases salinity levels in the water downstream. This is a cost for downstream users, but is it included in the analysis? It may be if the users are in the same political/administrative region; it may not if the users are on different sides of a political boundary. For years the United States passed along the external costs of increasing salinity to Mexican farmers with no compensation.

A second analysis–level problem is the need to assume a time stream of benefits and costs into an uncertain future. In planning a project a period of analysis (n) is chosen, but it is all but guaranteed that conditions will change in the future. Recall the Corps of Engineers' experience in the Everglades. One method to account for uncertain benefits and costs is to specify them as a function of some probability distribution, but such an approach can only account for natural variations, e.g., fluctuations in weather or streamflow. There is no way to account for the uncertainty associated with future changes in society's values.

A third problem arises in choosing the appropriate discount rate. Public spending uses a discount rate lower than that used by private markets. Low discount rates favor capital–intensive projects that produce benefits many years into the future; this is precisely the character of most water projects. High discount rates reduce future values so much that the benefits never offset the large initial construction costs, and the B/C ratio is less than 1.0. Until the government standardized the procedures for analysis, agencies treated the discount rate as a variable, manipulating it until the project gained a favorable B/C ratio. Figure 5.8 shows

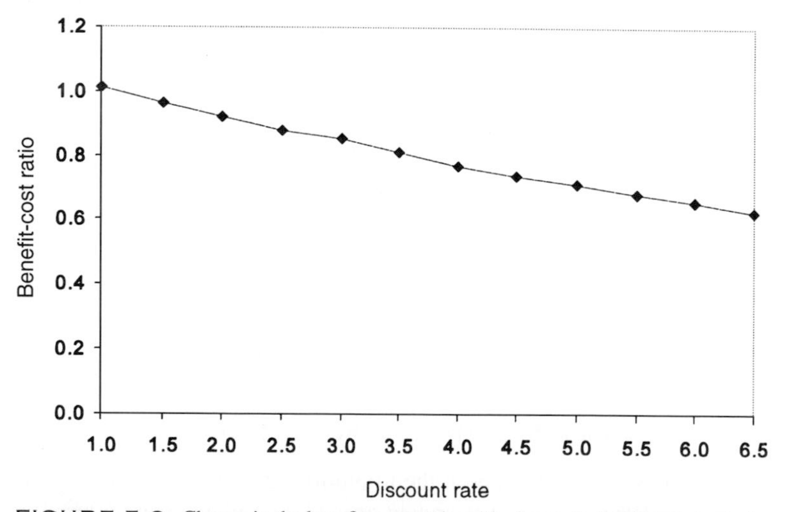

FIGURE 5.8 Change in the benefit–cost ratio with change in the discount rate for the hypothetical data in Table 5.4.

how changing the discount rate for the hypothetical data in Table 5.4 changes the B/C ratio. As the discount rate is lowered, the B/C ratio increases. For this particular set of data the discount rate would have to be a ridiculously low 1 percent before the project achieved a B/C ratio greater than 1.0.

The last problem is that the cost–benefit procedure gives the impression of analytical precision when in fact it is based on many assumptions and estimates. This may give quantifiable values an advantage over nonquantifiable values in the minds of decision makers. Decision makers may be more impressed with the presentation of "hard" numbers as opposed to more qualitative descriptions of intangible impacts. This was certainly a motivation for the U.S. Fish and Wildlife Service to quantify ecological impacts, and the National Park Service to get dollar estimates for recreational experiences using various indirect measures such as "willingness to pay" and "shadow prices." Without being equivalent in kind, theses values cannot be part of the formal cost–benefit analysis.

Theoretical Problems

Cost–benefit analysis is founded upon the concept of the *Kaldor–Hicks* (K–H) compensation test. The K–H test is a necessary condition for *Pareto optimality.* Briefly, the K–H test says that if a change in resource allocation would create a new economic surplus (net benefits) that would be sufficient to compensate any losers (those who would be made worse off by the change) and still leave some excess for the winners (those who would be made better off), then the change in resource use is economically efficient. In other words, the change is economically efficient if the winners would be able to fully compensate the losers and still be ahead. This is a theoretically sound concept, but that is just the problem, the K–H test is only theoretical. The test says *if* winners could compensate losers; compensation is not actually required. As a result, a project could have a benefit–cost ratio greater than 1.0, and yet there may be people who are worse off as a result of the project.

A second theoretical problem is that the analysis assumes a constant marginal utility of money between all individuals. In plain English this means one dollar is assumed to have the same value to everyone. This assumption is not true. One dollar to a homeless person is undoubtedly of greater marginal utility (value) than one dollar to a multimillionaire.

RANDOM NATURE OF FUTURE PAYMENTS AND "EXPECTED VALUE"

Since future benefits (and some costs) depend upon the availability of water, which is uncertain because of natural fluctuations in the hydrologic cycle, future benefits from water development are likewise random and uncertain. For example, the annual recreational benefits at a reservoir will vary with the amount of water in storage. When water levels are low, as might happen during a drought, fewer people go to the reservoir (see Example 5.5). In a similar fashion flood control benefits (damages avoided), hydropower production, and water supply all vary with water conditions. Randomly varying benefits and costs can be assigned a single annual value for inclusion in a cost–benefit analysis by calculating their *expected value.* Expected values are probability–weighted values. The expected value of a random time series of benefits $E(B)$ is calculated as

$$E(B) = \sum_{i=1}^{n} P_i b_i, \tag{5.8}$$

where P_i is the probability of occurrence of the ith event, and b_i is the benefit associated with the occurrence of the ith event. Equation (5.8) says sum all the products of event probability and the associated benefits. The sigma symbol is just shorthand notation for the summation operation.

EXAMPLE 5.5

Assume the recreational benefits at a reservoir increase with increasing water level. Given below are water level ranges, the associated recreational benefits, and the probability of the water level being within a certain range during a summer season.

Reservoir level (feet above datum)	*Recreation benefits*	*Annual probability*
>30	$250,000	0.15
25–30	$100,000	0.40
20–25	$ 65,000	0.25
<20	$ 40,000	0.20

The annual expected value $E(B)$ is calculated from Equation (5.8) as

$$E(B) = 0.15(\$250{,}000) + 0.40(\$100{,}000) + 0.25(\$65{,}000) + 0.20(\$40{,}000)$$
$$E(B) = \$101{,}750 \text{ per year.}$$

The value $101,750 would be treated as a uniform annual value (A) in a cost–benefit analysis. Note that the sum of the annual probabilities must equal 1.0.

EVOLUTION OF WATER RESOURCE PLANNING OBJECTIVES AND PROCEDURES

Box 5.1 lists major developments in the evolution of federal water–resource planning and shows how the national objectives for water planning and management have changed over time. Prior to the 1936 Flood Control Act federal objectives included national and regional economic development, national security, and social welfare. Nowhere were these objectives explicitly articulated. They were implied from the types of activities that were undertaken by the federal govern-

BOX 5.1 EVOLUTION OF FEDERAL WATER PROJECT OBJECTIVES AND EVALUATION PROCEDURES

DATE	PROCEDURAL GUIDELINES	NATIONAL OBJECTIVE(S)
Pre–1936	Informal agency evaluation	Implicit objectives (economic growth, national security, social welfare)
1936	1936 Flood Control Act	Economic development
1950	Circular A–47 (Green Book)	Economic development
1962	Senate Document 97	National and regional economic development; social well–being; preservation of resources
1971	Water Resources Council's (WRC) proposed *Principles and Standards*	National economic development; environmental quality; social well–being; regional development
1973	WRC's *Principles and Standards* (P&S) for planning water and related land resources	National economic development; environmental quality. (where adverse impacts occur to regional development and social–well being they should be displayed)
1980	WRC's P&S revised as *Procedures*	National economic development; environmental quality
1983	WRC's *Principles and Guidelines*	National economic development

ment. Similarly, there were no official guidelines for evaluating water projects, or progress toward these objectives. The 1936 Flood Control Act stated that the federal government should invest in flood control wherever and whenever the benefits exceeded the costs. This act gave primacy to economic development as the objective of federal water development, and it qualitatively stated the procedure for evaluation. The techniques of cost–benefit analysis were progressively refined, and Circular A–47 (the Green Book) (1950) marked the first formal standardization of cost–benefit analysis procedures. Senate Document 97 further refined the analysis, but embraced multiobjective planning for federal water development. The Water Resources Council developed and refined multiobjective planning and project evaluation procedures. The WRC's recommended principles and standards included four explicit objectives: national economic development, environmental quality, social well–being, and regional development. The WRC recommended using four separate accounts to display all beneficial and adverse effects toward each objective. The four–account approach was quickly considered too cumbersome by the agencies, and the four objectives were reduced to two (national economic development and environmental quality) in the final *Principles and Standards for Planning Water and Related Land Resources* (WRC, 1973). Where appropriate, beneficial and adverse impacts on regional development and social well–being were to be displayed. In order to reduce the burden on federal agencies, the Reagan administration repealed most of the established planning procedures and instituted the *Principles and Guidelines* (WRC, 1983). The *Principles and Guidelines* had only one objective—national economic development. The *Principles and Guidelines* are still the basic guidelines for federal agencies today. In nearly 50 years of project evaluation the national objectives went full circle from economic development, to multiple objectives, and back again to the single objective of economic development.

FINANCIAL ANALYSIS

Cost–benefit analysis addresses the economic efficiency of a project. Financial analysis deals with cost allocation for repayment of the project. Cost allocation determines which purposes of a government project are reimbursable and which are nonreimbursable costs. The reimbursable costs are then allocated to the respective beneficiaries.

RESOURCE AND ENVIRONMENTAL ECONOMICS

Resource economics is a well–established field within the discipline of economics. Environmental economics has evolved and matured in the last four decades, and can be considered an evolutionary extension of resource economics. Externalities are central to environmental economics, but they are not unique to environmental economics. Any cost not paid for, or any benefit captured as a windfall, is an externality. Environmental economics takes the externality concept and applies it to environmental issues.

It is a fundamental tenet of environmental economics that pollution problems are symptoms of an improperly functioning market. Producers and consumers are making incorrect decisions, because the market is sending the wrong signals. If in producing good *A* external costs are passed on to either specific individuals, society at large, or the environment, then the price for *A* is actually too low because it does not include the external pollution costs. Internalizing the externalities raises the price of good *A*. When the price rises, consumers demand less, the supply of good *A* decreases, and there is less pollution. Producers also respond to the rising price by becoming more efficient in their production, thus reducing their pollution and lowering their costs. But producers are unlikely to internalize external environmental costs on their own, since they would be at a competitive disadvantage (higher price) compared to other producers. This is where the government plays a critical role by requiring all producers to internalize costs. A key issue, however, is the type of government intervention into the market. The United States has favored the "command–and–control" approach to pollution control. This approach uses uniform regulations to achieve environmental goals. The command–and–control approach is usually economically inefficient because it ignores the variable marginal costs of pollution mitigation by different industries. The approach also suffers from inflexibility and may discourage innovation.

Alternatively, the government may use a more market–based approach. With a market approach the government determines the total quantity of pollution that a particular area can handle, and then allocates pollution credits equal to this total to the various industries. Individual industries are free to trade their credits in a market–like system. Those industries that can more cost–effectively reduce their pollution will do so and sell their surplus credits to those who cannot easily reduce their pollution. In theory the overall environmental goal is met, but in a more economically efficient manner. Tradable credits have been used for individual air pollutants like sulfur dioxide, lead, and even chloroflourocarbons (CFCs), but have been rarely used for water pollution. The potential for establishing water pollution markets appears to be more feasible for point sources than for nonpoint sources. In addition to using tradable credits, the government can eliminate subsidies and reduce market barriers as ways to protect the environment and encourage optimal resource use.

Another area where environmental economics has influenced water resources is the development of methods for quantifying hitherto intangible values like recreational benefits, and fish and wildlife values. These must be quantified if they are to be included in cost–benefit analyses. Economists have developed and refined different methods of *contingent valuation.* Contingent valuation methods derive monetary values indirectly. One of the most popular contingent valuation methods is the willingness to pay survey. People are asked what they would be willing to pay for a resource or a certain level of environmental quality. But even as economists have refined their methods for estimating these values, new issues have emerged. How, for example, do you account for the benefits to people who value the resource, but who do not use it in any tangible way? In other words, there can be both use and nonuse values associated with a resource (Madariaga and McConnell, 1987). Economists call this type of value *existence value,* and there are different kinds. An example is an individual willing to pay to clean up a polluted lake, knowing that she will never visit or actually use the lake in any way. The person may simply value having the lake cleaner (*intrinsic existence value*), or the fact that other people will use and enjoy the water (*altruistic existence value*). Defining existence values presents some interesting theoretical challenges. Not only are they difficult to measure, they are difficult to define.

The last area where environmental economics has informed the debate over water quality is by addressing the question of how clean we want our environment. Economists argue that we do not want or need to completely eliminate pollution for two reasons. One reason is that water can decompose and assimilate some types of biodegradable pollution. Using the water's capacity to break down degradable pollutants is a reasonable use of water. We must be careful to not exceed the water's assimilative capacity, and we should not discharge nondegradable pollutants like heavy metals into the environment. The second reason we do not want zero pollution discharge is because it is not economically efficient. In using resources we create *waste disposal costs.* Waste disposal costs are composed of the *costs of cleaning up* or *preventing* water pollution and the *welfare costs* associated with the damages from pollution. Building a wastewater treatment plant or pumping and treating contaminated groundwater are both cleanup costs. When an industry upgrades its production technology and prevents water pollution by recycling wastewater, this is a pollution–prevention cost. Welfare costs include impacts on public health, lost recreation opportunities, and aesthetic impairment. If we consider deleterious impacts on aquatic ecosystems as reducing our welfare, as they surely do, then all of the environmental impacts can be considered welfare costs.

According to economic theory we should only clean up and/or prevent pollution to the point where the marginal cleanup/prevention costs just equal the marginal damages (welfare costs) from pollution. This situation is shown graphically in Figure 5.9. The economically efficient level of pollution removal is where the two marginal cost curves intersect (point X on the graph). To the left of X the benefits (damages avoided) of removing one more unit of pollution are greater than the costs of removing that unit. To the right of X the cost of removing the

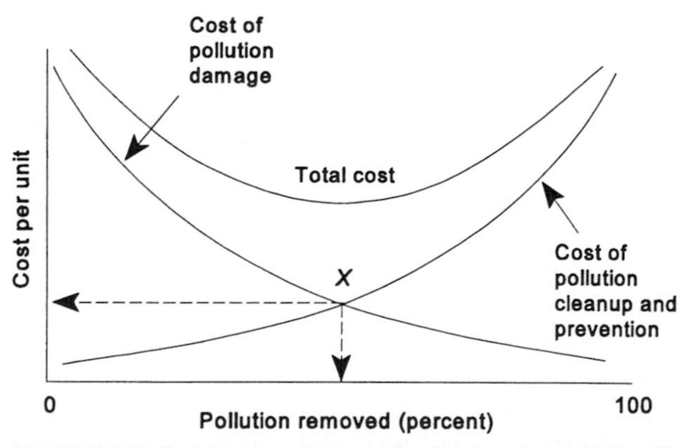

FIGURE 5.9 Schematic diagram showing the economically efficient level of pollution removal. The optimal level X is where the marginal cost of removal equals the marginal cost of damages. The total marginal cost is minimized at this point.

next unit of pollution exceeds the benefits received from doing so. Point X also indicates where the total cost (cleanup and prevention costs plus damages) is minimized. While this is good theory it is difficult to implement in the real world. It is hard to identify all the damages from pollution, and some costs are going to be intangible. Also, different groups will differ in how they value pollution damages. To some people a healthy aquatic ecosystem is extremely valuable; to others it may have little or no value. This highlights one of the fundamental limitations of environmental economics. We only assign value to environmental goods and services relative to a human reference—we do not know their "real" value apart from their relevance to the welfare of the human species.

6

WATER DEMAND AND SUPPLY

MANAGEMENT AND PLANNING

Water Demand

Urban Water Demand Planning

Water Supply

Dams and Reservoirs

Interbasin Transfers

Planning and Decision Making

Planning Entities

The Planning Process

Public Participation

Constraints to Water Management Decision Making

Water supply planning involves the duel components of estimating the future demand and assessing the potential sources of supply. Until the mid 1970s increasing water demand was considered inevitable (Figure 4.2), and planning activities focused almost exclusively on developing new sources of supply to meet the anticipated demand. As we came up against environmental limits in water availability, and negative environmental impact associated with increasing rates of water withdrawal, consumption, and pollution, it became imperative to embrace water–demand management as well. Managing water demand means using incentives and disincentives, new technologies, laws and regulations, and educational programs to promote water conservation and improved water–use efficiency. At the federal level the importance of water conservation was recognized as early as the Senate Select Committee Report (U.S. Congress, 1961), but it was the Carter administration that integrated water conservation into federal project planning in the late 1970s.

Satisfying the future demand for water supply will require expanding the range of choice of the options available. In some cases developing new supplies might still be the preferred option. More likely the option(s) will involve a creative mixture of improved water–use efficiency and water conservation, supplemented perhaps by some additional supply. In still other areas, especially in the West, demand management must be, and is becoming, the dominant approach to meet future water needs. In this chapter we consider some of the factors that influence water demand, examine some water supply alternatives, and finally consider the planning and decision–making process. Chapter 7 looks at water for urban and agricultural (irrigation) uses, and the role of water conservation and improved water–use efficiency within each sector.

WATER DEMAND

A variety of factors influence the demand for water regardless of the particular use. As we saw in Chapter 5, *price* is an effective tool for managing demand. As price goes up, demand goes down (Figure 5.2). For groundwater users it is not the price of water, but the price of the energy to pump the water which influences demand. In the plains of west Texas overlying the Ogallala Aquifer, the rising cost of energy during the 1970s was the instrumental factor in reducing irrigated acreage. The higher pumping costs facing the farmers could not be easily passed on to consumers in the form of higher prices for the food and fiber crops (largely cotton) being grown. Many irrigated areas have now reverted to dryland farming, even though there is still water in the aquifer. It just is not profitable to pump the water to grow low–value crops. For municipalities increased pumping costs are easily passed along to consumers in the form of higher water bills. Domestic water use, being more of a necessity, is much more inelastic than water for irrigation.

Technology is an important factor influencing water demand. On the one hand technological inventions have increased water use during the 20th century. The development of indoor plumbing dramatically increased per capita water use. Similarly, untold manufacturing and industrial processes require water as an integral component of production (Table 5.1). But technology can also reduce water use by increasing water–use efficiency. Low–flow toilets and showerheads and automatic timers on lawn sprinklers can save water in and around the home. Industrial wastewater recycling systems and redesigned production processes can reduce industrial raw–water withdrawals, as well as wastewater discharges. The potential water savings from new irrigation technologies like water scheduling and improved methods of water application are enormous. Urban water supply systems can easily lose 10 to 20 percent of their treated water supply through leaks in the distribution lines. Since the city already paid to treat this water, better leak detection saves not just water, but the money spent on water treatment.

A very simple technological device that affects water demand is the water meter. For example, the city of Boulder, Colorado, adopted universal metering in the early 1960s, and the impact on water use was immediate (Figure 6.1). Meters work in combination with price because people can be charged per unit of water used. With unmetered systems people pay a flat rate regardless of the amount of water actually used. For domestic water use, metering is most effective in reducing outdoor water use, since outdoor use is more elastic than indoor use. Still today most residential water users in New York City are unmetered. Farmers in some western states overlying the Ogallala Aquifer have been required to install meters on their irrigation wells as part of groundwater management programs.

Laws and regulations affect water demand in many ways. The Clean Water Act has provided the incentive for the adoption of industrial water recycling. Industries are finding that it is cheaper to recycle their effluent than to treat it for discharge directly to the environment. In 1992 Congress passed the National Energy Policy Act. One of the most significant sections of the act, as it effects water resources, was the requirement that in January 1994 all new toilets sold in the United States have a maximum flush capacity of 1.6 gallons. Most existing toilets use 3.5 gallons, and the oldest models can use as much as 5 gallons per flush. More than 4.8 billion gallons of water are flushed down toilets each day (U.S. Environmental Protections Agency (USEPA, 1995). As the existing stock of older

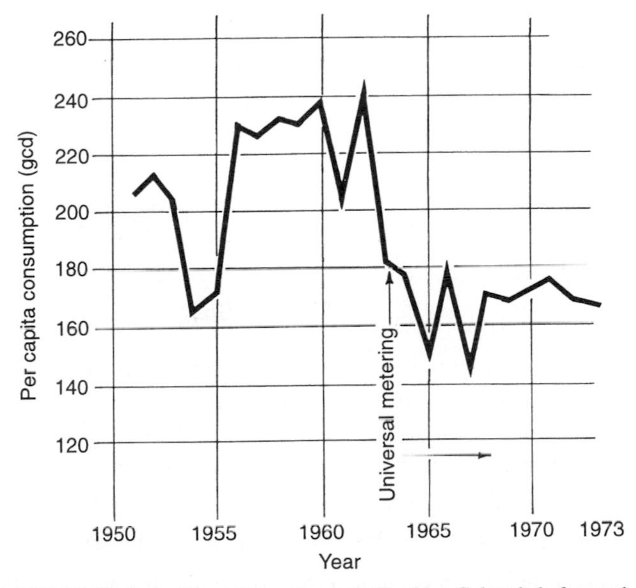

FIGURE 6.1 Per capita water use in Boulder, Colorado before and after the adoption of universal metering. (Source: Metropolitan Denver Water Study Committee, 1975)

toilets is eventually replaced this law could have a significant impact on both the demand for urban water supply and the need for future wastewater treatment capacity. Some cities have building codes that mandate water–conserving plumbing fixtures in new construction. The city of El Paso, Texas, started a program to replace water–guzzling toilets more than 10 years ago.

In the West, outdoor water use can compose 30 to 50 percent of the total domestic use. Lawn watering is one of the major culprits. Most lawn grasses are not adapted to the drier climates of the West, and must be irrigated regularly. This is also true for golf course turf, though some golf courses irrigate with treated sewage effluent. More and more cities are encouraging the use of climatically appropriate landscaping, which means using native or introduced plant species that are adapted to the local climate. Some cities, like Tucson, Arizona, do not make it optional; native landscaping is mandated by law.

During a drought, local or state governments may limit or prohibit certain types of water use. During the drought in California in the 1980s lawn watering was widely banned. Some residents just could not stand the sight of a brown lawn and actually spray–painted the dead grass green.

Western water laws can encourage inefficient water use. The use it or lose it provision in prior appropriation law encourages water diversions, even if the water is not needed so as to avoid any presumption of abandonment. The fact that an appropriator might not be able to claim ownership to any conserved water is another impediment to conservation.

Public education plays an important role in reducing water demand by increasing awareness of water conservation options. People may not know how much water a lawn needs, or may not realize that watering in the late evening saves water because evaporation is reduced. Likewise they may not know that water–efficient plumbing fixtures are available and can save not just water but energy and money as well. But one of the, if not the single, most important role of education is that of developing and instilling an ethic of respect for the resource. All of the factors listed above can help people use water more wisely, but it is when people truly believe in what they are doing that behavior changes and the best decisions and choices are made.

URBAN WATER DEMAND PLANNING

For municipal water supply future *population size* and *distribution* are important variables used in planning future demand. The change in population equals the sum of the *natural growth* and the *net migration*. Natural growth is the difference between births and deaths; net migration is the difference between immigration and emigration. Expressed as an equation, population change is

$$\text{population change} = (\text{births} - \text{deaths}) + (\text{immigration} - \text{emigration}). \quad (6.1)$$

Projecting population growth is an uncertain proposition since it means forecasting into the future. Various methods are used to project population, including simple (linear) trend extrapolation, exponential growth, or an "S–curve" growth model. Depending upon the length of time into the future for which the projection is being made, it may be prudent to develop high–, medium–, and low–range projections to accommodate the inherent uncertainty and avoid the appearance of precision that accompanies a single estimate.

The location of future growth is important because water must be supplied to the point of use. Municipalities need to plan not just for water supply but for treatment plants, pumping stations, storage facilities, and distribution lines. Population growth beyond the established service areas forces the extension of the water and wastewater infrastructure. Experience has shown that once the water and wastewater lines are extended, even more development will occur. This is where land–use planning and growth management should be coordinated with water–resource planning. Some local governments require proposed developments to demonstrate in advance that they have a sufficient water supply as a precondition for approval. Land–use planning affects water–resource planning in other ways too. Increasing impermeable surface area reduces groundwater recharge and promotes rapid storm runoff, which can make flooding more of a problem (Figure 1.13). Surface and subsurface runoff from the land strongly influences water quality as well.

The simplest method for planning future water demand is to multiply the projected population growth by the current (or a future) value for per capita water use. For example, if per capita use is currently 150 gallons per day, and population is projected to increase by 2000 people over the next 10 years, then the water system will need an additional 300,000 gallons per day at that time. If conservation can reduce per capita use to 145 gallons per day, then the future increase in demand drops to 290,000 gallons. If the water system provides water to a mix of uses such as residential, commercial, institutional, and industrial, this simple approach might assume that water use for each sector increases in direct proportion to the increase in population. The validity of this assumption is certainly open to challenge. At the other end of the planning spectrum are the *disaggregated* demand models. Water demand is disaggregated by water–use sector and representative values of water use by each sector are determined. Growth in each sector is projected, multiplied by the sector's water–use value, and summed to give the total water demand. The challenge is to accurately project not just population growth, but growth in industry, commercial development, tourism, etc.

Urban water system planning involves other issues besides future water needs. Urban water systems must plan for fire flow capacity, maintain water quality levels, have contingency plans in times of droughts, and be prepared with emergency response procedures in case of an accident that threatens to contaminate the supply.

WATER SUPPLY

There are many different options for obtaining water, some conventional and some that are more unconventional (Table 6.1). The conventional approaches include storage reservoirs, interbasin transfers, groundwater development, and reallocation of water between existing uses. Unconventional alternatives include wastewater recycling, desalination, weather modification, and towing icebergs.

TABLE 6.1 Estimated Costs for Various Water Supply Alternatives[a]

Source	Dollars/acre–foot
Reservoir storage	\$100–\$200[b]
Interbasin transfer	\$100–\$200[b]
Groundwater development	\$72[b]
Desalination	
Brackish water	\$200–\$800[c]
Seawater	\$1050–\$6500[c]
Reuse and recycling of wastewater (advanced treatment)	\$162–\$393[c]
Towing icebergs	\$16–\$690[c]
Weather modification	\$8[c]

[a] (Gleick, 1993); [b] 1980 dollars; [c] 1985 dollars

What is considered conventional and unconventional changes over time and between cultures. In the United States desalination is used, but its use is not widespread, whereas desalination is quite common in Saudi Arabia. Recycling wastewater was uncommon 30 years ago, but has gained in popularity in the last decade. Between 1950 and the mid–1980s there was a great deal of research on weather modification as a means of augmenting water supplies. This alternative is no longer seriously pursued in the United States though it is still considered an option elsewhere.

Table 6.1 includes cost estimates for the different supply alternatives. The conventional methods all cost roughly the same, though groundwater development is a little cheaper. The costs for interbasin transfers can be much higher than that shown in Table 6.1. It was estimated back in 1975 that new interbasin transfers from the Colorado River basin to municipalities along the Front Range in Colorado could run \$3000 to \$4000 per acre–foot or higher (Metropolitan Denver Water Study Committee, 1975). Of the unconventional methods weather modification (cloud seeding) is potentially the cheapest. Past research indicated that under ideal conditions cloud seeding could increase local precipitation perhaps 10 percent. Desalination is expensive because of the high energy costs, especially when seawater is used. The use of reservoirs, interbasin diversions, and desalination is discussed below.

DAMS AND RESERVOIRS

The most widely used method for increasing water supply is the water–storage reservoir. The cost is moderate compared to other alternatives and can provide large volumes of water. Streamflow is variable in time and reservoirs store water during periods of high flow for use during periods of low flow. Thousands of single–purpose water supply reservoirs have been built in the United States. The Corps of Engineers surveyed nearly 50,000 reservoirs and found 34 percent were

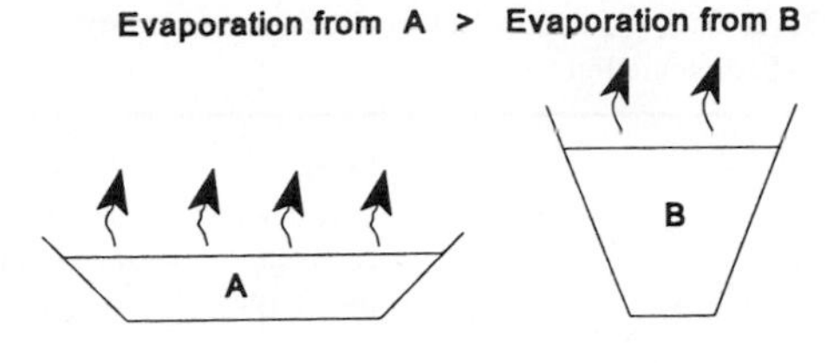

FIGURE 6.2 Two reservoirs with equal storage volumes. Reservoir A is wide and shallow which increases evaporation compared to narrower and deeper reservoir B.

used primarily for recreation, 16 percent for flood control, 15 percent for public water supply, 13 percent for irrigation water supply, 9 percent for farm ponds, 3 percent for hydroelectricity, and 10 percent for "other" uses (U.S. Army Corps of Engineers, 1977).

From a water supply point of view the best reservoir is narrow and deep because this profile minimizes the water surface area and promotes cooler water temperatures, both of which help reduce evaporation (Figure 6.2). This means reservoir sites in narrow, steep canyons are preferable to sites of low relief. Narrow canyons are also preferred from a economic point of view because the dam itself can usually be smaller (cheaper) per unit volume of water stored. But narrow canyons and gorges provide some of the most spectacular natural landscapes. This means reservoir construction is on a collision course with aesthetic considerations and wilderness preservation. The controversy at Hetch–Hetchy and Echo Park are two of the better–known examples from history. In the early 1990s the U.S. EPA and the courts disapproved the city of Denver's plan to construct the Twin Forks Dam and water supply reservoir in the foothills of the Rocky Mountains southeast of the city. The main objections were the potential negative impacts on bighorn sheep and destruction of the scenic landscape.

Reservoirs may also lose water through seepage into the surrounding rock. It has been estimated that as much as 0.5 million acre–feet of water seeps annually from Lake Powell behind Glen Canyon Dam on the Colorado River. To the extent that this water is stored underground it is not really lost; however, there are few opportunities to use the water on the overlying desert land, and, of course, it must be pumped to the surface. Seepage can raise water tables adjacent to and downstream from the reservoir. Seepage from Cochiti Reservoir (a Corps of Engineer flood control reservoir) on the Rio Grande west of Santa Fe, New Mexico, has waterlogged farmland in the Cochiti Pueblo immediately below the dam.

Reservoirs are ephemeral. River–transported sediment deposits behind the dam and reduces the reservoir storage capacity. The ultimate fate for most reservoirs is to be plugged with sediment, and the dam itself becomes a waterfall. The removal of sediment is physically possible but rarely economically justified. Table 6.2 gives average storage capacity reductions from sedimentation for different regions

TABLE 6.2 Rates of Reservoir Storage Capacity Loss Due to Sedimentation[a]

Region	Storage capacity lost (percent per year)	Percent of sediment originating on cropland
Corn Belt states	0.26	63
Lake states	0.27	64
Northern Plains states	0.23	36
Pacific states	0.49	9
Northeast	0.08	29
Appalachian	0.13	29

[a] Source: Gleick, 1993.

around the country. The highest rates of sedimentation are occurring in the Midwest (Corn Belt states), the states surrounding the Great Lakes, the Northern Great Plains, and the Pacific states. In the Great Lakes and Corn Belt states erosion from cropland is the primary source of the sediment. Rugged topography, logging, land development, and fire all contribute to the high rates of sedimentation in the Pacific states. The lowest rates of sedimentation are in the Northeast and the Appalachian regions.

Reservoir Storage Estimation

The maximum possible yield from a reservoir is equal to the annual average streamflow minus losses to evaporation and seepage. The *safe yield* from a reservoir is not a single value but depends upon the level of reliability users will accept. Reliability is the probability that the reservoir will deliver that yield. The relationship between yield and reliability is shown qualitatively in Figure 6.3. For the same storage capacity a lower yield is obtained with greater reliability than is a

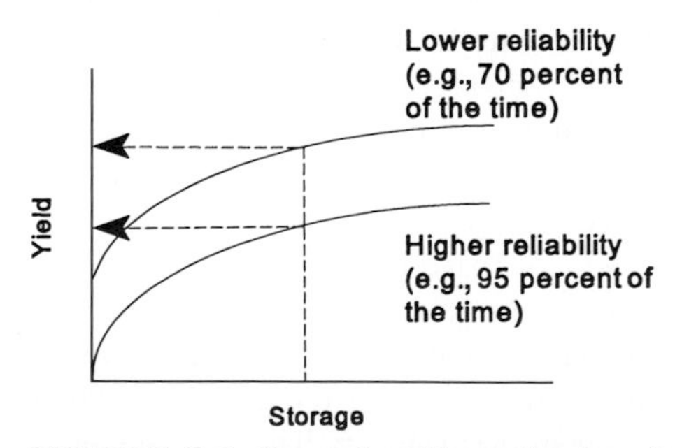

FIGURE 6.3 Reservoir yield as a function of reliability. Reliability is the probability of providing that yield. A lower yield can be obtained with higher reliability.

higher yield. Using the values in Figure 6.3 the reservoir can deliver the lower yield 95 percent of the time and the higher yield only 70 percent of the time. Urban water systems require a higher reliability than irrigation systems, because the consequences from failure are more severe for the urban system. The 95 percent reliability may be the safe yield for the urban system, while 70 percent reliability is safe enough for the farmer.

Sequent Peak Method for Storage Estimation

One of the easiest methods for estimating the amount of reservoir storage needed to meet a given demand is the *sequent peak* procedure. The procedure accumulates the difference between supply (streamflow) and demand (reservoir releases plus losses) over some period of time. A graph of the accumulated differences produces a series of peaks and valleys. The required storage equals the maximum difference between a peak and the intervening valley to the next higher sequent peak (see Example 6.1).

EXAMPLE 6.1

Sequent peak demonstration

Table 6.3 gives hypothetical streamflows (Q) and demand (D). The units for streamflow and demand can be either average flows (cfs) or volumes. The time interval on which the values are measured can be months or years depending on the problem. The sequence of streamflows should include a critical dry period, usually the driest period of record. For this example water demand is a constant annual average flow of 300 cfs.

TABLE 6.3 Hypothetical Supply and Demand Values for the Sequent Peak Storage Calculation[a]

Year	Q	D	$Q-D$	$\Sigma(Q-D)$	Year	Q	D	$Q-D$	$\Sigma(Q-D)$
1	400	300	100	100	14	348	300	130	508
2	338	300	38	138	15	505	300	205	713
3	461	300	161	299	16	450	300	150	863
4	394	300	94	393	17	365	300	65	928
5	248	300	-52	341	18	305	300	5	933
6	275	300	-85	256	19	337	300	37	970
7	376	300	137	393	20	438	300	138	1108
8	365	300	65	458	21	222	300	-78	1030
9	490	300	190	648	22	281	300	-19	1011
10	208	300	-92	556	23	325	300	25	1036
11	321	300	21	577	24	356	300	56	1092
12	171	300	-129	448	25	294	300	-6	1086
13	230	300	-70	378					

[a] Values are annual average flows (cfs).

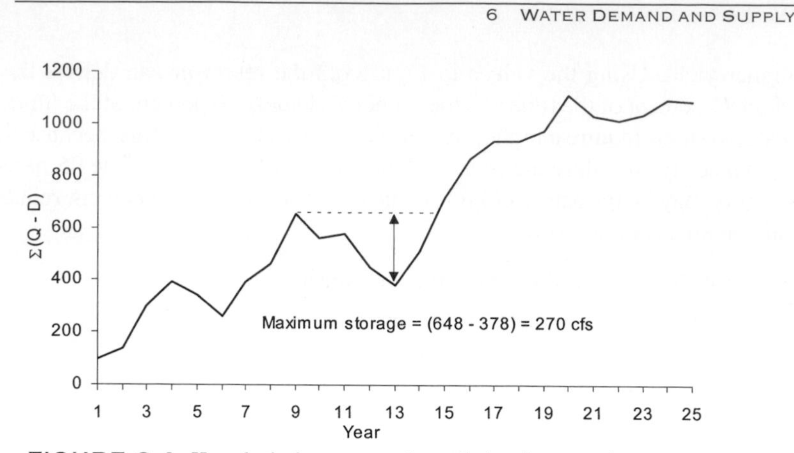

FIGURE 6.4 Hypothetical sequent peak graph showing a maximum storage of 270 cfs.

Column 4 (Table 6.3) is the difference between supply and demand ($Q - D$) for each year, and column 5 is the accumulation of these differences. Columns 6 through 10 repeat the calculations for years 14 to 25. The accumulated differences (columns 5 and 10) are graphed in Figure 6.4. A peak of 648 cfs occurs at year 9. The next higher sequent peak is 1108 at year 20. The required storage is the difference between the first peak and the intervening valley. For this example the required storage is $(648 - 378) = 270$ cfs.

Synthetic Streamflow Modeling

The following discussion draws extensively from Thompson (1998). The sequent peak method determines a storage value for a given demand and sequence of streamflows. But how representative of all possible streamflows sequences is the single, relatively short, historical record? There may have been much drier periods in the past before records were kept, and who knows what the future might hold? The value of storage determined from a single record of streamflows may prove inadequate in the future. Water supply planning must deal with such uncertainty. One way to address the uncertainty is through mathematical modeling. We can create a simple model to generate *synthetic streamflows* that are statistically identical to the historical record. In the parlance of synthetic streamflow modeling, an individual sequence of synthetic flows is called a *trace.* Statistically identical means that each trace (synthetic streamflow sequence) has the same mean, variance, and serial correlation as the historical data on which it is based. Serial correlation is a measure of the degree of association—dependence—between successive values in a sequence. If there is no dependence between successive values in a sequence, then the values are independent. Hydrologists call dependence between successive values *persistence* and it shows up in many hydrologic phenomena. Annual streamflow volumes frequently show persistence from one year to the next. Persistence is why we have droughts and wet spells. During a drought low streamflows tend to persist—low flows follow low flows—until the drought breaks.

The mathematical model introduced here is a *lag–1 Markov* model for annual average streamflow. The Markov model is also called an autoregressive (AR) model. Autoregressive is a mathematical term for the persistence between successive values. Lag–1 means that the dependence between values extends back only one time step. In other words streamflow for the current year t is serially correlated with streamflow one year earlier, i.e., year t-1. The model is therefore denoted as an AR(1) model. A major assumption with this type of model is that streamflows are *stationary* over time. Stationary means that the mean, variance, and serial correlation are not changing with time. If climate is changing then this assumption is wrong and the model should not be used. The lag–1 Markov model for annual average streamflows is

$$Q_t = \overline{Q} + r(Q_{t-1} - \overline{Q}) + e_t s\sqrt{1 - r^2}, \tag{6.2}$$

where Q_t is streamflow in year t, $\overline{Q}$ is the mean of the annual streamflow, s is the standard deviation of the annual streamflows, and r is the serial correlation coefficient between streamflows Q_{t-1} and Q_t. The mean, standard deviation, and serial correlation coefficient are estimated from historical data. The model (Equation (6.2)) states that streamflow in year t depends upon the average streamflow, the previous year's streamflow, the correlation between this year's and last year's streamflow, and a random component. The second term on the right–hand side of Equation (6.2) models the persistence between flows using the serial correlation. The third term generates random variations in the model. The random variate e_t has a mean of 0 and a standard deviation of 1, and is chosen from an appropriate probability distribution. Many investigators have found the normal distribution satisfactory, in which case the random variate is chosen from a standard normal distribution. Tables of random standard normal deviates are available in most statistics textbooks (e.g., Haan, 1977). Computer programs are also available for generating random variates from different probability distributions. There is no way to choose the probability distribution of the random variate *a priori,* and short historical records may not clearly define the underlying distribution. Any disparity between the assumed distribution and the true natural process is perpetuated in the simulated flow sequence, creating an operational bias. The accuracy of the model is limited by the accuracy of the estimated parameters. Whether the lag–1 dependence structure is adequate in replicating longer–term persistence in the natural process is subject to debate (Linsley et al., 1982). Once the model parameters are estimated, synthetic streamflow sequences of length n years can be generated. Given the uncertainty regarding the characteristics of long–term persistence it is recommended that n not exceed 100 years. This type of (stochastic) model does not attempt to replicate the actual pattern of historical flows, it simply generates statistically similar flows—flows that are statistically possible. Figure 6.5 shows mean annual flows for the Scott River in North Carolina

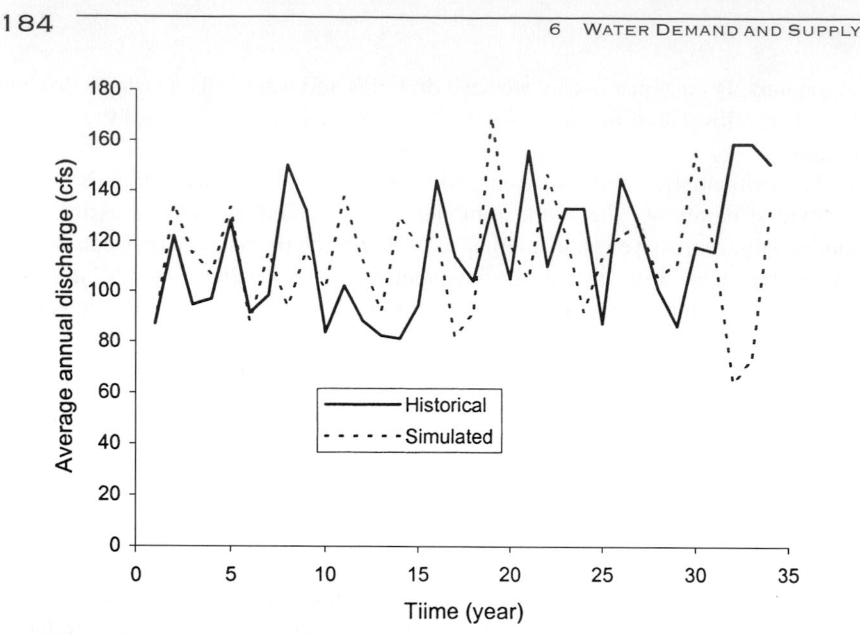

FIGURE 6.5 Graphs of the historical mean annual flow and a synthetically generated trace for the Scott River near Sylva, North Carolina.

and a synthetic trace generated by Equation (6.2). The model parameters for the Scott River are $\overline{Q} = 115.1$ cfs, $s = 24.8$ cfs, and $r = 0.132$. The model requires an initial discharge for Q_{t-1} to begin the simulation. The initial discharge was set equal to the observed discharge for the first year. Note that the trace has a higher maximum flow and a lower minimum flow than any found in the historical record. Box 6.1 contains a simple BASIC program from Thompson (1998) that generates synthetic streamflows using Equation (6.2). The random variate e_t is generated for a standard normal distribution in a two–step procedure. First, a random number between 0 and 1 is generated by the computer (line 160). This random number is then used to produce a standard normal variate using a

BOX 6.1 MARKOV STREAMFLOW SIMULATION PROGRAM

```
10 REM PROGRAM TO SIMULATE STREAMFLOW USING A LAG-1 MARKOV
   MODEL
20 REM
30 REM MAIN PROGRAM WRITTEN BY STEPHEN A. THOMPSON, 1995.
40 REM
50 CLS
60 DIM Q(500)
```

```
70 PRINT "FOR THE HISTORIC DATA SERIES ENTER FOLLOWING:": PRINT
80 INPUT "THE MEAN"; QBAR: PRINT
90 INPUT "THE STANDARD DEVIATION"; SD: PRINT
100 INPUT "THE LAG-1 SERIAL CORRELATION COEFFICIENT"; RHO: PRINT
110 INPUT "A STARTING FLOW VALUE TO 'SEED' THE MODEL"; Q(1): PRINT : PRINT
120 INPUT "HOW MANY YEARS DO YOU WANT TO SIMULATE?"; N
130 RANDOMIZE TIMER
140 PRINT "TIME"; TAB(15); "Q"
150 FOR I = 2 TO N
160 Y = RND(I)
170 IF Y > .5 THEN Y = 1 - Y:FLAG = 1
180 GOSUB 1000
190 Ei = X
200 IF FLAG = 1 THEN Ei = -1 * Ei: FLAG = 0
210 Q(I) = QBAR + RHO * (Q(I - 1) - QBAR) + Ei * SD * SQR((1 - RHO * RHO))
220 PRINT I; TAB(10); : PRINT USING "#####.#"; Q(I)
230 NEXT
240 END
250 REM
1000 REM ***** INVERSE NORMAL DISTRIBUTION SUBROUTINE *******
1010 REM
1020 REM CALCULATES AND APPROXIMATION TO THE INTEGRAL OF THE NORMAL
1030 REM DISTRIBUTION FUNCTION FROM X TO INFINITY
1040 REM USING A RATIONAL POLYNOMIAL.
1050 REM THE INPUT IS Y; THE RESULT IS RETURNED IN X
1060 REM ACCURACY IS BETTER THAN 0.0005 IN THE RANGE 0<Y<0.5.
1070 REM FROM RUCKDESCHEL (1981),
1075 BASIC SCIENTIFIC SUBROUTINES VOL. II,
1080 REM BYTE/MCGRAW-HILL, PAGE 164-166.
1090 CO = 2.515517
1100 C1 = 0.802853
1110 C2 = 0.010328
1120 D1 = 1.432788
1130 D2 = 0.189269
1140 D3 = 0.001308
1150 IF Y = 0 THEN X = 1
1160 IF Y = 0 THEN RETURN
1170 Z = SQR(-LOG(Y * Y))
1180 X = 1 + D1 * Z + D2 * Z * Z + D3 * Z * Z * Z
1190 X = (CO + C1 * Z + C2 * Z * Z) / X
1200 X = Z - X
1210 RETURN
```

subroutine (line 1000) that approximates the inverse transformation of the normal distribution (Ruckdeschel, 1981).

Frequency Analysis of Storage Requirements

Using the historical streamflow record we calculated a single estimate of reservoir storage for a demand of 300 cfs in Example 6.1. The synthetic streamflow generation model allows generation of multiple traces. For each trace a value of reservoir storage can be calculated. These multiple values of storage can be ranked from largest to smallest and assigned an *exceedence probability* using the formula

$$p = \frac{m}{n + 1}, \tag{6.3}$$

EXAMPLE 6.2

The hypothetical streamflow data from Table 6.3 are used as the "historical" streamflow data, and the mean, standard deviation, and serial correlation coefficient were calculated. Twenty traces were generated, each trace 50 years in length. Including the historical streamflows there are a total of $n = 21$ streamflow sequences for the analysis. For each

TABLE 6.4 Frequency Analysis of Reservoir Storage

Rank (m)	Storage (cfs)	Exceedence probability
1	655	0.045
2	650	0.091
3	611	0.136
4	585	0.182
5	548	0.227
6	529	0.273
7	502	0.318
8	493	0.364
9	480	0.409
10	435	0.455
11	431	0.500
12	430	0.545
13	385	0.591
14	315	0.636
15	270	0.682
16	253	0.727
17	247	0.773
18	237	0.818
19	218	0.864
20	148	0.909
21	117	0.955

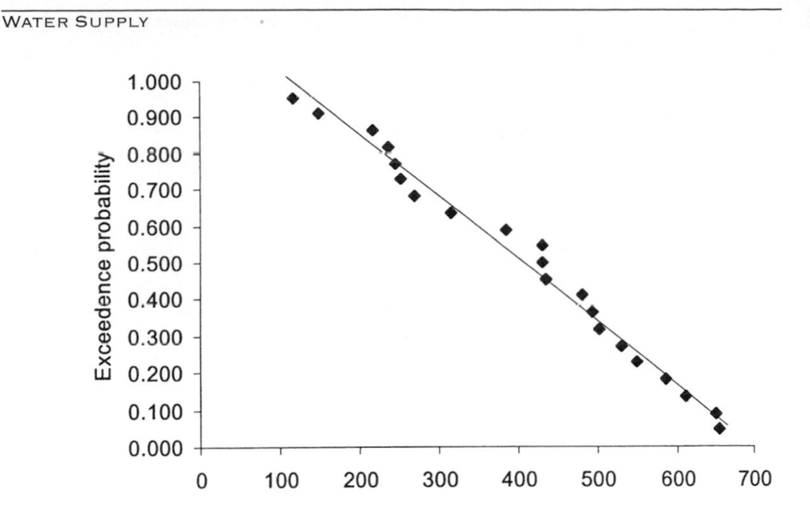

FIGURE 6.6 Plot of the storage–frequency analysis in Table 6.4.

trace the storage capacity needed to meet a demand of 300 cfs was determined by the sequent peak procedure. These storage values were ranked and assigned exceedence probabilities using Equation (6.3) (see Table 6.4 and Figure 6.6).

The original storage requirement of 270 cfs from Example 6.1 (Table 6.3) is ranked $m = 15$ with an exceedence probability of 0.682. The interpretation is that this storage value would be exceeded about 68 percent of the time. In fact, 50 percent of the time more than 431 cfs of storage is needed. It is up to the decision makers to choose the amount of storage based on the trade–off between cost and risk (reliability). The risk is that, if not enough storage is provided, the reservoir goes dry.

---

where p is the exceedence probability, m the rank position of the storage value, and n the total number of storage values in the analysis. Exceedence probability is the chance of equaling or exceeding that storage value. Equation (6.3) can be used to calculate exceedence probabilities for any set of independent data, and we use it again in Chapter 10 for flood frequency analysis. Example 6.2 demonstrates the combined use of the streamflow generation program, the sequent peak method, and frequency analysis to analyze reservoir storage requirements.

The amount of storage capacity required increases as the demand increases. Figure 6.7 shows the results of a series of simulations demonstrating how storage capacity increases with increasing demand. In this particular example no storage was required until demand exceeded 75 percent of the mean annual streamflow, after which the required storage capacity increased dramatically.

Storage must also increase as streamflows become more variable. This is one reason why reservoirs in the western United States have larger capacities than

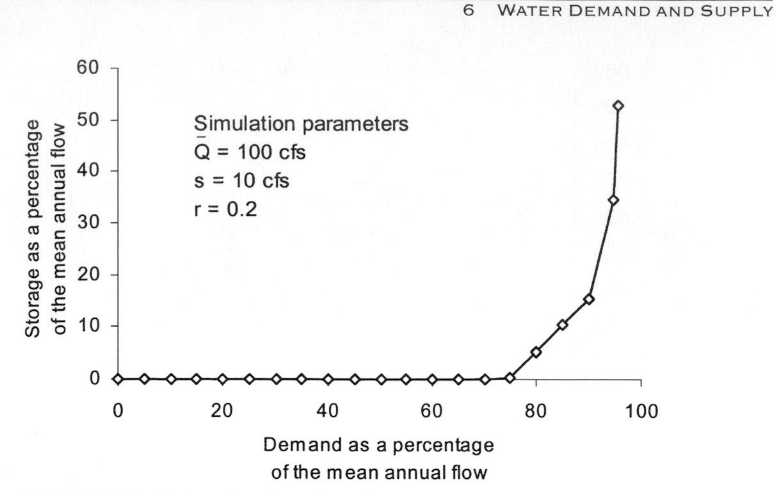

FIGURE 6.7 Results of a series of simulations that show the effect of increasing demand on the amount of storage required to meet that demand. For each simulation the mean is 100 cfs, the standard deviation 10 cfs, and the serial correlation coefficient 0.2. (Source: Thompson, 1998)

those in the East. To produce the same yield, reservoirs in the West need to be larger because of the greater variability in streamflow. Figure 6.8 shows how increasing streamflow variability increases the storage needed to meet a given demand.

Multipurpose Reservoirs

Multipurpose reservoirs control and release water for water supply, hydroelectric power generation, flood control, recreation, navigation, water quality maintenance, and various environmental purposes. A single reservoir cannot meet all of these purposes simultaneously. Operating a reservoir to meet certain purposes will

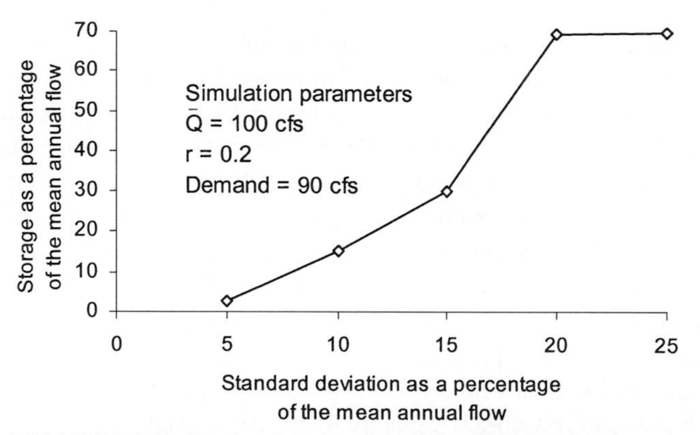

FIGURE 6.8 Results of a series of simulations that show how increasing streamflow variability increases the amount of storage required to meet a given demand. For each simulation the mean is 100 cfs, the serial correlation coefficient 0.2, and the demand 90 cfs. (Source: Thompson, 1998)

be in direct conflict with its operation for other purposes. For example, operating a reservoir for flood control means keeping the reservoir empty so as to have the largest available capacity to store flood flows. On the other hand the objective for water supply is to keep the reservoir as full as possible at all times. Considerable research has been conducted on optimizing the operation of multipurpose reservoirs to achieve the maximum benefits from the combined purposes.

Storage in a large multipurpose reservoir is divided into four zones (Figure 6.9). At the bottom of the reservoir is *dead storage*. Water in dead storage cannot be released from the reservoir since it lies below the lowest outlet structure and can only be used for recreation. Dead storage capacity is also allocated to sediment storage. Water in *active storage* is used to satisfy most of the purposes including water supply, river regulation, hydropower, and recreation. Water from the active pool is released in a controlled manner through the principal spillway. The *normal pool* level is the highest elevation of the reservoir under normal operating conditions. The *minimum pool* is the lowest level to which the reservoir is drawn down under normal operation. *Flood storage* is obviously used for storing flood flows. In addition to regulating discharge, flood storage is used for hydropower and water supply. *Surcharge storage* is water temporarily held above the emergency spillway. The difference between surcharge and flood storage is that surcharge storage cannot be controlled; the water flows uncontrollably over the spillway.

One final idea about using reservoirs to increase storage on a river is recognizing that the amount of new storage is the difference between the amount of natural storage before the reservoir existed and the amount of storage after the dam is built. Most or all of the active storage is likely to be new storage. The amount of new flood storage may be significantly less because floodplains provide a considerable amount of natural flood storage. The conflict, of course, is that people now want to use floodplains for purposes other than flood water storage.

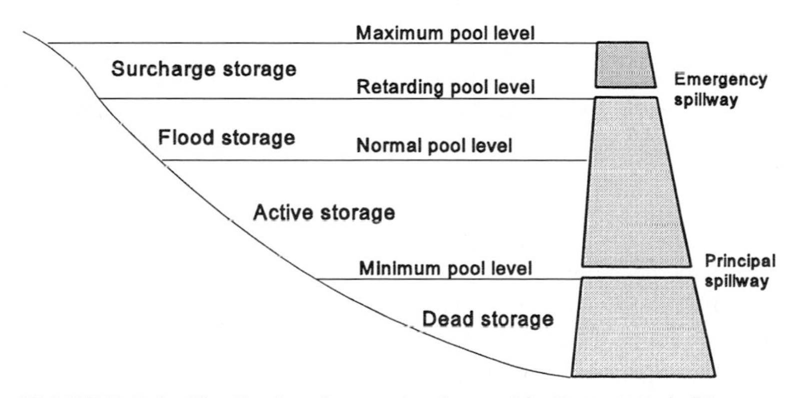

FIGURE 6.9 Classification of storage in a large multipurpose reservoir (Viessman and Welty, 1985; Dzurik, 1990.

Environmental Impacts of Dams

Dams and reservoirs provide many benefits but they create many environmental problems. The most obvious impact is the inundation of the valley and the natural stream channel. This destroys the existing flowing–water ecosystem and replaces it with a still–water ecosystem. Inundation may also displace human settlements. In the United States many small towns have been sacrificed in the name of increased water supply and flood control. The most significant episode occurred in connection with the TVA projects, where some 60,000 people were displaced over a period of years. In some countries the impacts on human settlements have been tremendous by comparison. The Sanmenxia Dam and reservoir in China displaced 870,000 people in the 1950s. China is now building the Three Gorges Project on the Chang Jiang River (Yantze). When completed this will be one of the largest dams in the world, and the reservoir will submerge farms and villages that are home to not fewer than 1.9 million people. The construction of the Kariba Dam in Africa displaced thousands of people from their ancestral tribal lands and generated social unrest because the displaced groups were forced onto lands of other tribes.

Dams are major obstacles to migrating fish. Anadromous fish live in the ocean but swim upstream to spawn. Fish ladders have been incorporated into some dams to facilitate fish migration, but in many cases the impacts on fish migration have been ignored. In the Pacific Northwest a number of salmon species have been designated as endangered, and restoration plans to improve salmon habitat and migration are being developed. Along the Susquehanna River in southcentral Pennsylvania the four hydroelectric dams situated between the Maryland–Pennsylvania border and the city of Harrisburg have installed or are installing fish lifts to help the native shad over the dams (Chapter 8).

Water released from large dams may be much colder than the natural streamflow. While this may be desirable from the viewpoint of lowering evaporation, it can have adverse impacts on downstream ecosystems. Other environmental impacts include small earthquakes induced by the weight of the water and the lubricating effect of water seeping into the ground, the raising of water tables and water–logging of soil near the reservoir, and changes in insect populations following the change from a *lotic* (flowing water) ecosystem to a *lentic* (still water) ecosystem, and, of course, dams can fail catastrophically. The National Inventory of Dams lists 74,053 dams, of which only 4 percent are owned or regulated by the federal government. The remainder are state regulated. The inventory classifies 10,400 dams as "high hazard," which does not refer to the condition of the dam, but to the potential for loss of life should failure occur. Another 13,000 are considered "significant hazard," indicating certain loss of property or infrastructure in the event of failure (Federal Emergency Management Agency (FEMA), 1996). The National Dam Safety program is aimed at reducing the risk of dam failures and increasing the development of state dam safety mitigation efforts. Currently 28 states have dam safety programs.

Lastly, dams disrupt the natural sediment budget of a river. Sediment carried by the river upstream is deposited in the reservoir, so the water released from the reservoir is clear. The clear water increases erosion of the stream bed below the dam resulting in "channel armoring." Armored channels have less fine sediment which may also adversely affect ecosystems. The impact on sediment transport was one of the key issues prompting an environmental reevaluation of the operation of Glen Canyon Dam in the 1980s. The evaluation lead to an historic event on March 7, 1996, when the spillway tubes of the dam were opened and a controlled experimental flood was allowed to roar through the Grand Canyon. The Secretary of the Interior has now officially changed the operation procedures for the dam in order to better protect the natural resources of Grand Canyon National Park and the Glen Canyon National Recreation Area. At a ceremony on October 9th in Phoenix Secretary Babbitt said, "We begin a new chapter in the fabled history of the Grand Canyon and Glen Canyon Dam. This marks a . . . change in the way we view the operation of large dams. We have shown that they can be operated for environmental purposes as well as water capture and power generation." Preliminary results indicated that the controlled flood increased the sandbar volume, widened the channel at several major rapids, increased backwater habitat for fish, and apparently had no adverse impact on endangered birds, fish, or archeological sites (Collier *et al.,* 1997). Instream flows for environmental purposes are covered at the end of Chapter 8.

INTERBASIN TRANSFERS

The practice of transferring water from areas of surplus to areas in need was a widely accepted in the United States until the environmental movement got rolling in the 1960s. Most interbasin transfers are found in the western United States for two reasons. First is the greater spatial imbalance between water supply and demand, and many cities and farmers have expanded their demand beyond the naturally available supply from the basin. The second reason is that western water law allows interbasin diversions. One of the precedent–setting cases in the development of western water law is *Coffin v. Left Hand Ditch Company* (1882). This case involved a relatively small interbasin diversion in the foothills of the Rocky Mountain north of the city of Boulder. The Left Hand Ditch Company established a water diversion from South St. Vrain Creek into the neighboring basin of Left Hand Creek. Subsequent in time, Mr. Coffin began farming along the banks of the St. Vrain downstream of the diversion. During a period of low flow in 1879 Mr. Coffin went up into the foothills and tore out the ditch company's diversion dam. He asserted that his was a common–law riparian right superior to the non-riparian diversion, and the ditch company's diversion was denying him water. In finding for the ditch company the Colorado Supreme Court proclaimed once and for all that prior appropriation was the law in Colorado, and that conveying water to a tract of land, regardless of where the land is located, was justified and legal, and necessary to make full use of the land and water resources.

Other important interbasin diversions include San Francisco's diversion from Hetch–Hetchy, diversions from the Colorado basin for urban and agricultural use in California, Colorado, New Mexico, and Utah (Figure 3.7), and the California State Water Project, which brings water from northern to southern California. Though somewhat less common, large interbasin diversions are found in the East as well. New York City progressively tapped the Croton River, streams in the Catskill Mountains, and finally the headwaters of the Delaware River. New York went in search of water not because demand exceeded the local supply, but because the local supply—the Hudson River—had become so polluted it was unusable. Not all interbasin transfers are for water supply. The Chicago Sanitary and Ship Canal diverts water from Lake Michigan south into the Illinois River, a tributary of the Mississippi. The canal serves the duel purposes of transportation and waste carriage away from the lake. The Grand Canal in China, once a major transportation artery which fell into disrepair for centuries, transferred water across the North China Plain from the Change Jiang to the Huang River. The Chinese government is considering refurbishing the canal and using it once again for transferring water for water supply.

One of the grandest interbasin diversion schemes ever conceived was the North American Water and Power Alliance (NAWAPA) proposed in the early 1960s. Proponents of NAWAPA envisioned diverting 110 million acre–feet of water annually from the Yukon and Tanana Rivers in Alaska and western Canada some 2000 miles south to the United States and Mexico. The plan also included a navigation canal across Canada that would carry water to the Great Lakes. NAWAPA never got beyond the conceptual stage, though for a time it sparked serious interest among western water–development groups. A recent updated version of this scheme was proposed by Governor Hickel of Alaska in the early 1990s. He proposed sending 4 million acre–feet of water south to California. The ballpark cost was $110 billion dollars, or about $3000 to $4000 per acre–foot. Needless to say this solution is far more expensive than just about any other alternative, especially water conservation.

One of the last large–scale interbasin schemes to receive serious attention by the federal government was a proposal to bring irrigation water to the High Plains from as far away as the Mississippi River. Recognizing that the Ogallala Aquifer is being mined, different studies have examined the feasibility of rescuing High Plains farmers with imported water. The most recent study (High Plains Associates, 1982) examined potential diversions from the Missouri River and/or other rivers near their junction with the Mississippi. Depending upon the particular scheme water would have to be pumped uphill from 1000 to 3000 feet vertically, and between 400 and 1100 miles. None of the alternatives were economically, politically, or legally feasible. The cost of water delivered to farmers in the 1982 study ranged from $227 to $569 per acre–foot (1977 dollars), which was well beyond the farmers' ability to pay. In 1982 the maximum that farmers could afford was estimated to be $120 per acre–foot (High Plains Associates, 1982). Even if the economics were more favorable taxpayers are unlikely to back such a project, and the areas from which the water would be taken are unlikely to allow the transfer in the first place. The 1982 study was careful to point out that it did *not* evaluate the *availability* of water. This plan demonstrates the problems with large interbasin diversions in the United States today. The areas of origin are unwilling to relinquish their resource, the projects are very expensive and rarely economically efficient, and the potential environmental impacts virtually preclude their ever being approved. While interbasin transfers must still be considered a water–supply option in the United States, like dams and other large construction projects they are more and more a relic of the recent past than a realistic alternative for the future. One of the last large–scale interbasin project was California's State Water Project (CSWP) built during the 1960s and 1970s. The project was never "finished" in the sense that it has never diverted and transferred as much water as was originally envisioned by its proponents. With each passing year it seems less likely that the project will ever be completed—that is, expanded—beyond its current configuration. The CSWP is described in the case study about Los Angeles below.

CASE STUDY: WATER FOR LOS ANGELES

Los Angeles is the second–largest city in the United States. By the year 2000 there could be 14 million people living in the greater Los Angeles area. The climate in southern California is dry in the summer and wet in the winter (see Figure 1.9). The average precipitation over the Los Angeles Basin is 14 inches per year. The city is surrounded by mountains on three sides and the Pacific Ocean to the west. Beyond the mountains lies desert. Los Angeles is one of the largest cities in the world located in a dry climate. Cairo, Egypt, is another, but unlike Cairo, which has the Nile, Los Angeles does not have a major local source of water. To satisfy the increasing demand Los Angeles has tapped sources as far as 400 miles away. The development of the city's water supply is a fascinating tale of engineering accomplishment surrounded by deception and intrigue.

The Naturally Available Supplies

The two naturally available sources of water include the Los Angeles River and groundwater. Los Angeles owns all of the water rights to the Los Angeles River. The City was granted "pueblo water rights" to the river after being confirmed the rightful successor to the original Spanish pueblo. Pueblo water rights are unrelated to the prior and paramount water rights of the Pueblo Indians in New Mexico. Only three cities have pueblo water rights—Los Angeles, San Diego, and Las Vegas, New Mexico.

In the late 1800s artesian springs were a common source of water. Overuse eventually reduced artesian pressure below the land surface and the wells thereafter had to be pumped. Continued overpumping led to salt water intrusion along the coastal areas. To limit salt water intrusion the city converted coastal well fields in the 1950s and 1960s into recharge zones by pumping treated wastewater back into the aquifer. Recharging the aquifer creates a groundwater mound within the aquifer that acts as a barrier to the inland movement of salt water. By 1900 the city's population exceeded 100,000 and it doubled again by 1904. In the same year the first publication by the newly created Los Angeles Department of Water and Power (LADWP) stated that a new source of water would have to be found. That new source was the Owens River. The man most responsible for bringing the Owens River to Los Angeles was William Mulholland, the visionary head of the LADWP.

The Owens River

The Owens River is one of the few rivers draining the eastern side of the Sierra Nevada. At the turn of the century the river supported a prosperous irrigation agriculture community in the Owens Valley. Starting around 1905 agents for Mulholland secretly began purchasing water rights from irrigators in the valley, eventually gaining control of most of the water.

Los Angeles received invaluable help from the federal government by being granted all of the necessary rights-of-way across public land. President Roosevelt and Gifford Pinchot both agreed that using the Owens River water in Los Angeles was more utilitarian—it provided a greater good to a greater number of people—than using it for irrigation agriculture. The federal government further aided the city by including most of the Owens Valley within the newly established Inyo National Forest, even though the only trees in the valley were the orchards planted by farmers. Reserving the land as national forest guaranteed that no homesteaders could come along and make appropriations, thus ensuring all of the water to Los Angeles. The 223-mile Los Angeles Aqueduct was constructed in just six years across some of the most hostile and desolate terrain in North America (Figures 6.10 and 6.11). Because the Owens Valley sits 4000 feet above sea level, the water travels all the way to Los Angeles under the force of gravity alone. The aqueduct began delivering water in November, 1913; however, the water was not delivered to the established city, but to the recently annexed San Fernando Valley. As it turns out, the city really did not need the Owens River water in 1913; the local supplies were still more than sufficient. Mulholland had greatly exaggerated the impending water shortage. To scare Los Angeles residents into passing the $23 million bond issue to finance the aqueduct, Mulholland ordered water be dumped from the city's reservoirs at night, lowering their level and making the city's water supply situation appear worse than it actually was. So Owens River water was used instead to create new irrigated farmland in the San Fernando Valley. A clutch of prominent city businessmen, who somehow received advanced notice that there would be surplus water available, bought up tens of thousands of acres in the San Fernando Valley "dirt cheap." When the water arrived, land values skyrocketed and they became multimillionaires virtually overnight. And capping off the entire project was a neat hydrological trick. Some of the irrigation water would seep down and recharge the aquifer under the San Fernando Valley. This groundwater then flowed east toward the city's well field and the Los Angeles River. The aquifer acted as an evaporation-free underground storage reservoir, and the city was able to sell the water twice—first to the irrigators and then to the residents of Los Angeles.

FIGURE 6.10 Construction of the Los Angeles Aqueduct. Mule teams moving segments of pipe used to construct the Jawbone Siphon. (Photo from the Los Angeles Department of Water and Power)

The city originally told the residents of the Owens Valley that it would only divert water into the aqueduct *below* existing farmland so that the established agricultural economy could continue. But explosive population growth and a drought in 1923 left the city little choice but to divert all the water, drying up the valley and destroying the community in the process. The community did not go quietly into oblivion. The aqueduct was dynamited

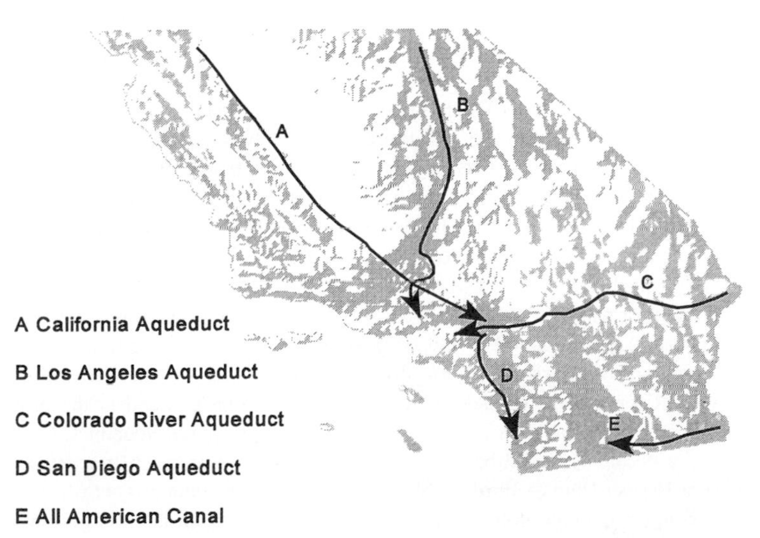

FIGURE 6.11 Major aqueducts bringing water to southern California.

numerous times, and a trainload of city detectives armed with rifles and machine guns were sent to patrol the aqueduct with orders to shoot–to–kill. The community had all but collapsed by the summer of 1927 when all of the banks in the valley were permanently closed. Ironically, Mulholland's illustrious career collapsed just a few months later.

Mulholland disliked reservoirs because he felt they wasted too much water by evaporation. The city's water system was thus short on water–storage capacity and had started building the Saint Francis Dam in San Francisquito Canyon north of the city. Dynamiting of the aqueduct, the drought, and the ever–increasing demand prompted Mulholland, against the advice of his engineers, to enlarge the dam beyond its original design. The reservoir was filled quickly and reached full capacity in early March, 1928. The dam immediately began to leak. As it turns out the dam was built astride an unknown lateral fault, and the pulverized rock in the fault zone (fault gouge) acted as a conduit for water to seep under the dam. Mulholland went up to look at the leak on March 12th, and after a perfunctory examination pronounced the dam solid and sound. Later that same day the leak became a gush under the center of the dam. Water swirling around the front of the dam began undermining the dam's east abutment, which was anchored in the poorly consolidated shale of the canyon's east wall. Just a few minutes before midnight the dam failed catastrophically, and 11.4 *billion* gallons of water formed a 200–foot–high wall of water roaring down the canyon. Five hours later when the flood finally reached the Pacific Ocean near Ventura between 400 and 500 people were dead, 1200 homes were destroyed, and erosion had completely stripped the topsoil from 8000 acres of farmland (Reisner, 1986). Corpses and debris were scattered along the beaches 200 miles south to San Diego. Mulholland's career was washed away in the flood.

In subsequent years Los Angeles began pumping groundwater in the Owens Valley and extended the aqueduct north to divert the streams feeding Mono Lake. The California Supreme Court drew the line at drying up Mono Lake and ordered Los Angeles to leave some of the lake's tributaries alone.

The Colorado River

Water from the Owens Valley satisfied the city's thirst for a time, but population growth meant finding still more water. Between 1900 and 1920 the city's population grew by 600 percent. Between 1920 and 1930 it was growing at a phenomenal 7 percent annually, a rate that would *double* the city's size every 10 years. Initially Los Angeles coveted the Colorado River more as a potential source of hydroelectricity than as a source of water. The drought cycle in the 1920s changed Mulholland's mind and he was now eyeing the Colorado for water as well as power. Tapping the Colorado for water would be a formidable task, and would be possible only if a large dam were first built to control the river's erratic flow. A second problem was that, unlike Owens River water, Colorado water would have to be pumped to the city. A power source was thus required along with the proposed aqueduct. Both of these problems would be satisfied by the federal government's construction of Hoover Dam. A third problem was that this aqueduct project was just too large for the city alone, and so in 1924 the city began negotiating with surrounding communities to create a Metropolitan Water District (MWD). Negotiations were successful and the new MWD was approved by the state legislature in 1927. In 1928 the Boulder Canyon Project Act was passed by Congress authorizing the construction of Hoover Dam. Passage of the act was in no small way influenced by the political clout of the MWD. With all the pieces in place construction began on the Colorado River Aqueduct and Parker Dam. Parker Dam was needed to create a reservoir (Lake Havasu) from which the aqueduct would draw water.

All did not go smoothly at first. Arizona had refused to ratify the Colorado River Compact and was suing California over rights to the river. When workmen from Los Angeles showed up at the river's edge, Arizona deployed its state militia to block construction. The shutdown was temporary but it further poisoned relations between the two states. The aqueduct and dam were completed in June 1941 and the MWD began delivering water to communities in southern California. The federal government leased the hydroelectric power facilities to the LADWP and the Southern California Edison Company.

The MWD was planning a second aqueduct to the Colorado to "borrow" the water which rightfully belonged to Arizona. Without the Central Arizona Project (CAP), Arizona had no way to use its allotted share and California's congressional delegation had successfully blocked the CAP for 20 years. In essence, MWD was betting that if it could establish Southern California's dependence on Arizona's water, the Supreme Court would be hard pressed to ever give it back to Arizona. First in time, first in right? Crucial to this plan, however, was that no other water was available from anywhere else. But in 1951 the California State Engineer, A. D. Edmunston, began a study for the transbasin diversion of water from the Feather River in northern California to farms and cities in southern California.

The California State Water Project

The Bureau of Reclamation published a study on the possibility of transbasin diversions of water from rivers in the Pacific Northwest to the arid Southwest in 1951. The Bureau considered transfers from the Klamath River to the Sacramento River and then south to be the most feasible scheme. The Bureau's report was the starting point for Edmunston's plan to bring water from northern to southern California, but instead of the Klamath he chose the Feather River. The Feather River Project required a new and expanded administrative agency, so the California Department of Water Resources was cobbled together from 52 existing water agencies. The MWD initially opposed the Feather River Project because it threatened the MWD's position with regard to claiming additional water from the Colorado River. Another problem was that half of the water from the Feather River Project was to go to farmers in the San Joaquin Valley, while urban residents in southern California would pay the lion's share of the costs—and this project was going to be a very costly.

The California State Legislature approved what was now called the California State Water Project in 1959. It then went to a state–wide referendum and was narrowly approved by 174,000 votes—48 of the 58 counties in the state voted against the project (Reisner, 1986). The California State Water

Project is currently composed of two massive dams (Oroville and San Luis) and a number of smaller reservoirs, two gigantic pumping plants to raise the water over the intervening mountains, and the California Aqueduct which snakes 444 miles to southern California. The California Aqueduct branches in southern California and the East Branch continues across the Mojave desert to Lake Parris, fully 600 miles downstream from Oroville. The California Aqueduct is the longest river in the state. The project was originally planned to transfer 4,230,000 acre–feet annually; however, the size of the project required it be built in stages. The project components as described above divert about 2,500,000 acre–feet per year. To reach the original goal of 4.23 million acre–feet new "Phase Two" components would have to be built. Action on Phase Two in the 1980s became nearly impossible for environmental and political reasons. The MWD has financed 70 percent of the State Water Project; however, the city of Los Angeles still gets over 90 percent of its water from the Owens River and Mono Basin. The residents of Los Angeles so far have paid a great deal of money for very little water.

Water Conservation

The one alternative Los Angeles (and the MWD) avoided like the plague was water conservation. The planners used technology to deny the climatic reality and perpetuate the myth that southern California was a water–rich oasis. By the 1980s two events forced the city and the MWD to seriously consider water conservation. One event was the completion of the Central Arizona Project which meant Colorado River water could now be delivered to Phoenix and Tucson. Los Angeles could no longer "borrow" Arizona's water. The second was the drought which began in 1986. The MWD began a program to pay member agencies $125 for each 1000 cubic meters of water they saved. The estimated savings through 1992 has been 33 million cubic meters, enough to supply 885,000 households (Postel, 1992). The LADWP too has instituted a water conservation program. The program includes a landscape management program, a water conservation advisory committee, a city ordinance mandating low–flow water fixtures in new construction, and free distribution of low–flow fixtures to residential customers. The department has spent $5 million (1990 dollars) on the purchase and distribution of fixtures. The programs reduced water use 4 percent from 1987 to 1990. (USEPA, 1995).

Desalination

There is plenty of water on Earth; however, most of it is too saline to be used directly for water supply. Desalination creates freshwater by removing ions of salt dissolved in the water. Removing the salt requires energy, which is why desalination is so expensive (Table 6.1). There is also the associated problem of disposing of the residual salt. There are at least five different desalination technologies currently available. The two most common are reverse osmosis and flash distillation. The reverse osmosis process essentially forces water through a porous membrane, filtering out the salt in the process. Distillation creates freshwater by evaporation. Water changing from the liquid state to the gaseous state leaves the dissolved salts behind. The theoretical minimum amount of energy required to remove salt from water is 0.6689 cal/g (2.8 kJ/liter). Current desalination technologies use anywhere from 5 (reverse osmosis) to 75 (distillation) times this amount of energy (Gleick, 1993). Distillation can use sunlight as a free energy source, though the rate of freshwater production is fairly low. Around the world there four times as many reverse osmosis plants as there are any other type, though almost twice as much water is produced worldwide using distillation technology. The reason is that distillation is preferred for large facilities.

The region of the world with the largest installed desalination capacity is the Middle East. Many Middle Eastern countries are energy rich, financially well off, and arid. Gleick (1993) lists 36 countries with desalination capacities greater than 20,000 m^3/d (0.053 bgd). These 36 countries account for 97 percent of the world's desalination capacity. The top 10 countries in terms of capacity are listed in Table 6.5. One–quarter of the world's desalination capacity in 1989 was in Saudi Arabia. The United States was second at 12 percent of the world's capacity in 1989.

Desalination in the Middle East costs between $1 and $8 per cubic meter ($1234 and $9870 per acre–foot). By comparison farmers in the western United States pay between $0.01 and $0.05 per cubic meter ($12 and $62 per acre–foot) for irrigation water, and urban residents pay $0.20 to $0.30 per cubic meter ($247 to $370 per acre–foot). Comparing Table 6.5 to Table 4.1, it is obvious that desalination is a very small component of freshwater use in the United States. A number of cities in California began emergency construction of desalination facilities in 1992 following six years of drought. One of those was Santa Barbara.

TABLE 6.5 Installed Desalination Capacity for the Leading 10 Countries in the World in 1989[a]

Country	Installed capacity		Predominant	Number of
	bgd	m^3/d	type[b]	units
Saudi Arabia	0.943	3,568,868	RO	1417
United States	0.420	1,588,972	RO	1354
Kuwait	0.367	1,390,238	MSFD	133
United Arab Emirates	0.352	1,332,477	MSFD	290
Lybia	0.164	619,354	RO	386
Japan	0.123	465,600	RO	615
Iraq	0.0816	323,925	RO	198
Qatar	0.0815	308,611	MSFD	59
Soviet Union	0.079	299,143	RO	53
Bahrain	0.073	275,767	RO	126

[a] Source: Gleick, 1993.
[b] RO, reverse osmosis; MSFD, multistage flash distillation.

In the 1960s the residents of Santa Barbara voted against joining the California State Water Project, choosing instead to rely on local supplies. After six years of drought the city's supplies were exhausted and the best option appeared to be desalination of seawater. Just as the plant was being finished the drought broke so the plant was mothballed. Millions of dollars later the plant sits on the beach having never produced a single drop of freshwater. Santa Barbara has also decided to connect to the State Water Project.

More important in the United States than desalination has been the increasing use of saline and brackish water for industrial uses where freshwater is not required. This is especially true for the generation of thermal electrical energy in coastal regions (Table 4.6).

PLANNING AND DECISION MAKING

Dzurik (1990) defines planning as "a logical and orderly way to think about the future." Water management involves the day–to–day actions and decisions that are made for using water and related resources. Since decisions made today (for example, issuing new building permits) have future consequences (increased water demand), planning and management go hand–in–hand. Dzurik identifies two general types of planning—*functional* planning and *comprehensive* planning. Functional planning is for a single purpose or need. Water supply planning is functional, as is planning for wastewater treatment, transportation, or prisons. Functional planning is carried out by the agency responsible for that particular service. Comprehensive planning is integrative in nature and comprehends multiple purposes. Comprehensive land–use planning relies on population projections to manage the spatial pattern and temporal changes in land use in a certain area. Zoning for different land uses, preservation of natural and historic areas, and regulations specifying where, when, and how new growth occurs are all parts of comprehensive plans. Comprehensive planning is carried out by planning departments. Where comprehensive plans exist, functional planning should coordinate with the comprehensive plan.

There is a third type of planning unique to water resources—basin–wide planning. Basin–wide planning, like comprehensive planning, is spatial in nature. It differs from comprehensive planning in that the boundaries of the planning area are physically determined (by the watershed divide), whereas the boundaries for comprehensive planning are political or administrative borders. Another similarity between comprehensive and basin–wide planning is that they both consider systematic interactions between land and water resources. Water–resource planning can and certainly has been done in spatial isolation. Numerous cities and towns have functionally planned their water supply system with little concern for neighboring areas. But the fluid nature of water generates spatial connections within, and beyond, the drainage basin. The early efforts at basin–wide planning in the Progressive and New Deal periods, focusing as they did mainly on water quantity issues, were motivated by the logic of this physical connectivity, though there may have been more tangible imperatives, like soil erosion, pushing the basin–wide approach. Today, growing population, increasing demands on land and water resources, the recognition of natural limits, environmental degradation, and questions about sustainable use are the imperatives, and we are seeing renewed interest in basin–wide approaches. One example is the federal EPA's adoption and promotion of basin–wide water quality management programs.

PLANNING ENTITIES

Functional and comprehensive planning for water resources is carried out daily by local governments. State governments vary in their commitment and capability to plan water resources. Until the 1960s the number of planning professionals working at the state level was quite small and there was little water–resource planning apart from the activities of the federal government. That picture changed drastically by the 1970s as state and local agencies became more skilled. Most states have legislation providing for water quantity management and planning. In some states this is combined with water quality management and planning. In addition a number of states have developed drought contingency plans following drought episodes in the 1970s, 1980s, and 1990s. Interstate water–resource planning occurs in states that are party to interbasin compacts (Table 3.2), or where river basin commissions still exist. In Chapter 2 we saw how federal water–resource planning has waxed and waned. The National Resources Planning Board and its predecessors during the 1930s accomplished a great deal. During the 1960s and 1970s the Water Resources Council and the federal–state river basin commissions made some important contributions. The decades in between were dominated by individual agencies planning and pursuing their own agenda, e.g., the iron triangle. There is no overall, coordinated federal planning for water resources today. Individual agencies still plan their own activities, and Congress passes legislation which initiates a cascade of action, some of which is planning, at state and local levels.

Two other entities involved in water–resources planning are special districts and the private sector. Special districts are created to accomplish either a single purpose, such as flood control, drainage, or irrigation, or multiple purposes. Special districts operate within a gray zone in the American political system. They are created as political subdivisions of states but are quasi–governmental in nature. Their powers may include taxation, eminent domain, and authority to issue tax–free bonds. But on the other hand their purpose and function are more appropriate for the private sector. The curious mixture of public and private attributes has led to confusion about their legal status as well as concern about adequate supervision by state authorities. The number of special districts in the United States has increased dramatically in recent decades. Leshy (1983) called them the new dark continent of American government. In 1969 special water districts distributed one–half of all the irrigation water used in western states (Leshy, 1983). The Middle Rio Grande Conservancy District (MRGCD) is one example of a multipurpose district (see Box 6.2).

BOX 6.2 THE MIDDLE RIO GRANDE CONSERVANCY DISTRICT

The Middle Rio Grande Conservancy District (MRGCD) in central New Mexico was created in 1925 to address the interrelated problems of drainage, flood control, sedimentation, and irrigation water supply. Between the years 1880 and 1920 irrigated acreage in the Middle Valley of the Rio Grande decreased 68 percent. The reason was that groundwater levels were rising, creating waterlogged soil, seeped areas, and alkali flats. Groundwater levels rose because of the massive increase in the amount of land under irrigation in the later half of the 19th century. Every irrigation project eventually becomes a drainage project. In order to rescue irrigation agriculture, drainage ditches were needed to lower the water table. While drainage was the most visible problem, the major force behind the creation of the MRGCD was the local Chamber of Commerce, which perceived flooding as the main threat to downtown Albuquerque. The MRGCD set about digging drains, building levees, improving the irrigation system, and, with the help of the Bureau of Reclamation, constructing El Vado Dam for water supply in the northern mountains. The district paid for the improvements with assessments levied against real property within the district. The MRGCD is centered on the floodplain of the middle Rio Grande and extends 150 miles from Cochiti Reservoir in the north to the Bosque del Apache Wildlife Refuge in the south, and varies from 0.5 to 1.25 miles in width. It includes parts of six Indian pueblos, four counties, and five municipalities. The district is divided into four divisions. The general structure of the water system in a single division is shown schematically in Figure 6.12.

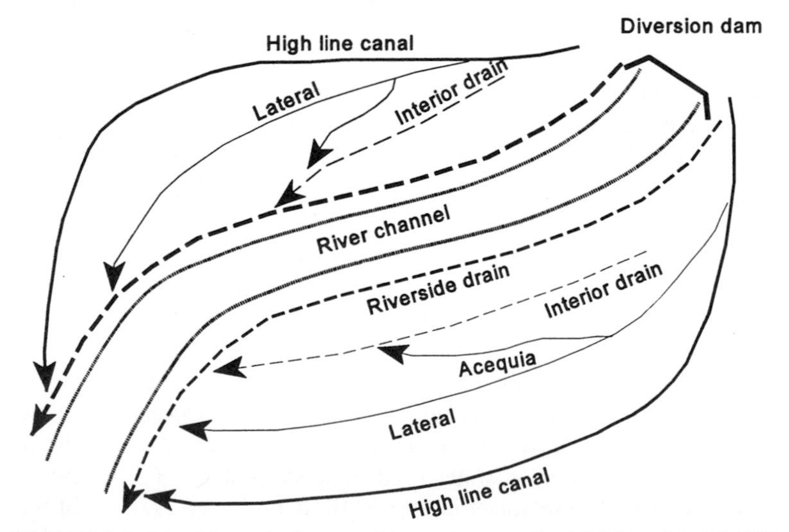

FIGURE 6.12 Schematic diagram of the water system for a division of the MRGCD

Urbanization has changed the MRGCD. While originally created to serve a rural (agricultural) population, the Albuquerque Division has undergone extensive urbanization. In parts of the Albuquerque Division the reduction in irrigation combined with groundwater pumping by the city has lowered the water table and allowed the MRGCD to close and fill in some interior drains. The urban-to-rural transformation has also generated considerable conflict within the district. Urban residents feel their assessments are used to subsidize irrigation in other parts of the district. There is some truth to the complaint because urban land is assessed at a much higher rate than is agricultural land. The urban area in and around Albuquerque paid something like 70 percent of the district's total assessments from 1959 and 1984. Another problem is that city residents see the irrigation canals as hazards to small children and want them fenced, something the MRGCD has vehemently resisted.

Private companies regularly plan and manage water resources. Private water supply companies plan to meet consumer demand just as public utilities do. Private power companies operate hydroelectric dams and thermal electric power plants. In planning for the delivery of power they must consider their water needs and how their operation affects the water resource. Water planning and management are also done by self–supplied industries, by large agribusiness corporations that rely on irrigation water, and by individual home owners.

THE PLANNING PROCESS

The Rational Model

The *rational planning model* is an idealized sequence of steps (see Box 6.3). The model describes a *decision–making process.* The first step is the identification and definition of the problem. The problem may already exist or it might

BOX 6.3 STEPS IN THE RATIONAL PLANNING MODEL

1. Problem identification
2. Data collection and analysis
3. Formulation of goals and objectives
4. Identification of alternatives
5. Evaluation of different alternatives
6. Selection of the "best" alternative
7. Implementing the alternative
8. Monitoring the alternative's performance

develop in the near future if existing trends continue. In defining the problem the viewpoints of all relevant public and private groups should be considered. Public participation (of which we have more to say later) should be encouraged at this step. Once the problem is adequately defined, data are needed to understand current conditions and as a basis for generating future projections (step 2). Depending on the problem an array of demographic, economic, geophysical, biological, cultural, land–use, and historical data may be required (Table 6.6). Accompanying the data themselves should be *metadata* describing data characteristics such as their source and quality.

Step 3 in the model is the formulation of goals and objectives. Goals are general statements about what is to be accomplished. Objectives are more specific statements as to how goals will be reached. For example, the problem might be water pollution and water quality degradation of a stream. The goal might be to restore the stream to its original condition. Specific objectives that help achieve the goal might include treating all municipal waste discharges, eliminating industrial discharges, and reforesting part of the watershed to reduce nonpoint pollution. Objectives are important in that we use them to measure progress toward the larger goal. We saw in Chapter 5 that economic objectives dominated federal water planning until the late 1960s when multiobjective planning began. The objectives in federal multiobjective planning are more like "goals" as the term is used here.

Steps 4 through 7 involve formulating different alternatives, evaluating the merits of each alternative, and finally selecting the best alternative to accomplish

TABLE 6.6 Examples of Data That May Be Required in Water–Resource Planning

Data category	Examples	Data category	Examples
Demographic data	Population	Geophysical	Climatological
	Size		Hydrological
	Distribution		Surface water (quantity and quality)
	Growth rate		Groundwater (quantity and quality)
Economic	Income levels		Geological
	Economic structure		Rock types and structure
	Manufacturing		Seismic risk
	Industry	Biological	Ecosystem type and structure
	Commercial		Threatened or endangered species
	Agriculture		Wetlands
Land use	Residential		
	Industrial		
	Commercial		
	Manufacturing		
	Agricultural		
Cultural/historical	Archeological sites		
	Sites of historical interest		

the stated objectives. For example, municipal wastes can be treated by conventional centralized treatment technology using aerated lagoons or trickling filters, or by nonconventional means such as land application of wastewater or natural purification methods (wetlands). The Carter administration in the late 1970s instituted the requirement that federal agencies include at least one nonstructural alternative in every federal water–resource plan. Carter forced federal agencies to broaden their range of choice in the identification of alternatives.

One of the most important tools for evaluating and comparing alternatives is economic analysis. But economics is just one dimension of a project and environmental and/or social impacts may be of equal or greater importance. The environmental impact statement is the primary tool for identifying intangible and nonmonetary environmental impacts from federal actions. The relative weight given to economic versus noneconomic factors waxes and wanes with larger social conditions and the mood of the citizenry. Once the alternatives are identified, evaluated, and compared, the best alternative is selected. The chosen alternative should be acceptable to the majority of people. Given the recent experience in the Everglades, *reversibility* of the alternative might be an important characteristic to consider. It might also be that the best course of action is the "no action" alternative. After the alternative is implemented, the final step is to monitor the alternative's performance to see how well it accomplishes the original objective(s) (step 8). This type of postaudit evaluation has rarely been done in the past. A greater emphasis on monitoring performance might be in the offing with the growing interest and emphasis on *outcomes–based assessment* for public programs.

The rational model is not inherently spatial by nature, that is to say it can be applied to a single project at a single location. But neither is the model adverse to broader spatial application. Applying the rational model in a spatially–extensive basin–wide plan requires expanding the scope of the individual steps to include systematic interactions between resources. The general result is that the model becomes more complicated. At (larger) basin scales the problems become more complex (step 1), which means more and different types of data are required (step 2), and additional alternatives become available that were not possible, or necessary, at smaller scales (step 4). With increasing complexity, though, comes increasing opportunities for pooling resources, tapping human creativity, and accessing a wider range of skills and expertise. This presumes, of course, that the people and agencies want to cooperate and work together.

Modifications to the Rational Model

The rational planning model is an idealized process for decision making. Ideally we first identify a problem, collect and analyze relevant data, identify and compare alternatives, and make a decision. The model is based on the assumption of rational economic behavior, which is to say rational people make decisions that maximize their welfare. The rational model sounds good, but it is probably the exception rather than the rule in real life. Many factors conspire to make the rational model difficult to follow.

At least two alternatives to the rational model have been proposed. One alternative proposed by Simon (1959) says that people are not rational but rather *bounded–rational.* Bounded rationality means simply that there are limits to our rationality—we are, after all, only human. According to Simon, people do not maximize their utility but *satisfice.* Satisficing means accepting alternatives that meet minimum expectations; we accept a course of action that is not optimal but is good enough. According to Simon our aspirations are adjusted downward until our goals reach levels that are practically attainable. Some of the other key components of the bounded–rational model are that alternatives are not fixed and well specified; goals may be vague and inconsistent; information is incomplete or lacking and expensive to obtain; knowledge is incomplete; and our perceptions distort the information we receive.

Another modification of the rational model is Charles Lindblom's (1964) *disjointed incrementalism.* Lindblom's model assumes the basic problem may not be well defined; goals and values may be in conflict; only a limited range of alternatives can be considered at any one time, and they differ from each other only marginally; the means–ends relationship may be reversed, which is to say the ends (goals) are adjusted to meet the means available; policy makers often do not know what the public wants, but know what should be avoided; and the decision–making process never ends—it is a continuous sequence of incremental decisions. In essence, Lindblom says we "muddle through." The rational model is prescriptive and tells us how we *should* make decisions, whereas the alternative models of Simon and Lindblom attempt to describe how decisions are actually made.

PUBLIC PARTICIPATION

The public must be involved in decisions that affect them. As a general rule the public should be brought into the planning process as early as possible. In the rational model the public was (ideally) involved at the very first step of problem identification. A good reason for early involvement is that public perceptions may differ from those of the decision makers and the professional resource managers. Simon's model recognizes the potential for information distortion. *Perception* is the filtering and cognitive processing of environmental information. Perception is influenced by past experience, knowledge, fears, desires, and personality characteristics. In a classic study of the perception of the flood hazard, Kates (1962) verified differences in the perception of the flood hazard between residents living in the floodplain, and between residents and professional resource managers. A simple example of perceptual divergence is the meaning of the "100–year flood." A common misperception is that it means a flood that occurs only once every 100 years. This interpretation is wrong. The 100–year flood is a probabilistic statement. It means a flood of that size has a 1 percent chance of occurring each and every year. A flood of that size occurs *on the average* once every 100 years. This misperception is why the Federal Emergency Management Agency (FEMA) now prefers to use the term "Special Flood Hazard Area."

Probability is a difficult concept to grasp, even by professionals. Differences in perception lead people to view the same problem differently. In the study by Kates differences in perception led different people to choose different adjustments to the flood hazard. In the planning process it is best to get these disparate views early on so they can be discussed and reconciled. Some important questions regarding public participation in resource decision making are

To what degree should the public be involved?
Is the public being adequately represented?
Where in the process should public input be sought?
How should public input be obtained?

The first question addresses a fundamental tension that may exist between professional resource managers and the public at large. Some managers may view the public as ignorant and apathetic, which could be true in some cases. The managers may feel they have a better understanding of the issues and have more expertise in dealing with these types of problems. The potential exists for managers to assume that they know what is best and to view public participation as a waste of time and effort. The managers may only want to give the public token involvement. On the other hand interested citizens may want real power over the decisions that are made.

Who exactly is the public? Ideally input should come from a representative cross section of the public. But there are people who make it their business to get involved whether they are directly affected or not. Special interest groups, through professional lobbyists, can influence decisions out of proportion to the size of their constituency. The fact is the average citizen is often too busy, too harried, or too tired to get involved even on issues that are important to him or her. How can we be sure we are incorporating the views of the silent majority?

Where in the planning process should the public be involved? We have already said public participation should occur early to aid in the clear definition of the problem. Early involvement is important for other reasons. Early in the planning process (steps 1, 2, and 3 of the rational model) the decisions that are being made are *normative* (Smith, 1982). Normative decisions determine *what should be done.* When the process moves on to steps 4 and 5 the decisions being made are *strategic.* Strategic decisions are those which determine *what can be done.* Finally, at steps 7 and 8 *operational* decisions are being made. Operational decisions are made to determine *what will be done.* Too often the public is not consulted until the operational stage. By then the planners and managers are discussing operational issues while the public may want to address normative issues. As a result conflict ensues, there is a lack of communication, and the professionals criticize the public's participation as irresponsible.

Public input can be obtained in different ways, and the effectiveness of a given mechanism depends upon the situation. Table 6.7 lists three mechanisms for soliciting public input. The body of the table gives a qualitative appraisal of the effectiveness of each mechanism on representativeness, the quality of information flow

TABLE 6.7 Effectiveness of Different Public Participation Mechanisms[a]

Mechanism	Representative of the public at large	Ability to get information in from the public	Ability to get information out to the public	Ability to make a decision
Public meetings	Poor	Poor	Good	Poor–Good
Advisory groups	Poor–Good	Poor–Good	Poor–Good	Fair
Questionnaire survey	Good	Poor	Fair	Poor

[a] Mitchell, 1989.

to and from the public, and the ability to come to a decision. For example, public meetings are widely used and are good at getting information out to the public. They are poor at reaching a representative cross section of people, and are not very good at getting information from the public to the planners, managers, and decision makers.

CONSTRAINTS ON WATER MANAGEMENT DECISION MAKING

In Chapter 1 we talked about the characteristics of the physical resource. Chapter 2 touched on how policy, or the lack thereof, and administrative arrangements through history have influenced water–resource decisions. Chapter 3 outlined the basic legal structure for water, and in Chapter 5 we examined the role of economics. These and other factors combined are constraints on water–resource planning, management, and decision making.

The *physical system* constrains water availability in time and space. These physical resource limits are both absolute and relative. Every water management decision includes an appraisal of the availability of the physical resource. The total amounts of precipitation and evapotranspiration (water balance) set an absolute limit on natural water availability. Water may be transferred from one basin to another, but still there is a finite quantity available. Water transferred to a basin is unavailable for use in the basin of origin. There are also relative limits to the physical resource as when new and expanding uses require established water users to reduce their withdrawals.

Existing *water laws,* treaties, compacts and other and legal institutions constrain options for water management. Water–resource plans and decisions must obey the existing legal framework; however, laws can be changed. *Economics* is an important constraint and often dominates water–resource decisions. There are limited funds available regardless of whether it is spending by government or by private individuals. Economic efficiency says we should get the most for every dollar spent. The method by which costs and benefits are evaluated is important. What are considered costs and what are considered benefits are different between different cultures, and for the same society they change over time.

Administrative constraints arise when coordination is required between different levels of government, between different agencies at the same level, or between all of the above. Effective coordination has always been a problem in water planning and management. There is no question that water–resource problems will become more complex, more interdisciplinary, and more multijurisidictional in the future. Effective administration will be critical to successful management. The United States has never had a single national *water policy.* In some ways this is regrettable, as when the iron triangle drove water development with no clear goal in sight. In reality, though, a single policy could have never maintained relevancy, because the social changes have been too fast and fundamental to have been comprehended by a single policy. State and local governments have adopted policies to guide their water–resource activities. In some cases the policies combine land–use and water–use goals, e.g., maintaining irrigation agriculture as a viable economic sector.

Perception plays an important role in resource decision making. Perception of the flood hazard influences the types of flood–mitigation adjustments used by floodplain residents. Perceptions of the reality and severity of a drought can make all the difference between whether a government's plea for water conservation is heeded or not. The perception, on the part of water utility engineers, that the public would respond negatively to water demand management has been an important force behind their (utilities') preference for water supply management.

Finally, *environmental linkages* place a requirement of sustainability on our actions. Historically we have not understood environmental linkages very well. Even in cases where we did we were often too arrogant to care and routinely made decisions without regard for the environmental impacts. Research and experience have greatly increased our understanding of the natural environment, and we will have do things very much differently in the future if our resource–use systems are to be sustainable.

7

Offstream Water Use

Urban and Agricultural Uses

Urban Water Systems
- Urban Water Supply
- Geographic Pattern of Public Water Supply
- Drinking Water Quality
- Wellhead Protection and Management

Irrigation
- Historical Development
- Irrigation in the World Today
- Irrigation in the United States
- Water Application Methods
- Estimating the Evapotranspiration Requirement (E_{tp})
- Estimating the Irrigation Water Requirement (*IRR*)
- Irrigation Efficiency
- Economically Efficient Water Application
- Water Conservation in Agriculture
- Environmental Impacts of Irrigation

This chapter examines offstream water use for urban purposes and irrigation agriculture. For each sector we look at how water is used, the spatial patterns of use, selected quantity and quality characteristics associated with the use, and the role of conservation and water–use efficiency. Even though from a systems viewpoint it would be sensible to study the entire urban water system—water supply input and wastewater output—we will only examine the supply side in this chapter. Wastewater treatment is covered in Chapter 9 along with other water quality topics.

URBAN WATER SYSTEMS

Urban water supply systems characterized by centralized treatment and distribution facilities providing water to indoor taps are largely a product of the late 19th and early 20th centuries. Some cities developed urban supply systems much earlier than this, thousands of years earlier in the case of Rome and other cities from antiquity. The provision of public water supply in the United States has been largely accomplished by private companies and local governments with minimal help from the federal government. There are approximately 54,000 public water systems in the United States and related territories. A public water system is defined as one with at least 15 service connections or that regularly serves an average of at least 25 individuals daily at least 60 days out of the year. Of the 54,000 systems 794 are large systems serving more than 50,000 people and 6800 systems are medium sized and serve between 3301 and 50,000, while the rest are small and serve fewer than 3301 people (USEPA, 1997).

The goal of providing *safe* drinking water is barely a 100 years old. One of the classic stories in water supply epidemiology is the case of the Broad Street Pump. In 1854 more than 500 deaths occurred during an outbreak of cholera in the vicinity of Broad Street in London. Dr. John Snow deduced that the epidemic was caused by polluted water coming from a single well on Broad Street. His was a remarkable deduction since it predated the germ theory of disease. Dr. Snow's solution was simple and effective—he removed the pump handle, and the epidemic was contained. The development of clean water supply in the 20th century stands as one of the crowning achievements of water supply engineering. In 1880 water–borne typhoid fever killed 1 person in a 1000 in the United States. Today typhoid kills fewer than 1 person in 2 million. The development of wastewater treatment lagged decades behind water supply treatment. Dealing with wastewater was considered the problem of the next downstream user, not to mention it was viewed as a rather inglorious profession. Who in their right mind would choose a career dealing with sewage? As a result, water supply and wastewater systems usually developed as separate utilities with little planning coordination.

URBAN WATER SUPPLY

Figure 7.1 shows a typical urban water supply system. Water is withdrawn from a surface or groundwater source. Groundwater has enjoyed some advantages over surface water. One advantage was groundwater's comparative purity; however, this advantage has disappeared in many areas because of pollution. Another advantage is that groundwater is stored naturally—evaporation free—near the point

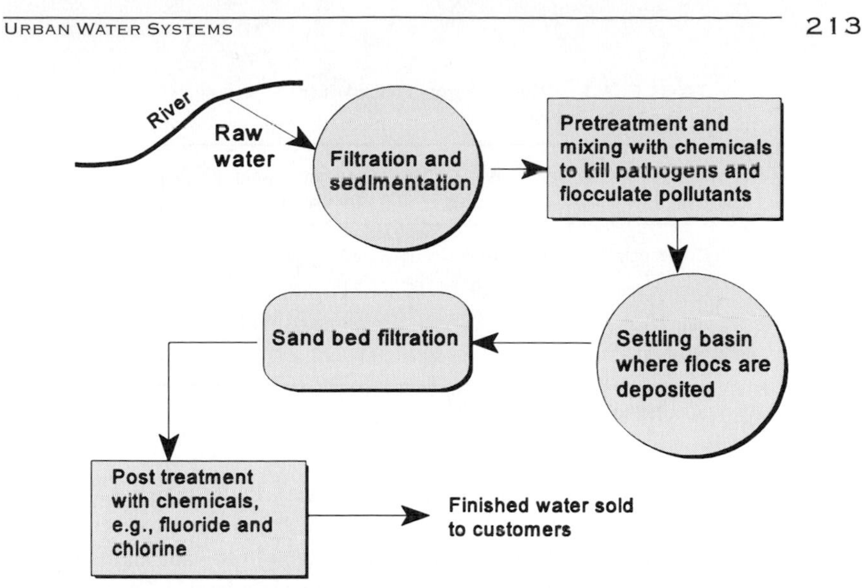

FIGURE 7.1 Treatment steps in an urban water supply system.

of use, obviating the need for reservoirs, aqueducts, and transmission lines. Raw water from the source is treated to remove contaminants or lower their concentration to acceptable levels. The initial quality of the raw water determines the number and types of treatment steps. The treatment process might include initial screening and sedimentation to remove suspended material, pretreatment with chemicals to kill pathogens (chlorine) and flocculate certain pollutants (alum), another period of settling where the coagulated "flocs" deposit out of the water, filtration through sand beds, and post–treatment with more chemicals (like fluoride to reduce tooth decay) and then the water off to the customers. Chlorine is the preferred method for disinfecting public water supplies in the United States. In France disinfection by ozone is widely used. Ultraviolet radiation is yet another disinfection alternative well suited to self–supplied domestic water systems.

Public water systems provide water for domestic use, commercial use, and industry. Domestic water use from public water systems in the United States averaged 105 gallons per capita per day (gpcd) in 1990. By contrast self–supplied domestic systems averaged 79 gpcd (Solley *et al.,* 1993). In 1990, 43 million people, or 17 percent of the population, supplied their own domestic water. When commercial uses are added to public water supply systems, the average use in 1990 increases to 124 gpcd. And when industrial uses are included, the average increases to 144 gpcd. Public water systems provided about 71 percent of the water used by commercial establishments (restaurants, hospitals, hotels, etc.), but only 27 percent of the water used by industry (Table 7.1).

Domestic water use is higher in the West than in the East. The principle reason is the greater outdoor water use for lawn sprinkling in the West. In eastern states domestic use typically ranges between 50 and 100 gpcd. In the West the range is

TABLE 7.1 Water Use from Public Water Supply Systems, 1990[a]

Use	Average (gpcd)	Percent of supply for use sector
Domestic	105	83
Commercial	19	71
Industrial	20	27
Total	144	

[a] Source: Solley *et al.,* 1993.

more like 100 to 150 gpcd, though in some states use exceeds 200 gpcd. Table 7.2 lists average values for the 10 states with the highest and the lowest per capita domestic use from public systems. The average resident in Nevada and Utah used more than four times as much water as the average resident in Ohio. Climate explains some of the difference in per capita use, but even between cities in the same region the differences can be dramatic depending upon the price of water, the use of meters, water conservation programs, the proportion of single–family versus multifamily dwellings (multifamily dwellings have less landscaping), and public attitudes.

The temporal demand for domestic water is highly variable. Figure 7.2 qualitatively shows the pattern of indoor and outdoor (mainly lawn watering) water use during the day. Indoor water use is lowest during the night and peaks in the morning as people prepare for the coming day. Water use peaks again in the evening as

TABLE 7.2 Domestic Water Use (gpcd) from Public Supplies for the 10 Highest and Lowest States and Washington, D.C., 1990[a]

Highest	gpcd	Lowest	gpcd
Nevada	218	Ohio	50
Utah	213	Wisconsin	52
Idaho	186	Maine	58
D.C.	179	Pennsylvania	62
Wyoming	162	Iowa	66
Arizona	150	Massachusetts	66
California	147	Rhode Island	67
Minnesota	147	North Carolina	67
Colorado	145	Connecticut	70
Texas	143	Kentucky	70

[a] Source: Solley *et al.,* 1993.

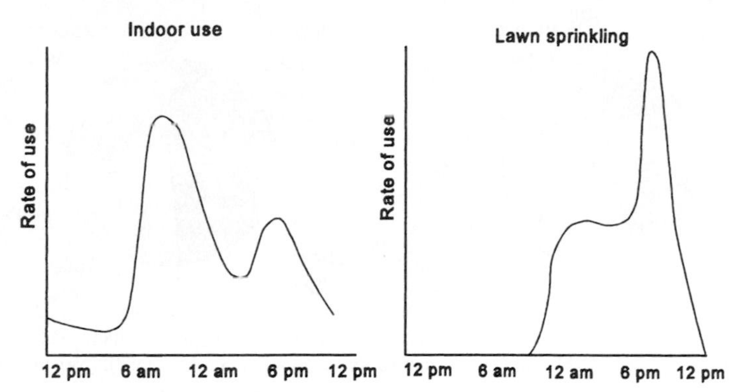

FIGURE 7.2 Qualitative representations of the temporal pattern of indoor water use and lawn sprinkling. Based on data for Kansas City, 1954 (after Savini and Kammerer, 1961).

people return home. Outdoor use for lawn watering is mainly an afternoon and evening phenomenon. With the use of more automatic sprinkling systems, lawn watering probably occurs later into the evening than shown in the figure. The maximum hour demand on public systems can be five times the average daily use, and the maximum day use can be twice the average daily use. Public water systems must be designed to accommodate these maximum use rates even though the excess pumping and transmission capacity sits idle much of the time. Public water systems must also have adequate pressure and flow for fire fighting.

Indoor Water Use

A sample–based study of residential water use sponsored by *WaterWiser* for the American Water Works Association (AWWA) has been underway in six western states and one Canadian province (Nelson, 1997). A total of 12 cities will be studied. In each city a sample of 100 single–family detached homes is chosen, and for two weeks in the summer and two weeks in the winter a small battery–operated data logger is attached to the water meter. The data logger records average water flow every 10 seconds. Using specially developed software it is possible to identify from the water–use record individual water–use events, i.e., a shower event, a toilet refill, and a clothes washer cycle. This is a significant advancement over older methods of determining end uses that were much more intrusive and necessarily relied on smaller samples. As of June 1997, logging had been completed in 6 cities in Colorado, Oregon, Arizona, Washington, and California. The database consisted of over 800,000 water–use events, from approximately 600 homes. The average indoor water use for these cities is calculated as 61.5 gpcd (Nelson, 1997). A 1968 study of domestic water use found a national average of 59.1 gpcd (American Society of Civil Engineers (ASCE), 1968). Even though the totals are very close there are some significant differences in indoor end–use categories. Figure 7.3 shows indoor water use without water conservation and with

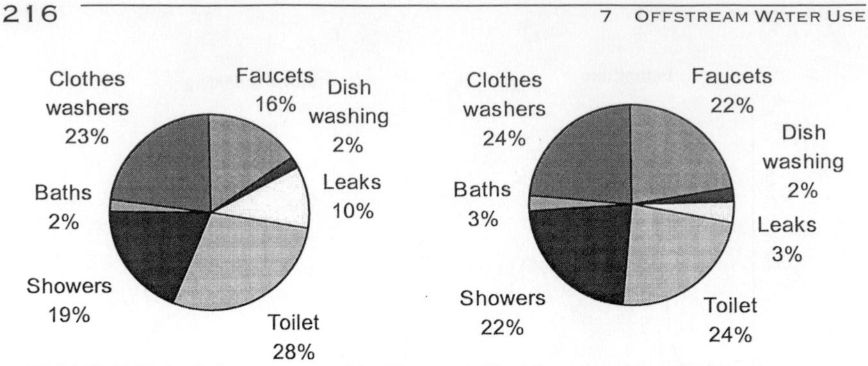

FIGURE 7.3 Indoor water use (a) without and (b) with commonly available water conservation in percent. See Table 7.3 for values in gallons per capita per day (gpcd). (Nelson, 1997)

water conservation. The without values are based on the actual sample data. Without conservation the average water use use is 64.6 gpcd (Figure 7.3a).

This value was calculated by taking the sample average of 61.5 gpcd and increasing it by 5 percent to account for existing water conservation practices in these cities. Water use by toilets is the single largest category at 18.3 gpcd (28.4 percent), followed by clothes washing, showers, faucets, and leaks. In the 1968 study, toilets were also the largest component of indoor use, but they used 24 gpcd, which represented 42 percent of total inside use. In a single–family home equipped with all of the commonly available water–conserving fixtures, overall per capita water use drops 31 percent to 44.7 gpcd (Figure 7.3b). The largest water savings is for toilets, which drops from 18.3 gpcd to 10.4 gpcd, a

TABLE 7.3 Single–family Indoor Water Use without Conservation and with Conservation[a]

	Without		With		Change due to conservation	
Use	gpcd	Percent	gpcd	Percent	gpcd	Percent
Faucets	10.3	15.9	10.0	22.4	−0.3	−2.9
Dishwashing	1.1	1.7	1.1	2.5	0.0	0.0
Leaks	6.6	10.2	1.5	3.4	−5.1	−77.3
Toilet	18.3	28.3	10.4	23.3	−7.9	−43.2
Showers	12.2	18.9	10.0	22.4	−2.2	−18.0
Baths	1.2	1.9	1.2	2.7	0.0	0.0
Clothes washers	14.9	23.1	10.5	23.5	−4.4	−29.5
Total indoor	64.6	100.0	44.7	100.0	−19.9	−30.8

[a] Source: Nelson, 1997.

decrease of 43 percent (Table 7.3). These are the 1.6 gallon per flush toilets mentioned in Chapter 6.

The average cost of water to residential consumers in 1997 was about $1.25/1000 gallons. Using 105 gpcd, a family of four would only pay about $192.00 per year for water. That is pretty cheap for something you literally cannot live without. It was common 20 to 30 years ago for public systems to use a *decreasing block rate* price structure. Users would pay one rate for the first few thousand gallons (first block), and then a lower rate for the next few thousand gallons (second block), and so on. This gave large water users a break in cost. However, with the acceptance of demand management many cities have switched to *increasing block rate* structures. With this structure the more water you use, the more you pay. Some water systems use *flat rate* structures. One of the arguments in support of a flat rate is that it provides a predictable income stream, which makes planning easier.

The United States is one of the few industrialized countries in the world where water–efficient plumbing technology is not widely used. Table 7.4 lists some plumbing fixtures and compares standard models to water–efficient models. The water-conserving fixtures can reduce water use from 40 to 70 percent. I have heard people complain about the performance of water–efficient fixtures. For example, one complaint is that low–flow showerheads do not deliver enough water. Another is that it takes too long to draw a glass of water. My experience is that there is no noticeable difference in performance between standard fixtures and high–quality, water–conserving fixtures. Water–conserving fixtures pay for themselves by saving both water and energy. For a family of four using 105 gpcd at a cost of $1.25/1000 gallons, reducing indoor water use 30 percent means a potential savings of about $57 per year, and more as the price of water goes up. Further, assuming it costs $150.00 per year to heat water, and that hot water use is reduced 30 percent as well, the potential savings in energy is another $45 per year. This is a total savings of about $102 per year. The annual savings are likely to be even greater since many cities tie the sewage utility bills to the amount of water used.

TABLE 7.4 Comparison of Standard and Water–Efficient Plumbing Fixtures[a]

	Water usage		
Fixture	Standard	Efficient	Median decrease (%)
Toilet	3.5–5 gal/flush	1.0–1.6 gal/flush	70
Shower heads	3.5–5 gal/min	1.3–2.5 gal/min	59
Residential faucets	2.6–5.5 gal/min	0.5–2.6 gal/min	66
Washing machines	35–50 gal/load	22–26 gal/load	44

[a] Sources: Gleick, 1993; EPA, 1995.

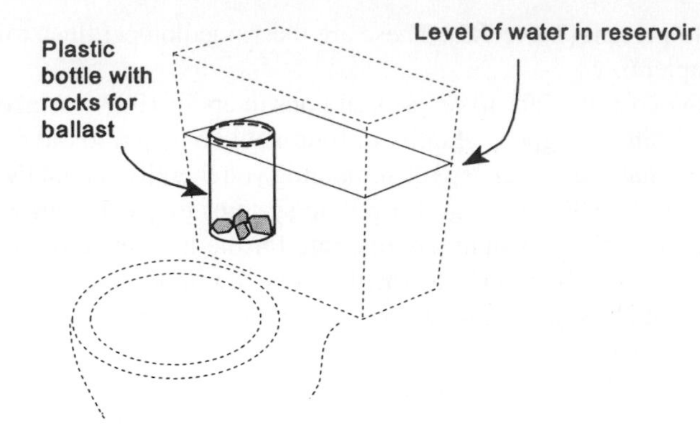

FIGURE 7.4 Plastic jar with ballast acting as a toilet dam water displacement device.

Even without buying a new water–efficient toilet people can reduce the water used by their existing model. Toilet dams are available in hardware stores or easily made using a plastic jar and some rocks for ballast (Figure 7.4). A plastic bottle is the best type of container, because brick will decompose, and glass can break. When fitting the jar into the reservoir one must be sure it does not interfere with the operation of the flushing mechanism. Box 7.1 lists a variety of water conservation measures applicable to both home owners and water system operators. Water conservation offers potential benefits for water quality improvement as well.

Outdoor Water Use

Outdoor water use is more variable than indoor use. The major component of outdoor use is landscape irrigation, especially turf (lawn) grass. Outdoor water use tends to be higher in the West than in the East, because the drier climate means turf lawns in the West need to be irrigated. The study by Nelson (1997) has found that in the western states where data logging has been done, outdoor use averages 65 percent of *total* residential use during the growing season. Some of the variables that affect landscape water use include natural rainfall, length of the growing season, size of the yard area, types of vegetation, the plants' evapotranspiration requirement, soil type, amount of shade, wind, and the method and efficiency of water application. Many of these factors are discussed later in the section on irrigation water use. The AWWA study has confirmed an earlier discovery that built–in (in ground) sprinkler systems tend to use 20–30 percent *more* water than "hose draggers." The reasons are that the built–in systems are usually set to the worst case conditions (hot dry weather) and left on that setting all season. This leads to overwatering in the spring and fall, and during cooler periods. The other reason is that hose draggers tend to irrigate less frequently. An automated sprinkler system that is adjusted periodically through the growing season to match the transpiration requirement of the vegetation could potentially save water.

BOX 7.1 URBAN WATER CONSERVATION

Methods for water users

Engineering methods

- faucet aerators
- water pressure reduction
- low-flow plumbing fixtures
- toilet dam in tank
- gray water use
- conservation landscaping
- drip irrigate shrubs and trees
- adjust automatic sprinkler systems
- develop deep turf grass roots

Behavioral methods

- shorter showers
- turn water off between uses
- flush toilet only periodically
- run appliances only when full
- water lawn when temperature is low
- allow grass to grow longer
- sweeping sidewalks rather than hosing

Methods for water system operators

- water meters
- leak detection
- water main rehabilitation
- water reuse and recycling
- capping unused artesian wells
- pricing strategies
- residential water audits
- public education

(Source: U.S. Environmental Protection Agency, 1995)

GEOGRAPHIC PATTERN OF PUBLIC WATER SUPPLY

Figure 7.5 show the pattern of water withdrawals for public supplies in 1990. California led all states at 5.83 bgd, followed by Texas (3.09 bgd), New York (2.91 bgd), Florida (1.93), and Illinois (1.86 bgd). The states with the lowest withdrawals were in the northern Great Plains, upper New England, and West Virginia.

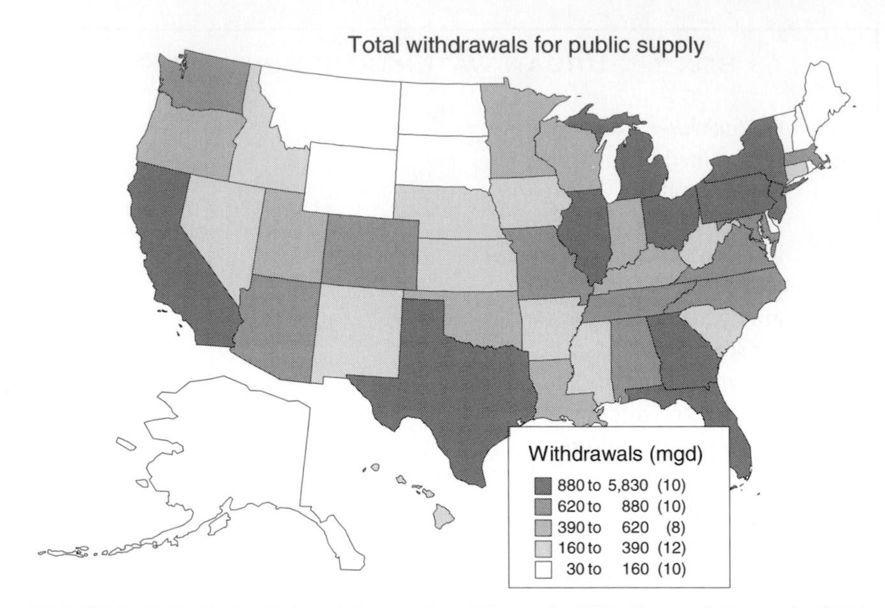

FIGURE 7.5 Total withdrawals by state for public supply, 1990. (Source: Solley *et al.*, 1993.)

DRINKING WATER QUALITY

The amount of water actually used to satisfy our metabolic requirements is a small fraction of domestic use (Figure 7.3). That notwithstanding, the public health implications of drinking water dictate water quality standards for public supplies. As with planning for quantity (yield and reliability), planning for water quality is viewed as nonnegotiable. This nonnegotiable approach " . . . is a guiding principle behind the conservative approaches, thought processes, and activities of planners, designers, and operators of municipal systems. All these activities, and the regulatory control of them, include the conscious realization that mechanical failures or negligence, even for short periods, can cause the illness and death of many people. It is not a responsibility to be taken lightly, and it isn't" (Lamb, 1985, p. 228).

The Safe Drinking Water Act (SDWA) (1974) established the basic guidelines for drinking water quality in the United States. The SDWA directed the EPA to set standards for contaminant concentrations in drinking water from public supplies. The regulations do not apply to small water systems and private wells. Individual states may set standards more stringent than the federal regulations. There are two categories of drinking water regulations. Primary standards are set for constituents that can affect the human health and are enforceable by the EPA. The primary standards are "interim" because research and revisions are ongoing. Secondary standards are guidelines for aesthetic characteristics (color, odor, etc.) that do not normally affect human health. By 1994 the EPA had identified 255 regulatory constituents, though it was thought the number could potentially be two or three times larger. By 1996, 83 constituents had final standards, and the EPA had either set, or was in the process of setting, primary standards for a total of 166 constituents. Of these, 119 were organic chemicals, 34 were inorganic chemicals, 6 were radionuclides, and 7 were microbiological contaminants (USEPA, 1994a).

Amendments to the SDWA in 1996 (PL 104–182) fundamentally changed how the EPA identifies regulated constituents, and how future standards will be set. The 1996 SDWA amendments are discussed below. Table 7.5 is a select list of constituents, their primary standard, and related health advisory data. For example, atrazine is a pesticide and a possible human carcinogen. The maximum contaminant level (MCL) is 0.003 mg/liter, and the status of this standard is final (F). For a 70–kg (155 lbs) adult the long–term exposure level is 0.2 mg/liter. This level of concentration is not expected to cause any adverse noncarcinogenic effects up to approximately 7 years of exposure, with a margin of safety. The lifetime exposure level is the concentration that is not expected to cause any adverse noncarcinogenic effects over a lifetime of exposure, with a margin of safety.

TABLE 7.5 Examples of Constituents, Standards, and Health Advisory Data for Drinking Water[a]

	Primary standards			Health advisories (70-kg adult)	
Constituent	Status of standard[b]	MCL standard (mg/liter)[c]	Status of health advisory	Long-term (mg/liter)	Lifetime (mg/l)
ORGANICS					
Atrazine	F	0.003	F	0.2	0.2
Benzene	F	0.005	F	–	–
Carbon tetrachloride	F	0.005	F	0.3	–
Chlordane	F	0.002	F	–	–
2,4-D	F	0.07	D	0.4	0.07
Diazinon	–	–	F	0.02	0.0006
Heptachlor	F	0.0004	F	0.005	–
Paraquat	–	–	F	0.2	0.03
Styrene	F	0.1	F	7	0.1
2,3,7,8–TCDD (Dioxin)	F	3E-08	F	4E-08	2E-08
Toulene	F	1	F	7	0.5
Vinyl chloride	F	0.02	F	100	10
INORGANICS					
Arsenic	(under review)	0.05	D	0.015	0.003
Boron	L	–	D	3	0.6

continues

TABLE 7.5 *Continued*

Primary standards					
				Health advisories (70-kg adult)	
Constituent	Status of standard[b]	MCL standard (mg/liter)[c]	Status of health advisory	Long-term (mg/liter)	Lifetime (mg/l)
Cadmium	F	0.005	F	0.02	0.005
Chromium	F	0.1	F	0.08	0.01
Fluoride	F	4	–	–	–
Lead	F	TT*	–	–	–
Mercury	F	0.002	F	0.002	0.002
Nitrate (as N)	F	10	F	–	–
Selenium	F	0.05	–	–	–
RADIONUCLIDES					
Beta particle and photon activity	P	4 mrem			
Gross alpha particle activity	P	15 pCi/liter			
Radon	P	300 pCi/liter†			
MICROBIOLOGY					
Cryptosporidium	L	–			
Giardia lamblia	F	TT			
Legionella	F′	TT			
Standard plate count	F′	TT			
Total coliform	F	**			
Viruses	F′	TT			
Secondary standards					
Aluminum	F	0.05–0.2			
Chloride	F	250			
Color	F	15 color units			
Corrosivity	F	non–corrosive			
Fluoride (under review)	F	2.0			
Iron	F	0.3			
Odor	F	3 threshold odor units			
pH	F	6.5–8.5			
Sulfate	F	250			
Total dissolved solids	F	500			

[a] Source: US EPA, 1994a.

[b] F, final; D, draft; L, listed; P, proposed; T, tentative; F′, final for surface water, considered for groundwater.

[c] TT, treatment technique; *action level = 0.015 mg/liter; **no more than 5 samples per month can be positive;

†the SDWA Amendments (1996) require the EPA to withdraw this MCL and established a complex multimedia (air and water) approach to setting the MCL.

There can be significant differences between long–term and lifetime advisories because of the EPA's conservative policies, especially with regard to carcinogenicity, and other factors that cause uncertainty. The health advisory data do not include any possible synergistic effects from multiple chemical exposure.

Amendments to the SDWA in 1996

Box 7.2 lists four themes characterizing areas of significant change in the Safe Drinking Water Act Amendments (1996). The emphasis on prevention of pollution of drinking water supplies reflects the general recognition that pollution prevention is more effective than after–the–fact pollution cleanup. The amendments direct the states to develop a program to delineate the source areas of public water supply systems and assess their susceptibility to contamination. This is the first step in developing locally based programs to protect water supplies. Additional support of this theme is provided by reauthorizing the wellhead protection program and a new state groundwater protection program (Section 1429). The EPA will also require states to assess the technical, financial, and managerial capacity of public water systems to meet new drinking water regulations.

Community water systems are now required to prepare and provide customers with an annual "consumer confidence report" describing the source of their drinking water and the levels of contaminants found in the water. In addition, customers must be notified of any violation of a national drinking water standard that could effect human health as a result of short–term exposure within 24 hours after the violation.

Contaminant regulation and standard setting is to be based on good science that considers the risks, costs, and benefits. The EPA is required to review at least five contaminants every five years, and use three criteria to decide whether or not to regulate a contaminant. The three criteria are the contaminant is harmful to human health; it occurs or is likely to occur in public water systems at harmful concentrations; and its regulation will lower the health risk. This approach

BOX 7.2 THEMES OF THE SAFE DRINKING WATER ACT AMENDMENTS (SDWAA) OF 1966

- New and stronger approaches to prevent contamination of drinking water.
- Better information for consumers including the "right to know" what is in your water.
- Regulatory improvements, including better science, prioritizing effort, and risk assessment.
- New funding for states and communities.

(USEPA, 1996a)

prioritizes the regulation of contaminants based on risk. The EPA must use the "best available peer–reviewed science and supporting studies" in carrying out standard–setting activities. The agency will also assemble and maintain a national occurrences database for both regulated and unregulated contaminants. Whenever the EPA proposes a new national primary drinking water regulation, it must conduct and publish a cost–benefit analysis. If the costs of an MCL standard exceed the benefits, then the EPA may adjust the MCL to a level that "maximizes health risk reduction benefits at a cost that is justified by the benefits." And the cost–benefit analysis must be comprehensible to the general public. Furthermore, where reducing the level of risk from one contaminant may increase the risk from another, the EPA is given the flexibility to minimize the overall combined risk. This risk–risk balancing can occur, for example, where chemicals added to the water to kill pathogenic organisms create disinfection by–products which themselves are dangerous. The inclusion of a cost–benefit analysis is a tangible outcome of the Republican control of Congress discussed at the end of Chapter 2.

The last theme deals with money. The SDWAA establish a Drinking Water State Revolving Fund. Congress authorized $9.6 billion for the period 1995–2003. The funds may be used to "facilitate compliance with national primary drinking water regulations" and "significantly further the health protection objectives" of the act.

Other aspects of the SDWAA include specific requirements regarding the potential regulation of arsenic, sulfate, radon, and disinfection by–products; new (reduced) monitoring requirements for regulated water systems; new requirements for water system operator certification; and variances for small water systems.

Is Your Water Safe?

Is your water safe? This simple question is infused with deeper considerations. What exactly is meant by "safe"? Safe implies that there is some *risk.* Risk is the probability of suffering an undesirable consequence. Risk is part and parcel of life, and nothing we do is completely risk free. We make decisions every day about risk, and we consciously or unconsciously make trade–offs between different risks. Driving a car is risky. Riding a motorcycle is riskier than driving a car. Is drinking tap water riskier than riding a motorcycle? People perceive, evaluate, and assess risks differently, something we touched upon at the end of Chapter 6. What is "safe," therefore, depends on the individual—it is a personal value judgement (Benarde, 1989).

Pursuant to the SDWA all public water supplies are to be tested regularly. If the water exceeds an MCL standard, or tests are not carried out, a Notice of Violation is issued. There were roughly 100,000 violations up through 1993, and a study by the National Wildlife Federation found that only 2 percent of violations were subjected to enforcement by the EPA (Miller, 1993). Which is more alarming, that there have been so many violations, or that so few have been acted upon? You should realize that a Notice of Violation does not necessarily mean the water is unsafe.

You might wonder whether drinking bottled water is safer than tap water. Bottled water is regulated by the Food and Drug Administration (FDA), not the EPA. The SDWAA requires the FDA to regulate the same contaminants in bottled water that the EPA regulates in public supplies, unless the FDA finds it is not necessary to protect public health. Often the source of bottled water is the same source as public supplies. Bottled water may be cleaner than tap water, but it should not be any dirtier. One thing is certain, bottled water is more expensive—about 700 times more expensive—than tap water.

The quality of water from private wells is rarely known. It can cost hundreds of dollars for a comprehensive chemical analysis and few home owners are willing to pay that much. The most common test performed on private water supplies is for fecal coliform bacteria. This test is usually voluntary and occurs most often during a real estate transaction. Coliform bacteria indicate water contamination by human or animal waste. This is a concern in rural areas with livestock, or with high–density home development using wells and septic systems.

WELLHEAD PROTECTION AND MANAGEMENT PROGRAMS

The amendments to the SDWA in 1986 included the first federal legislation aimed at groundwater management for public water supply. Groundwater is used in part or as the sole source of supply for 51 percent of the population. Additionally, 81 percent of community water systems rely on groundwater to some extent, and 74 percent of small community water systems (serving fewer than 3300 persons) depend entirely on groundwater (USEPA, 1996b). The 1986 amendments included a new section on state wellhead protection programs (WHPPs) for public water supply wells. The programs were voluntary though states were eligible for federal grants–in–aid to help get them established. The EPA has developed guidelines for WHPPs and the identification and delineation of wellhead protection areas (WHPA) (USEPA, 1987, 1993). Wellhead protection essentially involves land–use regulation around the well to prevent or limit contamination.

Job (1996) estimated benefit–cost ratios for wellhead protection programs in six regions around the country. The ratio of avoided contamination costs (benefits) to basic WHPP costs ranged from 5:1 to as high as 200:1, on a per well basis. The average for the six areas was 27:1 on a per well basis. Clearly the benefits of protecting groundwater resources exceed the costs of implementing the WHPPs.

Wellhead protection programs involve both administrative and technical issues. Administrative issues include budgeting for the program, developing guidelines and regulations that are understandable by the public and defensible in court, and intergovernmental negotiation where aquifers cross jurisdictional boundaries. Land–use regulation is an extremely sensitive issue. A successful WHPP must be supported by the public. It is imperative that the responsible agency cultivate public support and bring the public into the decision–making process early and in a meaningful way. If the public is convinced that wellhead protection is necessary to safeguard its water supply, new land–use regulations stand a better chance of passage.

Technical issues include identifying all potential sources of contamination and choosing and implementing a method for delineating the wellhead protection area. Identifying contaminant sources can be a major undertaking simply because of the number of potential sources, and in most communities these sources have never been catalogued and georeferenced. A geographic information system (GIS) is an excellent tool for storing, retrieving, and mapping contaminant–source information. Development of a WHPA essentially involves three steps: determining the objectives of the WHPA, choosing a delineation criterion or criteria, and finally choosing a method for mapping the WHPA based on the criterion/criteria.

WHPA Objectives

The EPA identified three general objectives for a WHPA: to provide a remedial action zone; to provide a zone for contaminant attenuation; or to provide a well–field management zone. A remedial action zone allows pollution cleanup before a contaminant reaches the well. An attenuation zone provides a buffer large enough so that contaminants are sufficiently degraded through chemical and biophysical processes before reaching the well. A well–field management zone is the most general of the objectives and aims to restrict the types of land uses and activities that occur within the zone.

WHPA Delineation Criteria

The term *criteria* as used by the EPA refers to the " . . . conceptual standards that form the technical basis for WHPA delineation." (USEPA, 1987, p. 3-1). Five criteria which can be used individually or in combination are (1) distance from the well, (2) the region of drawdown around the well, (3) the travel time of groundwater to the well, (4) groundwater flow boundaries, and (5) the assimilative capacity of the aquifer. The choice of criteria is not solely a technical issue, however, since the best criteria from a technical point of view may not be the best from an administrative viewpoint.

WHPA Delineation Methods

Wellhead delineation methods are the procedures for translating the WHPA criteria into boundaries on a map. Delineation methods range from simple and cheap to complex and expensive. The simplest method is the arbitrary fixed–radius method. The fixed–radius method draws a circle of specified radius around the well. The method uses distance as the sole criterion and does not consider physical groundwater processes, e.g., direction and velocity of flow, nor does it consider aquifer characteristics such as assimilative capacity. The method is inexpensive and quick, and is certainly appropriate as a "first step" toward a more comprehensive strategy. The other extreme is to collect geophysical data to characterize aquifer and groundwater flow properties, and to use sophisticated mathematical models to delineate the WHPA. This is certainly a much more expensive approach.

WELLHEAD DELINEATION IN PENNSYLVANIA

The Pennsylvania Department of Environmental Protection (DEP) has established a three-zone approach that local governments and water suppliers should use in delineating wellhead protection areas.

Zone 1 is an *arbitrary fixed-radius circle* around the well. The radius is 100 feet except in fractured carbonate rocks where the radius is 400 feet.

Zone 2 is the *zone of influence* of the well. This corresponds to the cone of depression (see Figure 1.14).

Zone 3 is the *area of contribution* to the well. This is the land area from which the well ultimately receives its water.

IRRIGATION

HISTORICAL DEVELOPMENT

With the exception of domestic use, irrigation is probably the oldest offstream water use of any consequence. Irrigation may have been practiced as far back as 9000 B.P. at Jerico in the Middle East. In ancient Mesopotamia archeological evidence indicates irrigation 6000 years ago. In Egypt and China irrigation was practiced 5400 and 3000 years ago, respectively. In the western hemisphere the oldest sites are in Peru and Mexico. In the American Southwest the Hohokam people were irrigating between 2200 B.P. and 700 B.P. in what is now Arizona. The Rio Grande Valley is the longest continuously settled region in the United States and it was founded on and maintained by irrigation agriculture.

Early Hydraulic Civilizations

The Egyptians likely had it the easiest in developing an irrigation system because the annual Nile flood and the growing season came in an orderly sequence. The Nile flood came in the fall; winter and spring were the growing seasons; and the fields were left fallow in the summer. With each flood sediment and nutrients were deposited on the fields in the floodplain. The left bank of the Nile was developed first. The right bank was left as an overflow safety valve for floodwater. Pharaohs of the 12th dynasty reclaimed the right bank and cut a flood diversion channel across the desert to the Faiyum Depression around 4000 B.P. (Teclaff and Teclaff, 1973). The irrigation system worked well, and was a good example of sustainable resource use. In the 19th century perennial irrigation commenced, and with it came problems of waterlogging and salinization.

In Mesopotamia the problems were more complex and the challenges more difficult than in Egypt. Sumerians had to deal with two rivers—the Tigris and the Euphrates—instead of one. Floods from melting snow in the spring came with crops already in the ground. The rivers meandered across much wider floodplains, so the location and timing of floods were more erratic. The two rivers were managed differently. The Euphrates was controlled by using canals to divert floodwaters out into the desert. The Pallacopas canal carried water 370 miles to the Chaldean Marshes. For the Tigris the flood diversion canals paralleled the river. The Nahrwan canal paralleled the Tigris for 250 miles (Teclaff and Teclaff, 1973). The Sumerians practiced perennial irrigation, which eventually created drainage problems leading to waterlogging and salinization of the soil. Deforestation in the mountains resulted in erosion and sedimentation of the canals which exacerbated the flood problem. Breakdown of the irrigation system, as well as of the centralized administrative system for water management, contributed to the collapse of Mesopotamian civilization.

Another ancient water supply system that developed in the Middle East was the karez system. The karez system is based on groundwater and uses a mother well, and a series of smaller access wells spaced 15 to 30 meters apart, to dig an underground tunnel (quanat). The quanat intersects the water table and conveys the intercepted groundwater to the surface downslope (Figure 7.6). Ancient karez systems are still used today for irrigation and urban water supply.

Irrigation in China developed along the Huang and Chang Rivers. Periods of political instability were interspersed with periods of efficient centralized water management. During China's feudal times water was used as a weapon between feuding states. States would build dikes along the river to force floodwaters onto lands of neighboring states (Teclaff and Teclaff, 1973). China's history is replete with stories of famine and flood, but the country has survived. Another ancient Chinese irrigation system based on groundwater was called the "Well Land" system. In this system the land was subdivided into eight privately owned parcels that surrounded a ninth public parcel. The well was placed on the public parcel. This ensured each parcel had access to water and prevented wells from being too closely spaced.

With the exception of the Rio Grande in central New Mexico, the earliest irrigation in the United States dates from the late 1840s and the Mormon settlers in Utah. Irrigation developed rapidly thereafter, commencing in central and southern California in the 1850s, the San Luis Valley of Colorado in the 1860s, and most other western states by the 1870s and 1880s. The federal government's involvement in irrigation followed the passage of the Reclamation Act in 1902.

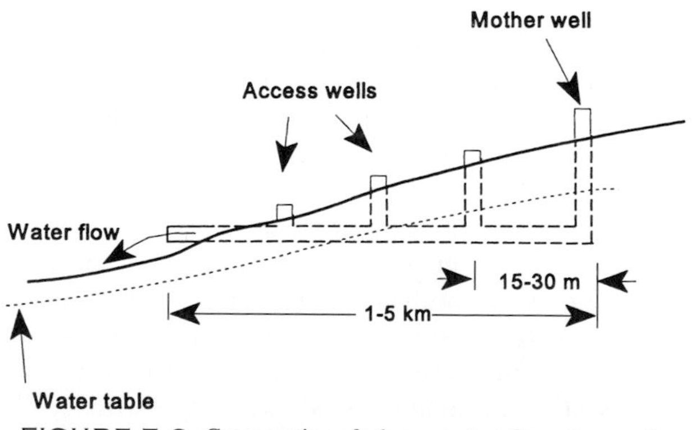

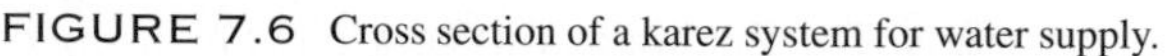
FIGURE 7.6 Cross section of a karez system for water supply.

TABLE 7.6 Irrigated Area for Five Countries and the World[a]

Country	Irrigated area (10⁶ hectares)	Irrigated area (10⁶ acres)	Percent of the region's total cropland
China	45,349	112,056	43
India	45,039	111,290	25
United States	23,200	57,326	11
Pakistan	16,220	40,079	78
Egypt	2,585	6,387	100
World	235,299	581,416	16

[a] Source: Food and Agricultural Organization (FAO), 1990.

IRRIGATION IN THE WORLD TODAY

China leads the world in irrigated land area with over 45 million hectares (Table 7.6) Forty–three percent of China's cropland is irrigated. India is a close second in total irrigated area, but irrigated cropland composes only 25 percent of India's total cropland. The United States has less than half the irrigated area of China and India, and it composes only 11 percent of the total cropland. Egypt is totally dependent on irrigation agriculture.

Between 1950 and 1990 irrigated cropland worldwide increased two–and–one–half times, going from 94 million irrigated hectares in 1950 to over 235 million in the 1990s. The development of irrigation agriculture has been a significant reason for the increase in world food production. Today one–third of all food and fiber comes from the world's irrigated cropland. The expansion of irrigation has slowed in the last two decades. From 1950 to 1979 worldwide irrigated area increased between 2 and 4 percent per year. The annual rate of increase since 1980 has averaged just over 1 percent. As a result, irrigated area per capita worldwide is now decreasing. The expansion has slowed because of low commodity prices, higher energy costs, and the costs of building irrigation facilities.

IRRIGATION IN THE UNITED STATES

Ninety–five percent of all the irrigated land in the United States is located in 16 western states and the 6 eastern states of Arkansas, Florida, Georgia, Louisiana, Mississippi, and Missouri (Table 7.7). The 16 western states alone contain 80 percent of the Nation's irrigated cropland. California leads all states with 9.48 million acres. Three–quarters of California's irrigated cropland uses some method of surface flooding, while one–quarter uses spray methods (sprinklers or drip systems). Sixty–two percent of the freshwater used for irrigation in California was withdrawn from surface sources, while 38 percent was groundwater. In addition to freshwater, about 148,000 acre–feet of reclaimed wastewater was used for irrigation in California (not shown in Table 7.7). Also not shown in the table was that 21.8 million acre–feet of the 31.3 million acre–feet of freshwater withdrawn in

TABLE 7.7 Irrigated Area and Water Withdrawals for the Top 22 States, 1990[a]

State	Irrigated area (10^6 acres) by water application method		Total area		Freshwater withdrawal in thousands of acre–feet per year		
	Spray	Flood	10^6 acres	10^6 hectares	Ground	Surface	Total
California	2230	7250	9480	3837	12000	19300	31300
Nebraska	3000	3860	6860	2776	4880	1950	6830
Texas	2160	4060	6220	2517	6270	3250	9520
Colorado	897	2660	3557	1439	2870	10100	12970
Idaho	1540	1870	3410	1380	7420	13500	20920
Kansas	1450	1660	3110	1259	4470	224	4694
Arkansas	297	2680	2977	1205	4820	1060	5880
Florida	1020	1130	2150	870	2180	2000	4180
Oregon	1070	965	2035	824	631	7060	7691
Washington	1510	472	1982	802	845	5920	6765
Wyoming	211	1730	1941	786	266	7760	8026
Montana	597	1340	1937	784	101	9990	10091
Arizona	409	940	1349	546	2300	3640	5940
Utah	457	837	1294	524	569	3460	4029
Georgia	1180	0	1180	478	294	199	493
Mississippi	449	728	1177	476	1970	146	2116
New Mexico	421	564	985	399	1540	1840	3380
Nevada	155	574	729	295	976	2190	3166
Louisiana	193	496	689	279	506	287	793
Missouri	228	323	551	223	376	40	416
Oklahoma	317	186	503	204	553	121	674
South Dakota	287	109	396	160	158	281	439

[a] Source: Solley *et al.*, 1993.

California in 1990 was consumed. Nebraska and Texas are second and third in irrigated cropland area, but these two states combined use less water than California.

A trend over the decade has been for irrigated area to decrease in some western states and increase in many eastern states. With the exception of the South, the amount of irrigated land in the eastern states is small by comparison to the West, and the water is mainly supplemental—a form of drought insurance. Table 7.8 shows the change in irrigated area for the top 10 states from 1975 to 1990. Five of the six western states lost irrigated acreage between 1985 and 1990. Irrigated area increased dramatically in Kansas and Nebraska from 1975 to 1985. In part this was a response to the loss of acreage in Texas. All three states overlie the Ogallala aquifer, and as wells dried up in Texas some farming operations moved north to

TABLE 7.8 Change in Irrigated Area for Select States, 1975–1990[a]

	Irrigated area (10^6 acres)			Percent change in acreage	
State	1975	1985	1990	1975–1985	1985–1990
California	8731	9580	9480	9.7	−1.0
Nebraska	3331	7480	6860	124.5	−8.3
Texas	6949	6750	6220	−2.9	−7.9
Colorado	2908	3355	3557	15.4	6.0
Idaho	2896	4100	3410	41.6	−16.8
Montana	1896	2302	1937	21.4	−15.9
Florida	1600	1914	2150	19.6	12.3
Oregon	1595	2041	2035	27.9	−0.3
Kansas	1589	2950	3110	85.7	5.4
Wyoming	1553	1811	1941	16.6	7.2
Arizona	1250	1317	1349	5.4	2.4
Washington	1219	1620	1941	32.9	19.8
Arkansas	1098	2022	2977	84.2	47.2

[a] Sources: Mather, 1984; Solley, *et al.*, 1993.

Nebraska and Kansas where water was still plentiful. In 1990 Nebraska and Texas ranked second and third in irrigated area; however, in Nebraska irrigated area is increasing while in Texas irrigated area is decreasing. Idaho is fifth on the list in area, yet it is second in total water use. The implication is that irrigation in Idaho is not very efficient. Figure 7.7 shows freshwater withdrawals for irrigation by state.

In 1994 corn occupied the largest amount of irrigated acreage (Table 7.9). Alfalfa hay accounted for 11.4 percent of all irrigated cropland. In addition to alfalfa, water is used to irrigate grass hay and pasture. Hay crops and pasture are relatively low–value crops. A large amount of water in the West is used on relatively low–value crops. The difference in value between using water to grow low–value crops and its value in urban uses is the driving economic force behind water markets.

WATER APPLICATION METHODS

Flood irrigation methods include wild flooding, furrow, and border strip irrigation (Figure 7.8). Spray methods include all sprinkler systems and microirrigation (drip and trickle) systems. With wild flooding a canvas dam is laid across the irrigation ditch, forcing the water to spill out over the field. The dam is then moved down the ditch and the process repeated. Wild flooding is common on field crops like hay and pasture. Furrow irrigation is used with row crops such as corn and vegetables. The crops are planted on the rows and the water flows between in the furrows (Figure 7.9). The water is released into the top of the furrow using siphon

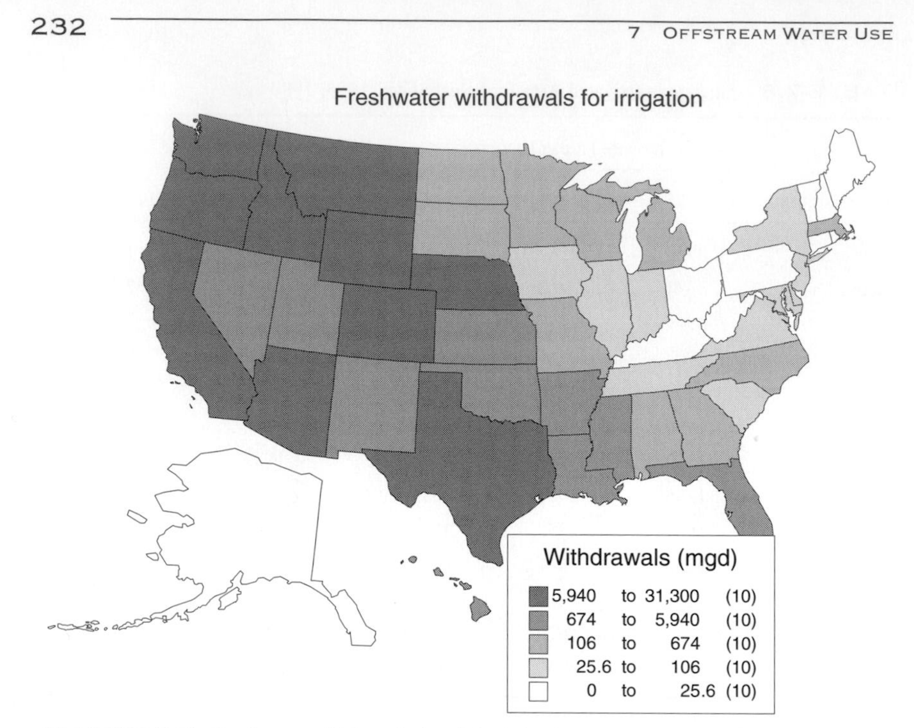

FIGURE 7.7 Freshwater withdrawals (mgd) for irrigation by state, 1990. (Solley *et al.*, 1993)

tubes or gated pipe. With border strip irrigation the field is contoured into strips, where each strip is separated from an adjacent strip by a small ridge (border). The water flows down the strip as in wild flooding, but is confined and controlled by the elevated borders on either side. Irrigation methods can use gravity, pumping, or a combination of gravity and pumping to apply the water. For example, groundwater can be pumped into gated pipe and then flows down the furrows by gravity. Table 7.10 lists the use of different irrigation application methods in the United

TABLE 7.9 Estimated Percent of all Irrigated Land in Different Crops, 1994[a]

Crop	Percent of all irrigated cropland	Cumulative percent
Corn	20.2	20.2
Alfalfa hay	11.4	31.4
Cotton	9.5	40.9
Orchards	7.7	48.6

[a] Source: U.S. Bureau of the Census, 1994.

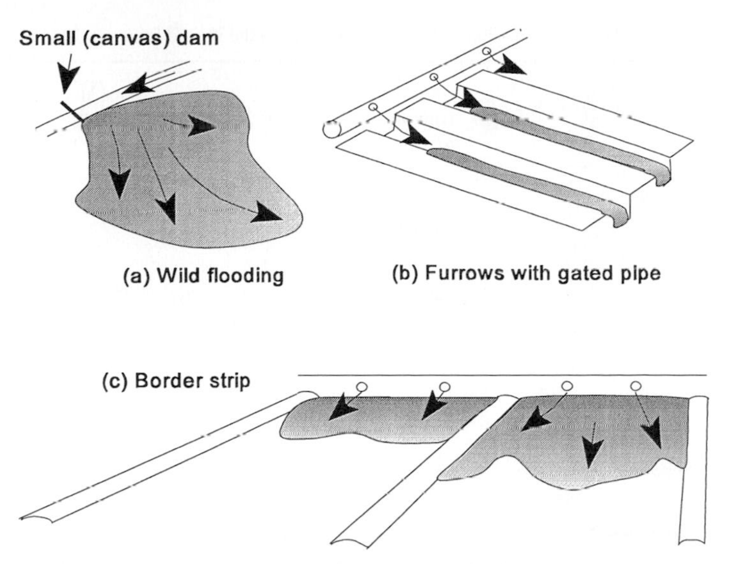

FIGURE 7.8 Some examples of irrigation application methods.

States in 1984. Flooding methods and spray methods were each responsible for irrigating half the farms. Among flooding methods the use of ditch with siphon tubes was the most common (see Figure 7.9). Hand–set sprinklers were the most common method for sprinkler systems. Drip and trickle were used on 4 percent of farms, and only on very high–value crops. A recent survey shows a continued

FIGURE 7.9 Corn being irrigated in northeast Colorado using siphon tubes drawing water from a ditch at the head of the field (photo by the author).

TABLE 7.10 Data on Irrigation Application Methods in the United States in 1984[a]

Method	Number of farms	Percent of farms	Area		
			10^6 acres	10^6 hectares	Percent
Flooding Methods					
Wild flooding	45,045	15.9	9,046	3,661	17.1
Furrow					
Gated pipe	42,826	15.1	8,367	3,386	15.8
Ditch with siphon tube	59,255	20.9	10,055	4,069	19.0
Spray Methods					
Sprinklers					
Center pivot sprinkler	32,442	11.4	16,884	6,833	32.0
Mechanical movement sprinklers	25,471	9.0	3,357	1,358	6.4
Hand moved sprinklers	46,885	16.6	2,918	1,181	5.5
Solid set/permanent sprinklers	19,694	6.9	1,233	499	2.3
Drip and trickle	11,651	4.1	986	339	1.9
Total	283,269	100.0	52,846	21,386	100.0

[a] Source: Bajwa *et al.*, 1987.

increase in the proportion of farms using sprinkler application methods (U.S. Bureau of the Census, 1994).

ESTIMATING THE EVAPOTRANSPIRATION REQUIREMENT (E_{tp})

One of the easiest formulas for calculating crop evapotranspiration is the Blaney–Criddle method. The method is based on correlations between crop consumptive–use data, monthly mean temperatures, and the percentage of daytime hours (Blaney, 1955). The original version of the Blaney–Criddle method has undergone numerous revisions, with the SCS (1970) version one of the more popular. A more recent revision of the Blaney–Criddle method was produced by the Food and Agricultural Organization (FAO) (Doorenbos and Pruitt, 1977). The original Blaney–Criddle method, and the SCS modification, should only be used for the western United States. The FAO version is intended to be used internationally. The original version of the Blaney–Criddle equation is

$$E_{tp} = k\frac{T_a p}{100}, \tag{7.1}$$

where E_{tp} is crop evapotranspiration (in/mo), T_a is mean monthly air temperature (°F), k is the crop consumptive use coefficient for the month, and p is monthly daytime hours given as a percent of the year.

The original method uses temperature in degrees Fahrenheit and calculates E_{tp} in inches per month. Values of p by month and latitude are given in Table 7.11.

TABLE 7.11 Percentage of Daytime Hours p for Each Month of the Year for Use with the Blaney–Criddle Method

Lat.	J	F	M	A	M	J	J	A	S	O	N	D
60 N	4.67	5.65	8.08	9.65	10.74	12.39	12.31	10.70	8.57	6.98	5.04	4.22
50	5.98	6.30	8.24	9.24	10.68	10.91	10.99	10.00	8.46	7.45	6.10	5.65
40	6.76	6.72	8.33	8.95	10.02	10.08	10.22	9.54	8.39	7.75	6.72	6.52
30	7.30	7.03	8.38	8.72	9.53	9.49	9.67	9.22	8.33	7.99	7.19	7.15
20	7.74	7.25	8.41	8.52	9.15	9.00	9.25	8.96	8.30	8.18	7.58	7.66
10	8.13	7.47	8.45	8.37	8.81	8.60	8.86	8.71	8.25	8.34	7.91	8.10
0	8.50	7.66	8.49	8.21	8.50	8.22	8.50	8.49	8.21	8.50	8.22	8.50
10	8.86	7.87	8.53	8.09	8.18	7.86	8.14	8.27	8.17	8.62	8.53	8.88
20	9.24	8.09	8.57	7.94	7.85	7.43	7.76	8.03	8.13	8.76	8.87	9.33
30	9.70	8.33	8.62	7.73	7.45	6.96	7.31	7.76	8.07	8.97	9.24	9.85
40 S	10.27	8.63	8.67	7.49	6.97	6.37	6.76	7.41	8.02	9.21	9.71	10.49

TABLE 7.12 Monthly Consumptive–Use Coefficients k for Use with the Original Blaney–Criddle Method[a]

Crop	Location	M	A	M	J	J	A	S	O	N
Alfalfa	CA, coastal	0.60	0.65	0.70	0.80	0.85	0.85	0.80	0.70	0.60
	CA, interior	0.65	0.70	0.80	0.90	1.10	1.00	0.85	0.80	0.70
	ND		0.84	0.89	1.00	0.86	0.78	0.72		
	UT, St. George		0.88	1.15	1.24	0.97	0.87	0.81		
	NM	0.70	0.75	0.80	0.90	1.00	1.00	0.80	0.70	0.65
Beans	NM			0.50	0.60	0.75	0.70			
Corn	ND			0.47	0.63	0.78	0.79	0.70		
	NM			0.50	0.70	0.80	0.80	0.70		
Cotton	AZ		0.27	0.30	0.49	0.86	1.04	1.03	0.81	
	NM		0.35	0.40	0.60	0.90	1.00	0.95	0.75	
Orchard	CA, coastal		0.40	0.42	0.52	0.55	0.55	0.55	0.50	0.45
Pasture	CA			0.84	0.84	0.77	0.82	1.09	0.07	
Potatoes	ND			0.45	0.74	0.87	0.75	0.54		
	SD			0.69	0.60	0.80	0.89	0.39		
Small grains	ND		0.19	0.55	1.13	0.77	0.30			
	NM		0.40	0.50	0.90	0.80				
Truck crops	CA, interior	0.19	0.26	0.38	0.55	0.71	0.82	0.69	0.37	0.35

[a] Blaney, 1959; Blaney *et al.*, 1960.

TABLE 7.13 Crop Evapotranspiration Requirements for Selected Crops in the Western States[a]

Crop	Southwest		Missouri and Arkansas basin	
	(in)	(mm)	(in)	(mm)
Alfalfa	42–61	1060–1550	23–32	591–799
Corn	17–24	439–607	14–22	375–558
Cotton	28–42	716–1070		
Wheat	17–27	445–683	16–22	415–549

[a] Source: Gleick, 1990.

Table 7.12 gives monthly crop coefficients (k) for different states. Table 7.13 gives some representative ranges for crop evapotranspiration (see Example 7.1).

EXAMPLE 7.1

Calculate the evapotranspiration requirement for alfalfa near Salt Lake City, Utah, using the Blaney–Criddle method. Alfalfa is a perennial crop and begins growing when the average air temperature is approximately 50°F (10°C), and stops with the first frost. The growing season for alfalfa is assumed to be April through September since coefficients have been determined for these months. The coefficients for St. George, Utah, are used here (Table 7.14), though they may not be the same as those for Salt Lake City.

TABLE 7.14 Calculation of E_{tp} by the Blaney–Criddle Method

	T_a			E_{tp}	
Month	(°F)	p	k	(in)	(cm)
April	49.2	8.95	0.88	3.9	9.8
May	58.8	10.02	1.15	6.8	17.2
June	68.3	10.08	1.24	8.5	21.7
July	77.5	10.22	0.97	7.7	19.5
Aug.	74.8	9.54	0.87	6.2	15.7
Sept.	65.0	8.39	0.81	4.4	11.2
			Total	37.5	95.1

The seasonal consumptive use for alfalfa in this area is estimated as 37.5 inches or about 3.1 acre–feet of water per acre. As we shall see later in the chapter, irrigation water application is less than 100 percent efficient, and a farmer would need more water than this to raise a crop.

ESTIMATING THE IRRIGATION WATER REQUIREMENT (*IRR*)

The crop evapotranspiration requirement (E_{tp}) is the largest component of the *irrigation water requirement* (*IRR*) of a crop. But there are two other components to consider, the *leaching requirement* (L_r) and *effective precipitation* (R_e). The leaching requirement is extra water needed to flush salts from the root zone. The leaching requirement may range from 0 to 15 percent of the evapotranspiration requirement, depending upon the salinity of the soil and the irrigation water. The effective precipitation is that portion of rainfall that is stored in the soil and helps satisfy the evapotranspiration requirement, thereby decreasing the amount of irrigation water needed. Effective precipitation is the total precipitation minus surface runoff and deep percolation. The *IRR* is calculated as

$$IRR = (E_{tp} + L_r) - R_e. \tag{7.2}$$

IRRIGATION EFFICIENCY

Irrigation using a surface water source usually involves two steps. First, the water is withdrawn from the source and conveyed to the farm in a canal, ditch, or pipe. The second step is applying the water to the field. Some sprinkler systems using groundwater pumped directly from beneath the field, e.g., center pivot systems, essentially skip the conveyance step. Each step has it own water–use efficiency. Open earthen ditches lose water by seepage and evaporation. An enclosed pipe, on the other, hand loses very little water. *Conveyance efficiency* E_c is defined as the volume of water output (V_o) from a conveyance facility divided by the volume of water put in to the structure at the upper end (V_i):

$$E_c = \frac{water\ volume\ output\ (V_o)}{water\ volume\ input\ (V_i)}. \tag{7.3}$$

In a similar fashion *field efficiency* E_f is defined as the volume of water beneficially used by the crop divided by the volume of water applied to the field. Water not beneficially consumed either runs off the end of the field as *tailwater* or percolates beneath the root zone, perhaps recharging groundwater. (Percolating water in excess of the leaching requirement is not beneficially used.) The overall *farm efficiency* E_s is the efficiency of the entire irrigation system and is the product of the conveyance and field efficiencies:

$$E_s = E_c \times E_f. \tag{7.4}$$

Efficiency can be defined in a similar manner for a reservoir, an irrigation district, or even an entire drainage basin. At the larger spatial scales conveyance losses, tailwater return flows, and deep percolation are not lost, but can be diverted and reused by farmers downstream. Multiple reuse thus increases the overall irrigation efficiency within a district or drainage basin (Jensen, 1984).

Irrigation application methods differ in their field efficiency. Field efficiency for wild flooding might be as low as 20 percent and as high as 50 percent. Sprinkler systems typically range between 70 to 80 percent. Trickle and drip systems can reach efficiencies in excess of 90 percent. Since irrigation uses such large volumes of water, small improvements in efficiency translate into large volumes of conserved water. The total amount of water (Q) a farmer would need to satisfy the irrigation water requirement of the crop and account for system efficiency losses is

$$Q = \frac{(E_{tp} + L_r)}{E_s} - R_e. \tag{7.5}$$

Example 7.2 shows the effect of improving irrigation efficiency on water supply.

EXAMPLE 7.2

This hypothetical example shows how improving irrigation efficiency can release potentially large volumes of water for reallocation to other (urban) uses. Assume a farmer irrigates 1000 acres of corn. The component values for the total seasonal water supply (Q), on a per acre basis, are listed below.

Crop evapotranspiration requirement is E_{tp} = 28 in = 2.33 ft.
Leaching requirement is taken to be 10 percent of E_{tp}, or L_r = (0.10 × 2.33 ft) = 0.23 ft.
Effective precipitation R_e = 10 in = 0.83 ft.
System efficiency E_s = 0.50 (50 percent).

From Equation (7.5) the water supply needed per acre is

$$Q = [(2.33 \text{ ft} + 0.23 \text{ ft})/0.5] - 0.83 \text{ ft} = 4.29 \text{ ft}.$$

The total volume of water needed to grow the corn crop is 1000 acres ×4.29 ft = 4290 acre–feet of water. If system efficiency were increased from 50 percent to 60 percent, Q would decrease to

$$Q = [(2.33 \text{ ft} + 0.23 \text{ ft})/0.6] - 0.83 \text{ ft} = 3.44 \text{ ft}.$$

The difference is (4.29 ft − 3.44 ft) = 0.85 ft of water. This represents a water volume of (1000 acres x 0.85 ft) = 850 acre–feet, or 277 million gallons. At a domestic use rate of 105 gpcd, this would provide enough water to supply 7226 people for an entire year.

Relatively modest improvements in irrigation efficiency can yield potentially large volumes of water. This is why agriculture is viewed an a potential source of new water supply for urban areas in the West.

ECONOMICALLY EFFICIENT WATER APPLICATION

The estimate of water supply from Equation 7.5 is physically based without any consideration of the cost of irrigation water. The general relationship between

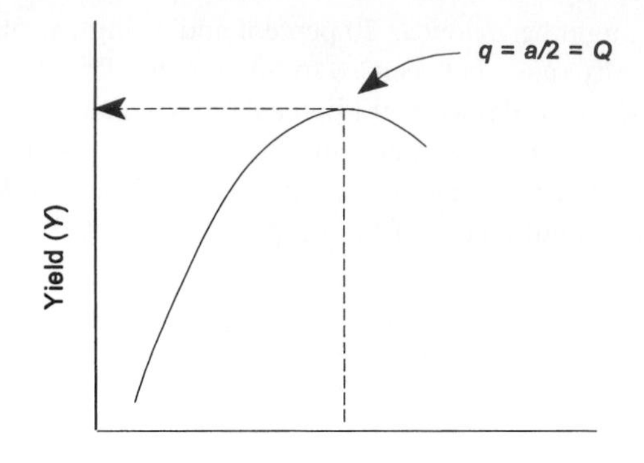

FIGURE 7.10 Generalized curve of crop yield as a function of applied irrigation water. Maximum yield occurs when the amount of applied water equals Q.

crop yield (Y) and the amount of applied irrigation water (q) is shown in Figure 7.10. The curve in Figure 7.10 is of the form

$$Y = aq - q^2. \tag{7.6}$$

Crop yield increases with applied water up to a maximum, but decreases thereafter. The maximum yield is found by differentiating Equation (7.6) and setting the result equal to zero. (Differentiation is a basic calculus procedure.) The peak of the crop yield–applied water function is $q = a/2$. This value is also the value of Q given by Equation 7.5, so $q = a/2 = Q$. A farmer would only apply Q if the water were free. When water must be purchased, the farmer should apply less than Q to maximize his income.

The net income (profit) from irrigation agriculture equals the gross income minus the cost of production (James and Lee, 1971). Gross income equals the crop yield Y times the price received for the crop P_c per unit of yield. The total costs of production include fixed costs F_c and variable costs V_c. Fixed costs include equipment, the land mortgage, and taxes. These are fixed costs because the farmer pays them whether or not he grows a crop. The variable costs depend on the level of production. Variable costs include the costs of fertilizer, seed, pesticides, fuel, and irrigation water. The cost for water equals the price per unit of water P_w times the amount of water used q. Putting these together into a single equation gives the net income I as

$$I = P_c Y - P_w q - V_c Y - F_c. \tag{7.7}$$

Since crop yield is a function of the applied water, the right side of Equation 7.6 can be substituted for Y in Equation (7.7) and rewritten as

$$I = P_c(aq - q^2) - P_w q - V_c(aq - q^2) - F_c. \tag{7.8}$$

Equation (7.8) gives net income I as a function of the applied irrigation water q. The objective is to find the amount of irrigation water that maximizes net income. Again, this is found by differentiating Equation (7.8), setting the result equal to zero, and solving for q. The result is

$$q = 0.5 \left[a - \frac{P_w}{(P_c - V_c)} \right]. \tag{7.9}$$

Equation 7.9 shows that fixed costs F_c play no role in the determination of the optimum amount of water to use. Fixed costs do, however, help determine whether crop production is profitable or not (Example 7.3).

EXAMPLE 7.3

Find the optimal amount of irrigation water to apply using the following data. The yield–water function is $Y = 9q - q^2$, where q is in acre–feet/acre.

The costs are

variable costs, V_c = \$10.50/ton
price of water, P_w = \$20.00/acre–foot
price received for the crop, P_c = \$25.00/ton
fixed costs, F_c = \$80.00/acre.

By Equation (7.9), the economically optimal amount of irrigation water to apply is

$$q = 0.5 \left[9 - \frac{20}{(25 - 10.5)} \right] = 3.81 \textit{ acre-feet/acre.}$$

By contrast, Q calculated as the peak of the function is $Q = a/2 = 9/2 = 4.5$ acre-feet/acre.

The net income per acre for this example is given by Equation (7.7) as

$$I = 25(19.78) - 20(3.81) - 10.5(19.78) - 80 = \$130.61 \text{ per acre.}$$

WATER CONSERVATION IN AGRICULTURE

Water conservation usually means improving water–use efficiency (Equation (7.4)). Conveyance efficiency can be increased by lining ditches and canals to reduce seepage. Field efficiency can be improved by switching to more efficient water–application systems. Sprinkler systems are more efficient than flooding, and drip irrigation is more efficient than sprinklers. Of course improving efficiency must make sense economically. Changing from flooding to sprinkler application saves water but requires a capital outlay for the equipment and has higher energy costs. And certain irrigation application methods are unsuitable for certain types of crops. As evidence of the increase in water conservation in agriculture,

GROWING RICE IN CALIFORNIA

The fourth largest water-using crop in California is rice (after alfalfa, cotton, and pasture, in that order). Rice is indigenous to tropical humid climates, not the dry mediterranean climate of the Sacramento Valley where it is grown. Rice can be grown in California with irrigation, but should rice be grown there? California produces 18 percent of the Nation's rice, making it the second largest producer. Rice production is worth $500 million dollars per year and uses enough water each year to supply one-fourth of California's urban residents (Botkin and Keller, 1995). Rice production has also caused water pollution from pesticides and fertilizers. In recent years rice growers have responded to public pressure by reducing fertilizer and pesticide use. They have also modified their practice of burning the rice stubble remaining in the fields after the harvest. The stubble is now left to rot and the fields are sometimes flooded to provide habitat for wildlife, especially migrating waterfowl.

sprinkler systems were estimated to irrigate 46 percent of the total land irrigated in 1994, compared to 40 percent in 1988, and approximately 38 percent in 1984 (U.S. Bureau of the Census, 1994). Tailwater recovery systems can improve efficiency by pumping tailwater back to the top of the field for reuse. Modifications to the operation of existing application methods have also yielded higher efficiencies. For example, *pulse* irrigation is a variation on traditional furrow and border strip application. Instead of running a continuous flow of water down the furrow or strip the water is released in pulses. With each successive pulse the water moves rapidly over the previously wetted area, reducing the amount of deep percolation at the upper end of the field. Laser leveling of fields improves the uniformity of water application. An important nonstructural approach to improving efficiency is irrigation scheduling to provide the correct amount of water when it is actually needed by the crop. Genetic research to develop crops that use water and fertilizer more efficiently could also reduce irrigation water demand. Laser leveling of fields and irrigation scheduling are some of the earliest attempts at what is now being called "precision agriculture." A final way to conserve water is not through improving efficiency, but by using excess water to recharge groundwater supplies. As water becomes more valuable new technologies will continued to be developed offering farmers more efficient methods of using water for irrigation.

ENVIRONMENTAL IMPACTS OF IRRIGATION

Irrigation has many impacts on the environment. Building the water supply infrastructure (dams, interbasin diversions, pumps, etc.) disrupts the natural flow regime of streams. As with all industrial agriculture, farmers use chemical pesticides and fertilizers. These chemicals are carried from the fields into water bodies

SAVING WATER AND REDUCING SALT LOADS

Motivated by the need to reduce salt levels in the Colorado River the U.S. Department of Agriculture began a voluntary, on-farm salinity control program. Farmers in the Big Sandy Watershed in Wyoming have been converting from border strip flood irrigation to low-pressure overhead sprinkler irrigation. The conversion began in 1988 and by 1994 about 42 percent of the farm land in the watershed was participating.

For the farmers involved crop yields have increased 100 percent, and water use has been reduced 50 percent. Total salt savings are estimated at 22,000 tons per year.

Farmers sign a 25-year operation and maintenance contract in exchange for 70 percent federal cost sharing. The program also authorizes cost sharing with farmers who agree to replace wildlife habitat which may disappear when the return flow from flood irrigation no longer forms seasonal saline ponds.

(American Water Resources Association (AWRA), 1994)

by surface and subsurface return flows. The large consumptive use by irrigation means there is less water available downstream, causing detrimental impacts on aquatic ecosystems.

Salinization of the soil and saline return flows are pervasive and abiding impacts of irrigation. When irrigation water is evapotranspired, dissolved salts are left behind in the soil. With each irrigation more salt is deposited. This is why farmers add a leaching fraction to their irrigation water requirement. This extra water flushes the salt from the root zone. But the salt must go somewhere, and that somewhere is eventually back to the stream, either slowly as groundwater flow or rapidly through a system of drains. Underground drains are installed under the most valuable irrigated farmland to rapidly and efficiently carry away the drainage water and salts.

In the soil, salt interferes with plant growth and has an adverse impact on soil structure. The concentration of total dissolved solids (TDS) is a commonly used indicator of the salt content of the water. TDS concentrations of less than 500 mg/liter have no detrimental effects on crop growth. Salt–sensitive crops may be adversely affected at TDS concentrations of between 500 and 1000 mg/liter. At concentrations between 1000 and 2000 mg/liter most crops will be adversely affected. Hansen *et al.* (1980) reported the downstream increase in TDS along some western rivers. Concentrations along the Colorado River went from 100 mg/liter upstream to 1178 mg/liter downstream, while concentrations on the Arkansas River went from a "trace" upstream to 2200 mg/liter downstream.

A relatively new environmental impact associated with irrigation has been the mobilization of selenium in the subsoil. Many soils in the western United States contain selenium, a toxic heavy metal. Under the naturally dry conditions selenium is relatively immobile in the soil. With irrigation and deep percolation it becomes mobile and is carried along in the drainage water. Selenium pollution has killed and caused deformities in wildlife at the Kesterson Wildlife Refuge in the San Joaquin Valley. The only solution in this case was to plug the subsurface drains. The inevitable result, should irrigation continue, will be progressive salinization and eventual destruction and abandonment of these fields.

8

INSTREAM WATER USE

HYDROELECTRIC POWER AND RECREATION

Hydropower

- Concepts and Definitions
- Types of Hydroelectric Power Plants
- Categories of Electrical Power
- Hydroelectric Power Potential
- Regional Water Use for Hydroelectric Power Generation
- State Water Use for Hydroelectric Power Generation
- Hydroelectric Dams
- Environmental Challenges for Hydroelectric Power

Water for Recreation

- Recreation on Federal Public Lands
- Federal Land Agencies
- Federal Water Resources Agencies

Environmental Values and Instream Flows

Instream water uses include hydroelectric power generation, outdoor recreation, transportation (navigation), water quality improvement, and instream flows for environmental purposes. This chapter focuses mainly on two instream uses—hydropower and recreation, though instream flows for environmental maintenance are discussed at the end of the chapter. Water quality and ecosystems are the topic of the next chapter. Instream water uses can be mutually compatible, as when providing sufficient flow for navigation satisfies the water requirements for recreational uses, ecosystems, and water quality. Offstream uses are less compatible with instream uses, at least in theory. Flat–water recreational opportunities are enhanced upstream of large hydroelectric dams by the presence of the reservoir. On the other hand water releases for generating hydroelectric power may cause large fluctuations in river flow, which might be detrimental to downstream uses. Environmental and aesthetic values supported by instream flows have become more and more important in the last few decades. Just 10 years ago it was taken for granted that environmental and recreational instream uses would have to be reconciled with the use of the river for hydroelectric power generation. That is not the case today, and these instream uses are gaining at least coequal status with hydroelectric power on many streams and rivers.

HYDROPOWER

CONCEPTS AND DEFINITIONS

The hydrologic cycle continuously replenishes rivers, streams, and lakes, making hydroelectric power a renewable source of energy. Falling water is the conversion of potential energy stored in the water into the kinetic energy of motion and heat energy due to friction. At a hydroelectric power plant falling water is directed through penstocks to the turbine generators, which convert the kinetic energy into electrical energy (Figure 8.1). The electrical energy is sold and distributed to customers through an electrical "grid" composed of power stations and transmission lines. Though the energy source itself is renewable, the power–generating infrastructure (dams, transmission lines, power stations) has a finite life span. The most interesting and serious questions revolve around what will be done with these dams after they reach the end of their useful life.

Elevated water possesses potential energy equal to the mass of water times its

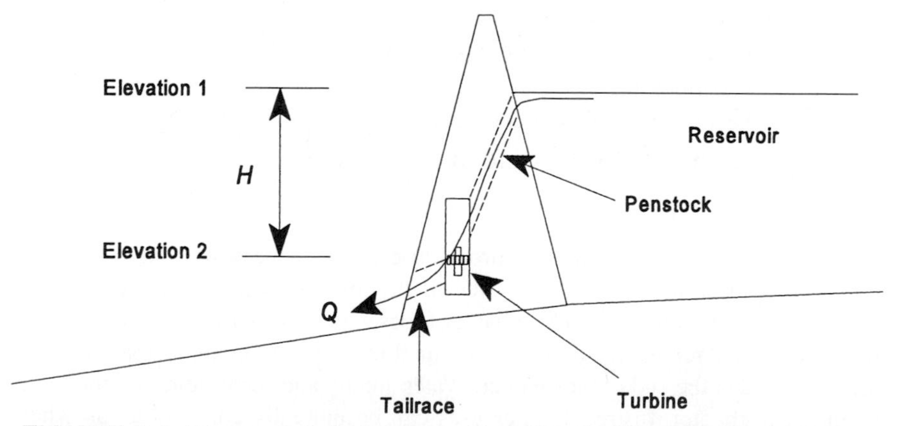

FIGURE 8.1 Diagram of a hydroelectric dam. Head (H) is the difference in elevation between two points.

elevation. Power is the rate at which work is being done. Using English units the equation for calculating the power from falling water is

$$P = \frac{(QH\gamma)}{737.26} \times e. \tag{8.1}$$

In Equation (8.1) P is power in kilowatts, Q is streamflow (ft^3/s), γ is the weight density of water (62.4 lbs/ft^3), H is head, which is the difference in elevation (feet) between two points (Figure 8.1), and e is the mechanical efficiency of the plant. One kilowatt (kW) is 1000 watts; 1 megawatt (MW) is 1 million watts. The basic unit for measuring a quantity of electrical energy is the kilowatt–hour (kWh). One kW of power generated for 1 hour is 1 kWh. The mechanical efficiency of hydroelectric plants is typically 75 to 85 percent. Example 8.1 gives a sample calculation for power.

EXAMPLE 8.1

Calculate the instantaneous power production (kW) for the following situation.

Streamflow $Q = 100$ ft^3/s
Head $H = 42.6$ ft
Density $\gamma = 62.4$ lbs/ft^3
Efficiency = 0.75.

$$P = \frac{(100\ ft^3/s)\ (42.6\ ft)\ (62.4\ lbs/ft^3)}{737.26\ ft \cdot lbs\ /\ s \cdot kW} \times 0.75 = 270.4\ kW.$$

Producing power at the rate of 270.4 kW for 1 hour generates 270.4 kWh of electricity.
Alternatively, a streamflow of 200 ft^3/s could produce the same power with only half the head.

$$P = \frac{(200\ ft^3/s)\ (21.3\ ft)\ (62.4\ lbs/ft^3)}{737.26\ ft \cdot lbs\ /\ s \cdot kW} \times 0.75 = 270.4\ kW.$$

As a very approximate rule of thumb, 1 kW of installed generating capacity is sufficient for the electrical demands of 1 household. The power generated in this example would provide enough power for approximately 270 homes.

TYPES OF HYDROELECTRIC POWER PLANTS

There are basically two types of hydroelectric power plants—*run–of–the–river plants* and *storage plants.* Run–of–the–river plants generate power relying on the available streamflow. Run–of–the–river plants have little or no water storage, and may or may not use a dam to increase head on the river. This type of plant is best suited to large perennial streams with fairly uniform flow throughout the year. Storage plants use a dam and reservoir to store runoff for controlled release at a later time (Figure 8.2). Storage plants are further distinguished based on the size of their reservoir. *Pondage plants* are the smallest and store water through the nighttime hours for peak electrical generation the following day. *Seasonal storage* plants store runoff from the wet season for use during the dry season. Many of the dams along the western slope of the Sierra Nevada in California are of this type. *Cyclical storage* plants are the largest and can store multiple years worth of streamflow (James and Lee, 1971). The *pumped–storage* plant is a variation on the storage plant. Pumped–storage plants pump water from a river or lower reservoir up to a higher storage reservoir. Water is pumped uphill (consuming electricity) at times when the demand for electricity, and therefore the price

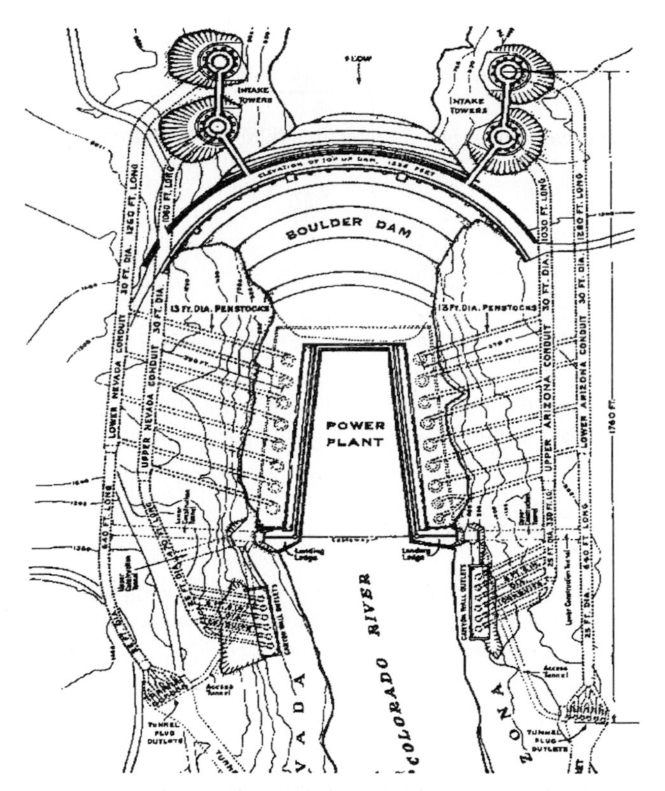

FIGURE 8.2 Plan view of Boulder (Hoover) Dam and power plant. The power plant is located below the dam on both sides of the river. Water is fed to the penstocks through conduits along both canyon walls. (Source: U.S. Bureau of Reclamation. Drawing downloaded from the Bureau of Reclamation's home page, http://www.hooverdam.com/workings/damplan.htm)

per kWh, is low. Water is released from the upper reservoir (generating electricity) when the demand, and the peak price, is high. In operation, pumped–storage plants consume more energy than they produce. They only make sense from an economic point of view. If the value of the peak power produced exceeds the value of the off–peak power used to pump the water, then the plant makes sense economically.

CATEGORIES OF ELECTRICAL POWER

Since streamflow is variable in time, electric power generation is also variable, though reservoir storage acts to smooth out the variability. Electrical power is categorized and priced on its availability. *Firm* or *prime power* is the maximum annual rate that energy can be generated without interruption during the critical dry period (James and Lee, 1971). Because firm power is guaranteed, it is the most expensive, and is sold to residential and commercial users. *Secondary power* is available more than 50 percent of the time, but cannot be guaranteed. Secondary power is cheaper than firm power and is marketed to industries and other users that can suspend operation during power outages. *Dump power* is available less than half the time, and is the least expensive. Figure 8.3 is a flow duration curve with the three power categories defined. A flow duration curve shows the percent of time (X axis) that streamflow equaled or exceeded a given flow rate (Y axis). The installed capacity (kW or MW) is the upper limit on power generation. Flood flows exceeding the installed capacity are spilled from the reservoir and do not generate power.

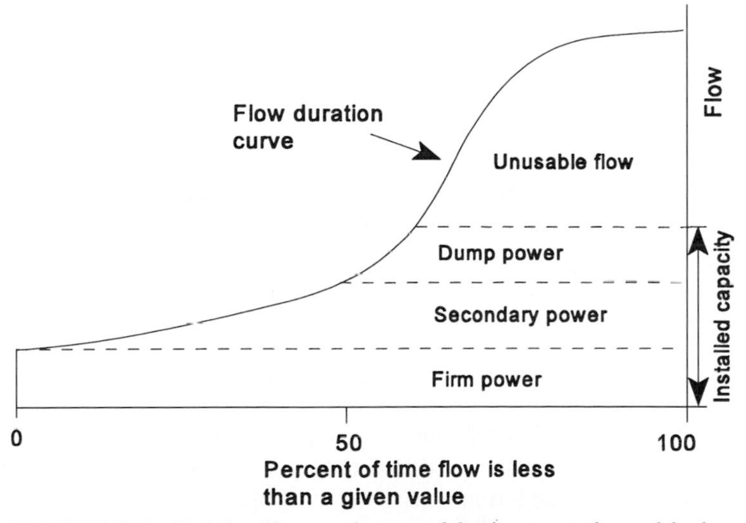

FIGURE 8.3 Flow duration curve for a run-of-the-river power plant and the three categories of power, Flows above the installed capacity are unusable.

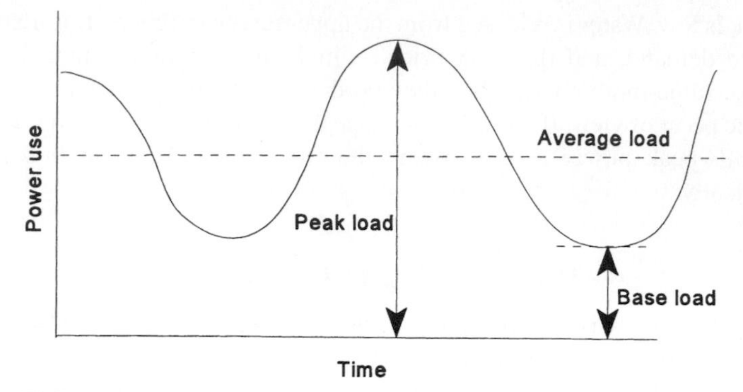

FIGURE 8.4 Qualitative representation of electricity demand (after James and Lee, 1971).

The demand (load) for electricity varies daily and annually (Figure 8.4). Demand is high during the day and lower at night. Depending upon climate, the demand for electricity can fluctuate seasonally and with the weather. Winter is the heating season and demand rises with the use of electrical resistance heaters. Days are shorter during winter so lights are on longer, and people spend more time indoors. In the summer, air conditioning is an important component of the peak load. Along the East Coast, peak electricity–use records are usually set on hot summer days. Warmer indoor temperatures in the summer cause refrigerators, which are a large–wattage appliance, to run longer. A truly amazing statistic is that in 1993 Americans used more electricity just for air conditioning than was used by all the 1.2 billion people in China for all purposes (Miller, 1995).

Hydroelectric facilities are well suited to meet peak demands. By opening and closing penstock gates, turbine generators can be brought on line, or taken off line, as demand rises and falls. This flexibility is a distinct advantage of hydroelectric plants over thermal (fossil fuel or nuclear) power plants, which cannot rapidly adjust their power output. Hydroelectric plants are often used to satisfy peak loads, while thermal power plants satisfy the base load. But again the rapid changes in river flow can be detrimental to downstream recreational uses and ecosystems.

The "load factor" measures the relative magnitude of the difference between the peak load and the average load. The load factor is calculated as

$$\textit{Load factor} = (\textit{average load/peak load}) \times 100. \tag{8.2}$$

For residential areas load factors run 30 to 40 percent. Load factors for industrial users are often twice this high. Since power companies must have enough capacity to satisfy the peak demand, from the power company's perspective the higher the load factor the better, because it means they have less idle generating capacity.

HYDROELECTRIC POWER POTENTIAL

The two main variables in hydroelectric power generation in Equation (8.1) are streamflow (Q) and head (H). Combine these and you can see that areas with both large streamflows and large topographic relief will have the largest hydroelectric potential. Likewise, the lowest potential is in dry climates with low relief. For a given area there is a *theoretical potential* for hydroelectric power (Mather, 1984). The theoretical potential is the product of the average annual streamflow and the overall basin relief. The theoretical potential represents a physically based upper limit for power generation. The theoretical potential can change, however, if the climate changes and causes runoff to increase or decrease. The theoretical potential can never be attained for a number of reasons. The second law of thermodynamics states that conversion of energy can never be 100 percent efficient; some is always lost as waste heat. Also, it may not be possible to develop all the necessary dam sites. We could never perfectly equalize seasonal and annual variations in streamflows. And when we impound water in reservoirs, seepage and evaporation reduce the amount of water available for power generation.

The *technical potential* for hydroelectric power is the theoretical potential reduced by all of the inefficiencies, losses to evaporation and seepage, and sites that can not be developed. This represents the hydroelectric potential that could be developed given the existing level of technology. Some sites, though technically capable of development, are not justifiable on economic grounds. The *economic potential* defines those sites that, in addition to being technically possible, are economically feasible. Both the technical and the economic potentials change with time as new technologies create new possibilities. Today's turbine generators have higher mechanical efficiencies than older models and have increased technical potential. Economic feasibility changes in response to the marginal economics of energy supply and demand, and is itself a function of technology. More efficient electrical transmission lines might make development of remote sites more economical. One factor that is not included in the calculation of either the technical or the economic potential is the social and political feasibility. From both a technical and an economic standpoint the Grand Canyon would be a superb location for a hydroelectric dam. Social and political considerations all but guarantee the canyon will never produce a watt of electrical energy. The value we place on free-flowing streams is a major obstacle to future hydroelectric power development in many areas, even though the technology exists and the economics are favorable. On the other hand, rising concern over climate change and air pollution from burning fossil fuels makes thermal power plants a less attractive option and the relatively "clean" hydropower facilities more attractive by comparison. Given the increasing importance of social considerations, perhaps the economic potential should be called the *socio*economic potential.

Table 8.1 shows the *known exploitable potential* and the installed hydroelectric capacity for the 10 countries in the world with the largest exploitable potential.

TABLE 8.1 Hydroelectric Power Potential and Installed Capacity for the 10 Countries With the Greatest Exploitable Potential, 1991[a]

Country	Known exploitable potential (MW)	Installed capacity (MW)	Installed capacity as a percentage of exploitable potential
China	2,168,304	30,100	1.4
Brazil	1,116,900	45,558	4.1
Indonesia	709,000	1,950	0.3
Canada	614,882	59,381	9.7
Zaire	530,000	2,772	0.5
Colombia	418,200	7,201	1.7
Peru	412,000	2,396	0.6
Argentina	390,038	6,499	1.7
United States	376,000	90,141	24.0
Zambia	309,009	2,245	0.7

[a] Source: World Resources Institute, 1994.

The known exploitable potential is defined as the hydroelectric energy exploitable under existing technical and economic constraints (Gleick, 1993). The term is essentially synonymous with economic potential as used here. The installed capacity represents the developed portion of the exploitable potential. Eight of the 10 ten countries are less developed countries (LDCs). Globally, China is the most favorably endowed in exploitable hydroelectric potential. As of 1991, when the data in Table 8.1 were collected, China had developed only 1.4 percent of its known exploitable potential. This will change when the Three Gorges hydroelectric project is completed in 2009. The installed capacity of that project alone will be 18,200 MW. Three Gorges will increase China's installed capacity by 60 percent. Of the top 10 countries, only the United States has developed a significant fraction (24 percent) of its potential. Many countries around the world have higher ratios of installed capacity to exploitable potential, but their known exploitable potential is small. Algeria has an installed capacity equal to 99.6 percent of its known exploitable potential, and Denmark has installed 86 percent of its known exploitable potential. Algeria's potential is 286 MW, and Denmark's is only 14 MW.

Hydroelectric power in the United States is a small fraction of our total commercial energy production. In 1991 hydroelectricity represented 9.3 percent of our total electricity production, and accounted for a paltry 1.5 percent of the total commercial energy production. Canada, which is fourth in known exploitable potential, has an installed capacity less than 10 percent of its potential. While these numbers indicate considerable room for further development, social, political, economic, and environmental constraints will make future development difficult.

The Canadian government has already faced considerable opposition to the staged development of the massive LaGrande hydroelectric project on the James Bay.

REGIONAL WATER USE FOR HYDROELECTRIC POWER GENERATION

The estimated water–use figures for hydroelectric power generation, in general, are based on better data and less extrapolation than the values for most other water–use categories (Solley *et al.,* 1993). For one–time–through hydroelectric plants, the use estimates are quite good. It is more difficult to estimate water use for pumped–storage facilities where water may be reused multiple times. Table 8.2 gives hydroelectric water use and power generation by water–resource region (Figure 4.6). By far the Pacific Northwest dominates in both water use and power generation. The Pacific Northwest used fully 38 percent of all the water

TABLE 8.2 Water Use and Hydroelectric Power Generation by Water–Resource Regions, 1990[a]

Region	Water use		Power generated in million Kwh	Acre–feet /kWh
	Millions of gal/d (mgd)	Thousands of acre–ft		
New England	168,000	188,000	8,080	23.3
Mid–Atlantic	192,000	215,000	11,700	18.4
S. Atlantic–Gulf	275,000	308,000	18,500	16.6
Great Lakes	506,000	567,000	30,100	18.8
Ohio	147,000	165,000	5,860	28.2
Tennesssee	294,000	330,000	19,777	16.7
Upper Mississippi	73,200	82,100	2,200	37.3
Lower Mississippi	26,600	29,800	1,250	23.8
Souris–Red–Rainy	1,280	1,430	45	31.8
Missouri Basin	109,000	122,000	12,600	9.7
Arkansas–White–Red	109,000	122,000	8,370	14.6
Texas–Gulf	12,100	13,600	953	14.3
Rio Grande	3,520	3,950	569	6.9
Upper Colorado	11,900	13,300	4,760	2.8
Lower Colorado	34,700	38,900	6,640	5.9
Great Basin	2,360	2,650	284	9.3
Pacific Northwest	1,250,000	1,400,000	142,000	9.9
California	69,000	77,300	23,700	3.3
Alaska	1,790	2,010	980	2.1
Hawaii	264	296	89	3.3
Caribbean	362	406	108	3.8
Total	3,287,076	3,682,742	298,565	median = 12.3

[a] Source: Solley *et al.,* 1993.

and generated 47 percent of all the hydroelectric power in the United States. The Columbia–Snake River system is a hydropower workhorse. To achieve this dominance there are some 20 dams on the main stems of the Columbia and Snake Rivers, including some of the world's largest, and 100 more dams on tributaries throughout the basins. The second–largest region in both water use and energy production was the Great Lakes. The Tennessee Valley was third in water use but fourth in power generation. California was third in total power generation but 12th in water use. Power plants in California generate more electricity per unit of water used than those in the Tennessee Region. In fact, if you divide water use (column 3) in Table 8.2 by power generated (column 4), the result is acre–feet of water used per kWh of electricity generated (column 5). The East versus West pattern for the regions based on the values in column 5 is clearly evident. The eight eastern water–resource regions (New England through Souris–Red–Rainy) used an average of 23.3 acre–feet/kWh. The 12 western regions, excluding the Caribbean, used an average of 7.5 acre–feet/kWh. The pattern can largely be explained by referring once again to Equation (8.1) and Example 8.1. In the East the bulk of the power is generated by large volumes of water flowing at relatively low head. Out West the flows are smaller but the potential energy stored in the water is much greater due to the rugged topography and higher relief.

STATE WATER USE FOR HYDROELECTRIC POWER GENERATION

The state water–use pattern is similar to the regional pattern. Table 8.3 lists the top 10 states in water use for hydroelectric power in 1990. The top 3 in water use are Washington, Oregon, and New York. These are also the top 3 in power pro-

TABLE 8.3 Water Use and Hydroelectric Power Generation by State, 1990

State	Millions of gal/d (mgd)	Thousands of acre–feet	Power generated in millions Kwh	Acre–feet /kWh
Oregon	481,000	539,000	40,800	13.2
New York	459,000	515,000	29,400	17.5
Alabama	218,000	244,000	10,300	23.7
Tennessee	160,000	170,000	11,800	14.4
Michigan	110,000	123,000	3,040	40.4
Kentucky	83,000	93,000	2,880	32.3
Maine	82,700	92,700	3,960	23.4
California	75,000	84,100	23,900	3.5
Pennsylvania	68,000	76,200	3,190	23.9

[a] Source: Solley *et al.*, 1993.

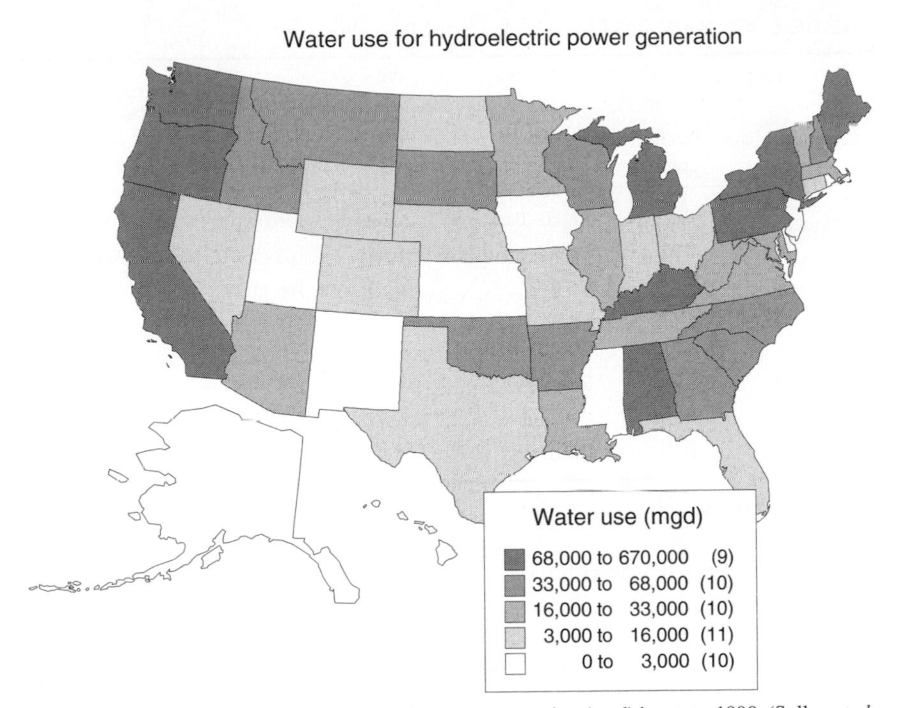

FIGURE 8.5 Water use for hydroelectric power generation (mgd) by state, 1990. (Solley *et al.*, 1993)

duction. Again the relative influence of flow versus head on power generation is evident. Oregon and New York used about the same amount of water, but Oregon generated 39 percent more power. California was eighth in water use, but fourth in power generation. The combined water use of Alabama and Tennessee (fourth and fifth in water use, respectively), was five–and–one–half times that of California, but their combined power output was eight percent less than California's. The 5 states with the lowest water use for hydroelectric power in 1990 were Delaware, Mississippi, New Jersey, Hawaii, and Rhode Island. Delaware and Mississippi used no water, and the three others used 380,000 acre–feet or less. New Mexico and Utah (aridity), and Iowa and Nebraska (low head and few sites), are also low in both water use and power production (Figure 8.5).

HYDROELECTRIC DAMS

What discussion of hydroelectric power would be complete without a comparative look at some of the world's largest dams? There are different ways to define large—dam height, length, mass, reservoir volume, or installed hydroelectric capacity. Here we use installed capacity as the variable for ranking and comparison (Table 8.4). The Itaipú Dam shared between Brazil and Paraguay on the Rio Paraná has the largest installed capacity at 12,600 MW. Itaipú has been

TABLE 8.4 The World's Largest Hydroelectric Power Plants, 1991[a]

Name	Year completed	Country	Installed capacity (MW)	Height in meters (feet)	Maximum reservoir volume in 10^6 m^3 (10^6)
Itaipú	1982	Brazil–Paraguay	12,600	196 (643)	29,000 (23,508)
Guri	1986	Venezuela	10,300	162 (531)	138,000 (111,866)
Grand Coulee	1942	USA	6,494	168 (551)	11,795 (9,561)
Sayano–Shus–hensk	1989	Russia	6,400	245 (804)	31,300 (25,372)
Krasnoyarska	1968	Russia	6,000	124 (407)	73,300 (59,419)
La Grande 2	1979	Canada	5,328	168 (551)	61,715 (50,027)
Churchill Falls	1971	Canada	5,225	32 (105)	32,640 (26,458)

[a] Source: Gleick, 1993.

number one since it was complete in 1982, but it will displaced from the top position by China's Three Gorges Dam. The Three Gorges Dam will stand 185 meters (607 feet) high and have an installed capacity of 18,200 MW, the power equivalent of 18 nuclear power plants (Zich and Sacha, 1997). Grand Coulee Dam on the Columbia River in northeastern Washington state is the largest hydroelectric plant in the United States, and currently third largest in the world. Grand Coulee was conceived during the New Deal era, and the reservoir is appropriately named Franklin D. Roosevelt Lake. Table 8.5 list some of the largest hydroelectric

TABLE 8.5 Largest Hydroelectric Power Plants in the United States, 1991[a]

Name	Year completed	River or basin	Installed capacity (MW)	Height in meters	Maximum reservoir volume in 10^6 m^3 (10^6 acre–ft)
Grand Coulee	1942	Columbia	6,494	168 (551)	11,795 (9,561)
John Day	1968	Columbia	2,160	70 (230)	3,256 (2,639)
Bath County	1985	Back Creek	2,100	146 (479)	44 (0.0356)
Chief Joseph	1955	Columbia	2,069	70 (230)	731 (0.592)
Saunders/Moses	1958	St. Lawrence	2,028	47 (154)	808 (0.654)
Ludington	1973	Lake Michigan	1,979	52 (170)	102 (0.0826)
Hoover	1936	Colorado	1,951	221 (726)	34,852 (28,251)
Raccoon Mtn	1979	Tennessee	1,530	37 (121)	n/a
Glen Canyon	1966	Colorado	1,288	216 (708)	33,304 (26,997)

[a] Source: Gleick, 1993.

plants in the United States. The two reservoirs on the Colorado River (Lake Mead behind Hoover Dam and Lake Powell behind Glen Canyon Dam) are enormous by U.S. standards, but pale in comparison to the truly awesome reservoirs worldwide (Table 8.4). Lake Mead is the largest man–made lake in the United States. Lake Mead and Lake Powell have a combined storage capacity large enough to hold more than 4 years worth of Colorado River flow.

ENVIRONMENTAL CHALLENGES FOR HYDROELECTRIC POWER

In Chapter 6 we discussed some of the negative environmental impacts caused by large dams. It is unlikely that we will see very many new large–scale hydroelectric projects in the United States in the future. The social and political climate is not favorable for large projects having potentially large environmental impacts. Hydroelectric power development in the last two decades has shifted focus with an emphasis on smaller–scale, low–head hydro projects, and improving efficiency at existing power plants by upgrading generators. The situation is different in other countries and China is just one example of where new large–scale hydroelectric power development is continuing.

In the United States some of the most important issues facing hydroelectric dams are environmental. Hydroelectric facilities are being forced to redefine their operating procedures to be more cognizant of instream environmental values. The following three examples demonstrate the range and magnitude of the issues that lie ahead.

Glen Canyon Dam, Colorado River

The instream values of the Colorado River include recreational use for white–water rafting and fishing, channel maintenance related to the movement of sediment, and riparian and aquatic habitat. With closure of the dam in the 1960s the seasonal flow regime of the river was dramatically altered. No longer do massive floods periodically roar down the canyon scouring sediment upstream and redepositing it downstream. In addition, sediment entrained in the river above the dam is trapped in the reservoir. Human modification of the river's flow regime thus changed the sediment budget. Some of the resulting impacts were progressive beach erosion, persistent blockages in the main channel from debris deposited by tributaries, and changes in the composition of the native riparian flora and fauna. All of these impacts were of increasing concern to recreational users, river ecologists, and environmentalists generally. The geomorphological and biological changes coincided almost precisely with an the explosion in recreational use of the river for white–water rafting. While the flood control function of the dam reduced the river's seasonal flow variability, the operation of the dam for hydropower generation increased the variability of daily flows. Under normal operating procedures, water releases for hydropower could cause river elevations to fluctuate as much as 13 feet in a matter of hours. Rapid daily variation in discharge posed

NIAGARA FALLS

The Niagara River forms a 32-mile-long boiundary between New York and Ontario, Canada, connecting Lake Erie to Lake Ontario. In this distance the river drops 326 feet, with nearly half of the drop occurring at Niagara Falls. The hydroelectric potential of Niagara Falls was first tapped in 1889, making it one of the earliest hydropower facilities in the United States. The river is wide and tranquil as it begins its journey, but narrows to about a mile in width as it approaches the falls. At Goat Island just upstream of the falls the river divides, creating two separate falls. The American falls is 1000 feet wide with a drop of 165 feet, and the Canadian or Horseshoe falls is 2600 feet wide and drops 158 feet. From the base of the falls the river flows in a steep gorge for 6 miles until it reaches the edge of the Niagara Escarpment. The gorge was created by the headward retreat of the falls, due to erosion of the soft shale, and undermining of the harder dolomite caprock. Horseshoe falls is retreating at rates of 2–5 feet/year; retreat of the American falls is less than 1 foot/year (Bolsenga and Herdendorf, 1993).

The average discharge of the Niagara River is 202,000 cfs. Because of the tremendous volume of storage in Lake Eric upstream, the interannual variability in discharge is quite low. The maximum recorded flow was 256,000 cfs in 1929; the minimum was 129,000 cfs in 1936. Compared to other large rivers the Niagara's 2:1 maximum-to-minimum ratio is remarkably small. The maximum-to-minimum ratio for the Mississippi is 25:1, while for the Columbia River the ratio is 35:1 (Bolsenga and Herdendorf, 1993).

Substituting the above values into Equation (8.1) we can estimate the theoretical hydropower potential of the Niagara River. The average discharge is $Q = 202{,}000$ cfs, the total head is $H = 326$ feet, and the efficiency is 100 percent. Inserting these values into Equation (8.1) gives the theoretical potential as

$$P = \frac{(202{,}000\ ft^3/s)(326\ ft)(62.4\ lbs/ft^3)}{737.26\ ft \cdot lbs\ /\ s \cdot kW} \times 1.0$$
$$= 5{,}573{,}563\ kW = 5{,}573\ MW.$$

problems for safe rafting and fishing. After a comprehensive environmental study, changes to the dam's operation policy were instituted. The dam is to be operated to fully accommodate environmental and recreational uses of the river. Glen Canyon Power Plant is currently restricted to a 20,000 cfs maximum discharge, which limits power generation to 767 MW. These restrictions will most likely become permanent upon completion of the Record of Decision on the Glen Canyon Dam Final Environmental Impact Statement.

The Columbia and Snake River Basins

The listing of a number of salmon species as endangered in the early 1990s portends profound changes in the operation of the large main–stem dams on the Columbia and Snake Rivers. Salmon are anadromous, which means they are born in freshwater, swim downstream to live in the ocean, and return to spawn upstream in the same freshwater tributaries where they hatched. Studies show 70–90 percent of juvenile fish are killed by turbines, predators, and other accidents as they make their way to the sea. For years the Corps of Engineers has barged the salmon around the dams. Barging has not stopped the precipitous decline in fish numbers. According to the Corps, the number of adult Chinook salmon on the Snake River dropped to less than 4000, down from 28,000, over the last 20 years (*New York Times,* 1997). Opponents of barging maintain the salmon need to be able to freely swim up and down the river. In order to successfully navigate the rivers the fish need higher flows during migration periods than are available under the present operating conditions. One proposed alternative is to release additional water during migration periods. This would be good for the fish but will reduce power revenues and the amount of water available for irrigation. Another alternative is to dig channels around the dams and install fish ladders. Even more radical proposals, heavily resisted by the region's industry and congressional representatives, call for removing some dams altogether. The Corps takes the position that it does the will of Congress. Right now the Corps operates the Columbia–Snake River system primarily for flood control, water supply, and power generation. If the public wants that mission changed, then it needs to persuade Congress to pass the appropriate legislation.

Shad Restoration on the Susquehanna River

The Susquehanna River begins in southcentral New York, flows south through Pennsylvania, and empties into the Chesapeake Bay in Maryland. The Susquehanna is the largest freshwater tributary to the bay. Until the 20th century, American shad, a type of anadromous herring, along with other herring species, freely migrated up and down the river. In the 1800s mill dams on the tributaries and diversion dams to support the riverside canal system temporarily interfered with the shad's migration. Overfishing and water pollution from coal mining and logging also caused shad numbers to decrease. But the death knell for the shad came when the Holtwood hydroelectric dam was built in 1910, 25 miles above the river's mouth. By 1932 three more hydroelectric dams had been built, the Conowingo, 15 miles downstream from Holtwood, and two more above Holtwood—Safe Harbor and York Haven. All four dams are privately owned. Starting in the 1950s shad restoration studies were underway. Restoration efforts first involved a combination of fish stocking, a temporary fish lift at Conowingo, and transporting shad around the dams in tanker trucks. An agreement was eventually reached with the owners of the dams to build permanent fish lifts. The first lift was built at Conowingo Dam in 1990. Fish lifts at the Holtwood and Safe Harbor Dams were

FIGURE 8.6 A low–head dam being removed from the Conestoga River in Pennsylvania to open the river to shad. The Conestoga is a major tributary to the Susquehanna (photograph by the author).

finished in the spring of 1997. The lift at the York Haven Dam is scheduled to be built in the year 2000. When that lift is operational shad will be able to swim all the way to New York for the first time since 1910. An additional component of the restoration program is to breach many of the of small mill and low–head hydro dams throughout the basin to reopen the tributaries (Figure 8.6).

WATER FOR RECREATION

Recreation is participating in an enjoyable activity during one's leisure time. The three factors that explain the increased demand for recreational resources in the United States are increased *leisure time, mobility,* and *income.* Leisure time is the time we have to ourselves after meeting all of our subsistent needs. Leisure time comes in the form of a few hours every day after work or school, as a couple of days on the weekend, and as a few weeks of vacation every year. Once we retire leisure time dominates our lives. The size of the block of leisure time available strongly influences how we use it. In the hours after work we might go for a walk, or to a neighborhood park. On the weekends we might drive a few hours to a reach regional recreation area, like a state park or a reservoir. A summer vacation might take us to another state, across the country, or abroad to another continent. The total amount of leisure time available to Americans continued to increase through the 1970s, but according to some studies has decreased since then (Zinser, 1995).

The second factor affecting recreation demand is mobility. After the second World War private automobile ownership skyrocketed, as did new roads to drive on. The interstate highway system, inaugurated by President Eisenhower in the 1950s, dramatically reduced travel times to distant areas. The third and last factor is rising income. The money used for recreation comes from our discretionary income, which, analogous to leisure time, is that portion of income left over after paying taxes and bills. Different types of water–based recreation have vastly different participation costs. Fishing in a nearby stream requires only a fishing pole and perhaps a license. Water skiing might involve an investment of tens of thousands of dollars in equipment (boat and skis), plus the cost for fuel, transportation, registration, and insurance.

There are other factors that affect recreation. *Physical geography,* expressed as combinations of climate, landforms, and vegetation, creates natural–resources regions having inherent physical limitations and recreational opportunities. Southern states with warmer climates have more months during the year when water–based recreation is possible compared to more northerly states. Water is itself a component of the physical environment. In the West, water is scarce; consequently water–based recreation tends to be concentrated at a fewer number of sites. And while our focus here is on water–based outdoor recreation, the presence of water enhances other recreation activities from camping to hiking to picnicking.

Technology expands recreational opportunities by creating new types of recreation and changing existing activities. Boats for water skiing, fishing, and pleasure sailing are constantly being redesigned, jet skis now offer a new way to enjoy the water, and, for better or worse, four–wheel drive vehicles allow people to reach ever more remote areas.

Demographic factors influencing recreation include the size, distribution, and age structure of the population. A larger population means more people using recreation resources. More people means the facilities are more crowded, and the recreational experience, perhaps, is diminished. Table 8.6 gives some examples of facility development standards used by New York state.

TABLE 8.6 Recreation Facility Development Standards—New York State[a]

Facility	Maximum user density	Standard per 1000 population
Boating	6–8 acres per boat	n/a
Boating access	40 boats per launch ramp	1 ramp per 2500
Fishing (stream)	5 users per mile	0.5 miles of stream per 1000
Swimming pool	1 user per 25 square feet	750 square feet per 1000
Camping	20 users per acre	n/a
Picnicking	35 users per acre	n/a

[a] Source: New York State Office of Parks, Recreation and Historic Preservation, 1989.

Recreation planning is done at all spatial scales from the national level, to the state level, and all the way down to the individual recreation site. Recreation planning and management are done both by government agencies and by the private sector. Government planning spans all spatial scales, while private industry is limited to the smallest scales. Our discussion of recreation is restricted to federal lands managed by seven agencies—the Forest Service, National Park Service, Fish and Wildlife Service, Bureau of Land Management, Corps of Engineers, Bureau of Reclamation, and the Tennessee Valley Authority. According to Cordell *et al.* (1990) about 60 percent of all outdoor recreation takes place at local neighborhood sites like parks and playgrounds. Unfortunately, there is very little systematic information about local–scale recreation. Recreation at state and private facilities each compose about 14 percent of the total outdoor activity. Recreation on federal land represented only 12 percent of the total participation in outdoor recreation nationwide, but it is the best documented in terms of types of activities and use levels.

RECREATION ON FEDERAL PUBLIC LAND

The seven federal agencies, along with the number of recreation visits measured in *recreation visitor days* (RVDs), are listed in Table 8.7. A RVD is equal to 12 *visitor hours,* and a visitor hour is defined by the National Park Service as one person on an area of land or water for the purpose of engaging in recreation for a period or periods totaling of 60 minutes. People visit the federal lands for many different types of recreation, including hiking, camping, picnicking, scenic driving, cultural enrichment, and directly water–related activities like boating, swimming, and fishing. Evaluating the role of water in the total recreation experience is complex because even land–based activities are augmented by the presence of

TABLE 8.7 Federal Agency and Recreation Visitor Days (RVDs), 1991[a]

Agency	Lands	RVDs, 1991	Percent of total RVDs
LAND AGENCIES			
National Forest Service	National Forest System	278,849,000	42
National Parks Service	National Park System	111,998,600	17
Bureau of Land Management	BLM Lands	44,981,600	7
Fish and Wildlife Service	National Wildlife Refuges	4,410,300	4
WATER AGENCIES			
Corps of Engineers		192,166,500	29
Bureau of Reclamation		23,365,200	0.7
Tennessee Valley Authority		1,069,800	0.2
	Total	656,841,000	

[a] Based on data from the various federal agencies and complied in Zinser (1995).

water. Of the seven federal agencies, the lands managed by the National Forest Service are highest in total RVDs. National forest lands made up fully 42 percent of all RVDs on federal land in 1991. Surprisingly, facilities operated by the Corps of Engineers ranked second in RVDs. Since all of the Corps' facilities are water based, this statistic gives some insight into the importance of water in outdoor recreation. Visitations to Bureau of Recreation facilities were less than 1 percent of the total RVDs, however, the reason for the apparently small figure is due to the method of reporting RVDs. Most (86 percent) of the Bureau of Reclamation's recreation facilities are managed by some other federal, state, or local agency. RVDs are reported under the agency directly responsible for managing the recreation facilities. Actual RVDs at Bureau of Reclamations projects are about two–and–one–half times those shown in Table 8.6, but those additional 33 million RVDs are reported and tallied under other agencies.

The following discussion relies heavily on the outstanding work *Outdoor Recreation* by Zinser (1995). Zinser compiled a great deal of statistical information gleaned from a wide variety of public sources to describe how our federal public lands are planned, managed, and used for outdoor recreation. The discussion here is organized by dividing the seven federal agencies into two groups. The first group consists of those that are primarily land–resource agencies, i.e., the Forest Service, Park Service, Bureau of Land Management, and Fish and Wildlife Service. The second group includes the three agencies that are primarily water–resource agencies, i.e., the Corps of Engineers, Bureau of Reclamation, and Tennessee Valley Authority. To further expedite the discussion only one agency from each group is examined in detail. For the land–resource group this is the Forest Service; for the water agency group this is the Corps of Engineers. The remaining agencies are covered in a more general fashion.

FEDERAL LAND RESOURCE AGENCIES

National Forest System Lands

The 191 million acres of land in the National Forest System includes 156 national forests, 20 national grasslands, and assorted miscellaneous holdings. The national forests compose 97 percent of the total (gross) acreage in the National Forest System. Most of the National Forest System is in the western states (69 percent) and Alaska (11 percent). Only 20 percent of the National Forest lands are found in the East. These lands are managed by the National Forest Service on a *multiple use, sustainable yield* basis. The statutorily recognized multiple uses are timber, range, watershed values, recreation, and fish and wildlife. Recreational uses officially received coequal status with the other four uses when Congress passed the Multiple–Use Sustained Yield Act (1960). The multiple use concept is intuitively appealing, but difficult to put into practice on the ground. As a result many areas are managed for a *dominant use* rather than on a multiple use basis. For years clear–cut timber harvesting was the dominant use on many national forest acres, and was clearly not compatible with any of the other uses. One of the recognized multiple uses is watershed values. The federal public lands are the major source of the annual renewable water supply for the western states. Aggregated over the nine western water–resources regions, 61 percent of the average annual water yield originates as runoff from federal public lands (Public Land Law Review Committee, 1970).

In terms of recreation the National Forest Service plans and manages lands for camping, picnicking, trails, rivers and other low–density uses, interpretive services, and wilderness. The recreational objective for rivers is to provide river and similar water–based recreation opportunities that are appropriate to the national forest's recreational role, that are within the carrying capacity of the resource base, and that protect the free–flowing condition of designated Wild and Scenic rivers. In 1992, 4316 miles of the National Wild and Scenic River System were within National Forest System lands (U.S. Forest Service, 1993).

California registered more RVDs on National Forest System lands than any other state in 1991 (Table 8.8). This high level of use reflects the combination of a large, mobile population and the large amount of National Forest System land located within the state. California, with 20,618,936 acres, is second only to Alaska (22,219,636 acres) in acreage managed by the National Forest Service. Nearly three–quarters of the RVDs on National Forest System lands, and 32 percent of all the RVDs on all federal lands, occurred in the 10 states listed in Table 8.8.

Figure 8.7 shows a national breakdown of RVDs on National Forest System

TABLE 8.8 Recreation Visitor Days (RVDs) to National Forest System Lands for the Top 10 States, 1991[a]

State	RVDs, 1991	Percent of National Forest System total RVDs	Percent of national total RVDs (Table 8.6)
California	65,220,800	23.4	9.9
Colorado	25,988,000	9.3	3.9
Washington	22,458,000	8.1	3.4
Arizona	21,548,800	7.7	3.3
Oregon	21,036,500	7.5	3.2
Utah	13,336,700	4.8	2.0
Idaho	12,908,500	4.6	1.9
Montana	10,595,300	3.8	1.6
Michigan	8,153,000	2.9	1.2
New Mexico	8,065,300	2.8	1.2
	Total	74.9	31.6

[a] Based on data from the NFS and contained in Zinser (1995).

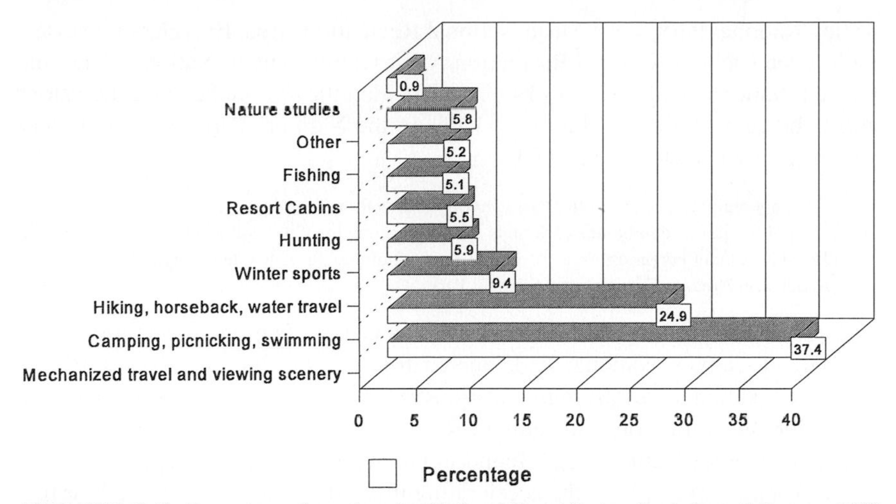

FIGURE 8.7 Percentage of people using National Forest System lands by activity type, 1992 (*Statistical Abstract of the United States,* 1997).

lands by activity group. Mechanized travel and scenic viewing were the activities people engaged in most frequently (37.4 percent). Camping, picnicking, and swimming were the second–most popular activities on National Forests System lands (24.9 percent). California ranked first in RVDs in all activity categories except winter sports and hunting. Colorado registered the most RVDs in those two categories.

National Parks, BLM Lands, and National Wildlife Refuges

The three remaining federal land agencies accounted for 28 percent of the total RVDs to federal lands in 1991 (Table 8.7). Lands in the National Park System are managed by the National Park Service. The National Park System totals some 357 "units," of which 50 are major parks; the other units include, among other categories, national recreation areas, national battlefields, historical sites, memorials, seashores, parkways, recreation areas, and trails. The total acreage in the National Parks System in 1991 was 80.1 million acres (National Parks Service, 1992). The 50 national parks made up 59 percent of the total acreage.

The types of uses allowed on these lands are more restricted than those allowed on the National Forest System lands. The management objective for the National Park System is the preservation of the natural and cultural resources in their unimpaired condition for the enjoyment of present and future generations. It is becoming more difficult to meet the preservation–without–deterioration objective. The intensity of use for the National Park System (1.40 RVDs/acre) is nearly as high as for the National Forest System lands (1.46 RVDs/acre).

In terms of land area Alaska dominates the National Park System. Of the 80.1 million acres in the entire system, 53.7 million (67 percent) are in Alaska. The seven largest units in the lower 48 states are Yellowstone National Park, Death

Valley National Park, Lake Mead National Recreation Area, Everglades National Park, Glen Canyon National Recreation Area, Grand Canyon National Park, and Glacier National Park. Zinser (1995, p. 89) identifies a number of conclusions about the geographic distribution of lands in the National Parks System, two of which are particularly relevant here.

> The only really large park in the East is Everglades National Park.
> Over 500 miles of continuous park units are found along the Colorado River—from Lake Mead National Recreation Area, to Grand Canyon National Park, to Glen Canyon National Recreation Area, to Canyonlands National Park.

Lake Mead National Recreation Area, Glen Canyon National Recreation Area, and Grand Canyon National Park ranked fourth, fifth, and sixth, respectively, among all National Park System units in RVDs in 1991. Among just the national recreation areas, the top four units in RVDs—Lake Mead, Glen Canyon, Ross Lake, and Coulee Dam—are all found in the West, and are all focused on reservoirs. A telling example of the accelerating interest in water–based outdoor recreation was the exponential rise in the number of people who went white–water rafting down the Colorado River in Grand Canyon National Park from the 1950s to the early 1970s (Table 8.9). In 1974 the National Park Service instituted a quota which limited the number of rafters to 14,253 per year. According to Hammitt and Cole (1987, quoted in Zinser, 1995), the waiting period to raft through the Grand Canyon can be as long as 10 years.

With nearly 269 million acres, the Bureau of Land Management (BLM) manages more land than any other federal agency. The BLM was formed in 1946 when Congress merged the old General Land Office, which administered the various homesteading programs, and the Grazing Service, which had been created by the Taylor Grazing Act in 1934. A great deal of the land under the BLM's control is rangeland and desert scrub, though there are also some forested areas. These lands were perceived to hold little potential for agriculture and were skipped over by the homesteaders. The planning and management goal for the BLM lands is multiple use, the same as with the National Forest System.

The wide variety of natural environments found on BLM lands means that virtually every type of outdoor recreation activity is possible. Table 8.10 lists

TABLE 8.9 Visitors per Year Floating/Rafting the Colorado River through the Grand Canyon[a]

Year	Visitors per year
1950	9
1960	205
1970	9,935
1972	16,432

[a] Source: Hammitt and Cole, 1987.

TABLE 8.10 BLM Water–Related Recreational Resources and Facilities[a]

Number	Description
44,981,600	RVDs in FY91
72,541,000	Visits in FY91
9,203	Miles of floatable rivers
4,130,078	Acres of lakes and reservoirs
156,328	Miles of fishable streams
533	Boating access points
28,390,000	Acres of waterfowl habitat
2,000	Miles of rivers in the Wild and Scenic River System

[a] Based on data from the BLM and contained in Zinser (1995).

BLM recreation facilities that are directly related to water recreation. The National Wildlife Refuge System is managed by a division of the U.S. Fish and Wildlife Service (USFWS). These lands are moderately restricted in terms of the types of uses that are allowed. Management of the Wildlife Refuge System is first and foremost for the protection and conservation of fish and wildlife, including endangered species, for their own benefit and for the benefit of people. This is done by a combination of habitat management and direct management of the wildlife populations. Habitat management seeks to provide adequate nesting and breeding sites, food, water, and protected environments for animal and fish species. Management of wildlife is accomplished through fish hatchery and stream stocking programs, and controlled hunting and fishing. A major function of the Wildlife Refuge System is to provide wildlife for recreational hunting and fishing. Other benefits from these lands and waters include nonconsumptive wildlife recreation, environmental education, and the ecological values provided by these ecosystems. The USFWS's major responsibilities are for freshwater fish, anadromous fish, endangered species, migratory birds, and certain marine mammals. The Wildlife Refuge System includes 485 national wildlife refuges, which compose over 97 percent of the system's total acreage, 17 fishery research stations, 84 fish hatcheries, 6 wildlife research centers, 164 waterfowl production areas, and administrative sites (U.S. Fish and Wildlife Sercie, 1992). Florida leads the nation in recreational visits to National Wildlife Refuge sites, followed by Oklahoma, Virginia, North Carolina, Illinois, California, and Alabama.

FEDERAL WATER RESOURCE AGENCIES

Of the three federal water agencies providing water–based recreation, the Corps of Engineers clearly dominates as measured by RVDs (Table 8.7). The Corps is second only to the National Forest System in RVDs on all federal lands. Even if we include under the Bureau of Reclamation the 33 million RVDs that occur at Bureau of Reclamation projects but are counted by other agencies, the

Corps still has more than a 3–to–1 margin over the Bureau of Reclamation in terms of RVDs. The Corps' dominance is explained by the fact that it has more water projects than the other agencies, it has more projects located near large urban areas, and it operates throughout the entire United States, whereas the both the Bureau of Reclamation and the Tennessee Valley Authority (TVA) are geographically restricted. The Bureau of Reclamation operates only in the 17 western states, and the TVA is limited to the drainage basin of the Tennessee River. For all three agencies the bulk of the recreation activities take place at their reservoirs. But in addition to reservoirs, the Corps operates waterways, locks, beach nourishment programs, and harbor projects that are also used for recreation.

Corps of Engineers

The spontaneous popularity of Corps projects for recreation caused Congress to broaden the agency's mission by giving it authority to plan for recreation in the 1944 Flood Control Act. Legislation in the 1960s and 1970s continued to expand the Corps' recreational planning role. Recreation is a recognized use of Corps projects and is included in project planning, but recreation cannot be the primary reason for building a project. The primary purposes are still flood control, hydropower, navigation, and water supply. Recreational components, however, consistently provide the greatest ratio of benefits to costs of all the project purposes. In a 1990 economic impact study the Corps estimated that visitors spent $33 for every dollar the Corps spent on recreational operation and maintenance (Zinser, 1995, p. 586).

Most Corps reservoirs were originally constructed in rural areas. Urban and suburban growth has combined with expanded road networks to bring urban development closer and closer to the reservoirs. About 80 percent of Corps reservoirs are now within 50 miles of a major metropolitan area. In the 1950s and 1960s individuals purchased property around the reservoirs and built private recreation facilities (boat houses, docks) (Zinser, 1995). The Corps stopped uncontrolled private development in the 1970s by requiring all existing facilities to be covered by a Shoreline Use Permit and banning new private development. There are some 50,000 private facilities under permit at Corps reservoirs.

Figure 8.8 Shows the number of Corps of Engineer projects by state. The two areas of concentration are (1) from the Ohio River basin through the middle Mississippi River basin to Texas, and (2) Oregon and California on the West Coast. The top seven states in number of Corps projects are Pennsylvania (39), Illinois (37), Ohio (33), Texas (29), Oklahoma (28), Kentucky (27), Arkansas (23), and California (22). As a region the northern Great Plains through the central Rockies into the Great Basin have the least number of projects. The states of Delaware, New Jersey, Maine, Utah, Nevada, and Wyoming have no Corps projects. The three western states (Utah, Nevada, and Wyoming) do have Bureau of Reclamation Projects.

Table 8.11 gives recreation–related information about Corps of Engineer projects. The Corps operates 57 percent of the recreation areas at Corps projects. The

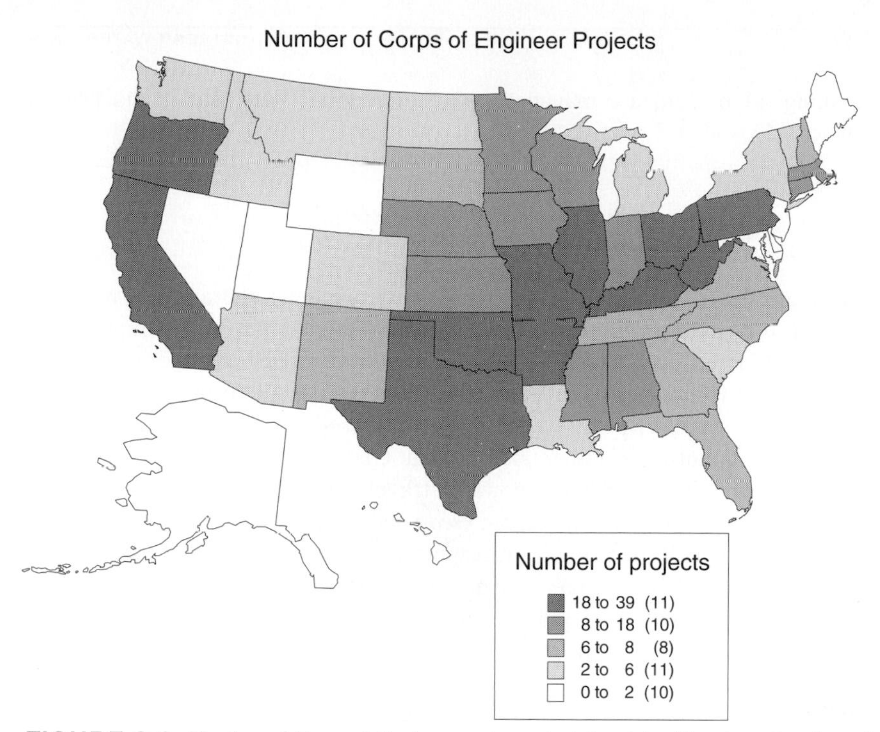

FIGURE 8.8 Number of Corps of Engineer water projects by state. (Based on data from the Corps of Engineers and contained in Zinser, 1995).

TABLE 8.11 Corps of Engineers Water–Related Recreational Resources and Facilities, 1992[a]

Project and facilities data	
Total number of projects	461
Total project area	11,897,449 acres
Total shoreline	40,647 miles
Number of recreation areas	4,382
Number managed by the Corps	2,500 (57 percent)
Total Lake Shore Use Permits	53,504
Number of projects with	
Boat launch facilities	395
Restrooms	457
Drinking water, picnicking facilities	430
Camping with electricity	250

Recreational use of Corps projects

Activity	Percent of visitors	Activity	Percent of visitors
Sightseeing	34	Camping	11
Fishing	26	Swimming	10
Boating	18	Water skiing	4
Picnicking	16	Hunting	3

[a] Based on data from the Corps of Engineers and contained in Zinser (1995).

remaining 43 percent are managed by other federal, state, local, and private agencies.

ENVIRONMENTAL VALUES AND INSTREAM FLOWS

Instream flows support a wide range of environmental and amenity resource values (Table 8.12). *Biological values* include habitat for fish, mammals, e.g., beavers, and other species that live in both the water and the benthic environment. In the western United States instream flows may be needed to maintain groundwater levels to support riparian vegetation. *Water quality values* include sufficient flow to dilute pollution and maintain acceptable levels of dissolved oxygen. *Recreational instream values* include water travel (kayaking, canoeing, rafting), swimming, fishing, and bird watching. *Geomorphological values* include the fluvial processes of sediment erosion, transport, and deposition for the maintenance of landforms (beaches, bars) and channel characteristics (width, depth), and for flushing fine sediment from gravel spawning beds. Fluvial processes, channel characteristics, and landforms are related to instream flows through the velocity, volume, and frequency of streamflows. Many of the environmental values in Table 8.12 are complementary. Improved water quality improves aquatic habitat and makes recreation more enjoyable. Beaches and bars are used by boaters as rest areas and campgrounds, and their presence (or absence) influences the amount and location of riparian vegetation and habitat.

As a generalization, instream flow issues in the eastern United States are concerned mainly with how water quality affects environmental and amenity resource values such as ecosystem health, recreation, and aesthetic enjoyment of the water. In the West, where prior appropriation water law encourages water diversions from streams, instream flow issues focus first on achieving and maintaining a sufficient quantity of flow. You cannot be concerned about water quality unless you have water in the stream to begin with. In areas with irrigation return flows that are contaminated by agricultural chemicals, water quality is a primary issue.

TABLE 8.12 Environmental Values Associated with Instream Flows

Biological	Geomorphological
Fish and mammal habitat	Sediment erosion, transport, deposition
Riparian vegetation	Channel bed sediment composition
Recreational	Landform maintenance (beaches, bars)
Water travel	Clearing channel obstructions
Fishing	Aesthetic
Swimming	Scenic view
Bird watching	Sound of running water
Water quality	
Dilution of pollution	
Dissolved oxygen levels	

BOX 8.1 PROCEDURE FOR PROTECTING INSTREAM FLOWS

STEP	COMMENTS/QUESTIONS
1. ASSESS AND IDENTIFY RELEVANT FLOW-DEPENDENT RESOURCE VALUES	Determine what environmental and amenity resource values are important for the stream. This may be done through resource planning or by government proclamation, e.g., a Wild and Scenic River designation.
2. DESCRIBE HOW RESOURCE VALUES DEPEND ON FLOWS	This is a preliminary assessment/description of how these resource values are related to flows.
3. QUANTIFY HYDROLOGIC AND GEOMORPHOLOGIC CONDITIONS	Determine the hydrologic, hydraulic, and geomorphic regimes for the stream. For example, what are the maximum, minimum, mean, and dominant "channel forming" flows?
4. DESCRIBE AND EVALUATE EFFECTS OF ALTERNATIVE FLOW REGIMES ON VALUES	How do incremental changes in flow affect the resource values? For example, at what flow level does whitewater rafting become uninteresting?
5. IDENTIFY INSTREAM FLOW REQUIREMENTS TO PROTECT VALUES	Identify the optimum and minimum flow conditions (quantity and timing) to support the values.
6. DEVELOP A FLOW-PROTECTION STRATEGY	Develop a realistic strategy for protecting flows that blends technical, legal, and administrative alternatives.

(Jackson *et al.,* 1989)

Jackson *et al.,* 1989 outline an interdisciplinary procedure for protecting instream flows in the West (see Box 8.1). The procedure begins with the assessment and identification of the environmental and amenity values associated with a stream (Table 8.12). The next step is a preliminary description of how the values are related to instream flows. The value–flow relationship is further defined in step 4. Step 3 in the process is the quantification of the stream's hydrologic, hydraulic, and geomorphologic regimes. This might include a flow duration analysis (see Figure 8.3), a flow frequency analysis (discussed in Chapter 10), or an analysis of how stream geometry (channel width and depth) changes with flow (e.g.,

Leopold and Maddock, 1953). Quantitative analysis defines the minimum, maximum, and mean flows for the stream. It may also try to determine which range of flows are most important in controlling physical channel characteristics. Step 4 is the further description and evaluation of how incremental changes in flow, channel, and vegetation characteristics affect the resources values. The next step is to merge resource values with the quantified hydrologic analysis, and recommend an instream flow regime that supports these resource values. Ideally both an optimum flow regime and a minimally acceptable regime are identified. These regimes are identified through an evaluation process done by individuals representing the various water–dependent resource values (Jackson *et al.,* 1989). Finally, an administratively, legally, and technically feasible strategy is developed that protects the instream flows. In the West this will likely mean establishing an instream water right, though there may be alternative ways of protecting instream flows. The increasing value that society places on instream flows for environmental and recreational uses is one of the most powerful forces changing traditional water management institutions.

9

Water Quality and Ecosystem Health

The Water Molecule
- Acidity/Basicity
- Pollution Concentration
- Categories of Pollutants

Water Quality Management
- Point Source Pollution
- Nonpoint Source Pollution
- Water Quality Monitoring

The Quality of Our Nation's Waters
- Rivers and Streams
- Lakes, Ponds, and Reservoirs
- Estuaries

The Watershed Approach to Water Quality

Ecosystems and Water Quality

There is no such thing as pure water in the environment. Human–caused water pollution is a change in the physical and/or chemical character of the water relative to its natural background condition. The cartoon in Figure 9.1 ran in the *Albuquerque Journal* after high concentrations of mercury were found in a number of state lakes. As far as scientists could determine, the mercury was from entirely natural sources.

FIGURE 9.1 Cartoon from the *Albuquerque Journal* (1991) reflecting local concern over high mercury concentrations found in some New Mexico lakes.

THE WATER MOLECULE

Water has a number of unique physical and chemical properties, one of which makes it highly susceptible to pollution. The molecule's one oxygen and two hydrogen atoms create a negative electric charge at the oxygen end and a positive electrical charge at the hydrogen end (Figure 9.2). This polar structure allows water to dissociate a wide variety of compounds, especially ionic compounds. This is why water is a nearly universal solvent. Water's dissolving power is beneficial in that it makes life possible, since cell membranes are permeable only to dissolved nutrients. Conversely this property means water is easily polluted.

ACIDITY–BASICITY

In water a small number of the molecules dissociate into positively charged hydrogen ions (H^+) and negatively charged hydroxide ions (OH^-):

$$H_2O \Leftrightarrow H^+ + OH^-. \tag{9.1}$$

In 1 liter of pure water, dissociation produces only about 10^{-7} moles each of H^+ and OH^+ ions. A mole of hydrogen ions weighs 1.008 grams, and a mole of hy-

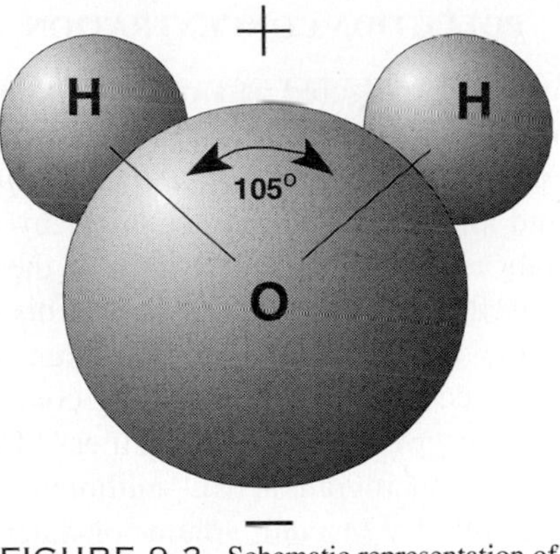

FIGURE 9.2 Schematic representation of a water molecule showing the polar charge structure.

droxide ions weighs 17.008 grams. The acidity of a substance is measured by the concentration of H^+ ions. If the concentration of H^+ ions were 10^{-6} moles per liter, this would be a 10–fold increase in acidity relative to pure water at 10^{-7}. The "pH" scale is a simplified way of expressing the acidity or basicity of a substance. The pH (potential hydrogen) number is the logarithm of the reciprocal of the hydrogen ion concentration:

$$pH = -\log [H^+] = \log \frac{1}{[H^+]}. \tag{9.2}$$

Pure water has a pH of 7, while a pH of 6 is 10 times more acidic than a pH of 7 (Figure 9.3).

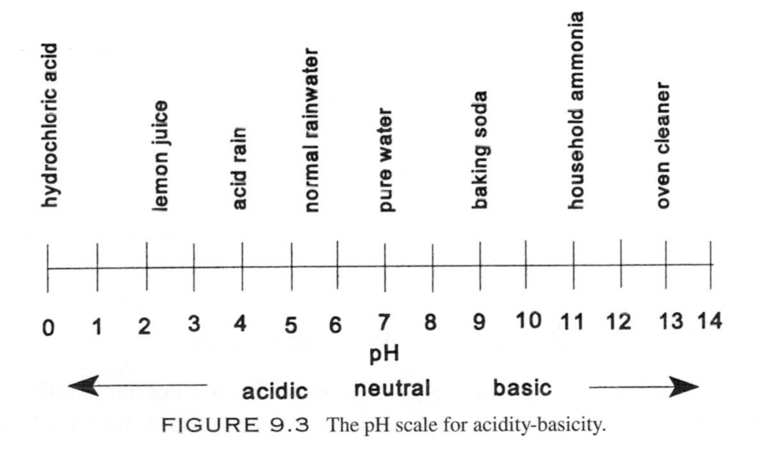

FIGURE 9.3 The pH scale for acidity-basicity.

POLLUTION CONCENTRATION

Many water pollutants are measured as a concentration of the substance dissolved in water. A common way of measuring and reporting concentration is as a dimensionless number, such as *parts per million* (ppm) or *parts per billion* (ppb). Concentration measured in ppm or ppb may be either by weight or by volume. Concentration by weight is calculated as the ratio of the mass (weight) of the pollutant to the mass (weight) of water in which it is dissolved. Only when all values are measured in the same units, for example, grams, can the concentration be dimensionless. Another common way of reporting concentration is *milligrams per liter* (mg/liter) or *micrograms per liter* (μg/liter). (One milligram is one–thousandth of a gram, and 1 microgram is one–millionth of a gram.) These units measure the mass of the pollutant in a unit volume of water, rather than as a mass to mass ratio, and the concentration value is not dimensionless (See Example 9.1).

EXAMPLE 9.1

Find the concentration by weight in ppm of 2 grams of a substance dissolved in 100 cubic feet of water. Assume a water temperature of 15°C.

First get all measurements into the same units, which for this example is grams. Convert the volume of water into its gram weight equivalent:

$$(100\ \text{ft}^3)(62.36\ \text{lbs/ft}^3) = 6236\ \text{lbs}$$
$$(6236\ \text{lbs})(453.6\ \text{g/lbs}) = 2828{,}650\ \text{g}.$$

The concentration is

$$(2\ \text{g})/(2{,}828{,}650\ \text{g}) = 0.000000707 = 7.07 \times 10^{-7} = 0.707\ \text{ppm}.$$

As a second example, convert "1 milligram per liter" into an equivalent "ppm" concentration. One milligram is one–thousandth of a gram, or 0.001 grams. One liter of water is 1000 grams. Therefore,

$$(0.001\ \text{g})/(1000\ \text{g}) = 0.000001 = 1.0 \times 10^{-6} = 1\ \text{ppm}.$$

CATEGORIES OF POLLUTANTS

Water pollutants can be grouped into one of eight categories, although many pollutants fall into more than one category.

DISEASE–CAUSING ORGANISMS

Water–borne infectious organisms are responsible for more human illness and suffering than any other environmental factor. Cholera, typhoid, and dysentery are all water–borne infectious diseases, and they infect millions of people around the

CATEGORIES OF WATER POLLUTANTS

Disease-causing organisms
Oxygen-demanding wastes
Sediment
Nutrients
Organic chemicals
Inorganic chemicals
Heat
Radioactive substances

world. Infectious diseases are largely controlled in the United States by water supply and wastewater treatment, but outbreaks still occur leading to illness and death. In 1993 an outbreak of cryptosporidium in Minneapolis infected 370,000 people and killed 80. The source of contamination was traced to livestock wastes flushed into the water system by surface runoff. In the period 1972–1990 the Centers for Disease Control identified 544 disease outbreaks caused by contaminated water (Alder, 1993). Because there are so many different pathogens it is impractical and expensive to test for every one. Instead, counts are made of "indicator organisms"—specifically, fecal coliform bacteria. Fecal coliform are found in the digestive tract of warm-blooded animals. Their presence in water indicates the possibility of contamination by other infectious organisms. In the United States water with more than 2 coliform colonies per 100 ml is considered unsafe for drinking. The EPA's standard for water used primarily for swimming is 100 coliform per 100 ml. Chlorination, exposure to ultraviolet radiation, ozonation, and advanced filtration are some treatment methods for controlling pathogens. In emergency contamination situations people are advised to boil their water to kill pathogens.

OXYGEN-DEMANDING WASTES

As a class of water pollutants, oxygen–demanding wastes are of singular significance in that they are at the heart of the federal regulatory programs for water quality management. The amount of dissolved oxygen (DO) in water has no discernable impact on human health, but it is crucial for the health of aquatic ecosystems. The bulk of our water quality regulations are therefore driven by a concern for the health of ecosystems rather than the health of human beings directly.

There are at least three groups of oxygen–demanding substances. The groups differ in terms of the types of reactions and materials involved. The first group includes inorganic chemicals—like sulfides—that react rapidly with DO in water. These reactions are wholly of a chemical nature, and the amount of dissolved oxygen consumed depends upon the types and quantities of the chemicals in the

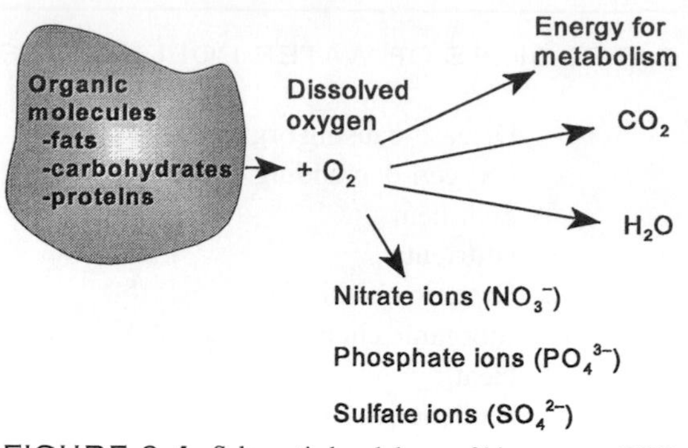

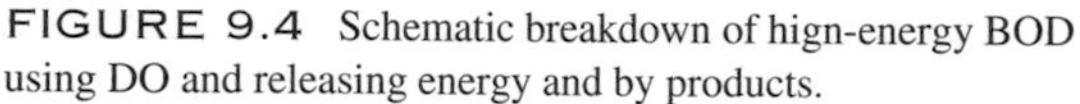

FIGURE 9.4 Schematic breakdown of hign-energy BOD using DO and releasing energy and by products.

water and on the amount of oxygen required for the specific reactions (Lamb, 1985). A second group of oxygen–demanding wastes are nitrogen compounds. Dissolved oxygen is used by certain types of bacteria in biochemical reactions that convert organic nitrogen to ammonia, ammonia to nitrite, and nitrite to nitrate. The ammonia–to–nitrite, and the nitrite–to–nitrate reactions are collectively called "nitrification." The DO required for nitrification is sometimes referred to as the "nitrogenous oxygen demand" or NOD. The last group of oxygen–demanding substances are those that exert a biochemical (or biological) oxygen demand (BOD). BOD represents the DO that is used for the metabolic processes of living organisms. Microorganisms eat organic matter as their energy source, breaking it down into simpler by–products (Figure 9.4). As part of their respiration they release water and carbon dioxide. There can be multiple sequential reactions (a chain reaction) involving different microorganisms to achieve complete decomposition of the organics.

Quantifying and distinguishing the relative contribution to DO depletion from inorganic chemical reactions, NOD, and BOD is not easy. Various laboratory test are performed to estimate these for natural watercourses. The standard test for BOD concentration is a five–day BOD test. For this test two water sample are taken and the DO is determined immediately in one. The second sample is incubated for five days in total darkness at 20° C, after which time its DO is measured. The difference in DO between the two samples is the amount of oxygen consumed by microorganisms in five days and is designated BOD_5. In practice many factors complicate the test and a major inconvenience is having to wait five days for the result. An alternative is a chemical oxygen demand (COD) test, which uses a strong oxidizing agent to react with the organics in the sample. The COD test takes only a few hours. By developing an empirical correlation between COD and BOD, the relatively rapid COD test can be conducted and BOD predicted using the correlation relationship (Lamb, 1985).

Figure 9.5 shows saturation for DO in water as a function of water temperature,

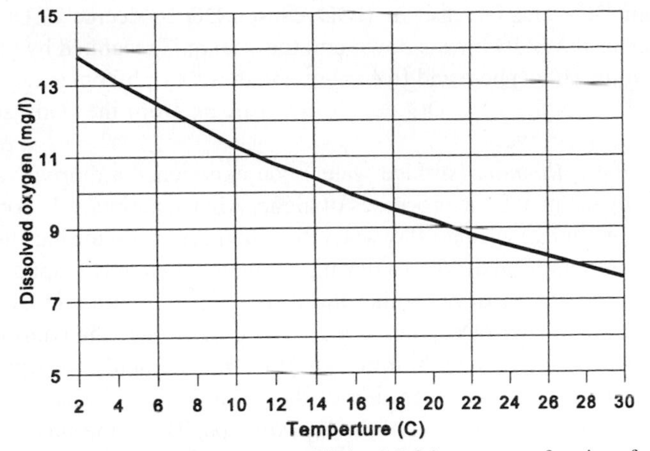

FIGURE 9.5 Saturation (equilibrium) DO in water as a function of water temperature.

at an air pressure of 1 atmosphere. The saturation DO value represents a condition of dynamic equilibrium for oxygen exchange between the water and the atmosphere. Oxygen is constantly entering the water from the atmosphere and leaving the water surface back to the atmosphere. At saturation—equilibrium—the amount of oxygen entering the water equals the amount leaving. Oxygen-demanding wastes reduce the amount of DO in the water, and thus the amount of oxygen going to the atmosphere. The DO *deficit* is the difference between the saturation DO and the actual DO in the water. For temperatures normally encountered in natural waters, saturation DO ranges from 8 to 14 mg/liter.

Figure 9.6 qualitatively shows the effect of a point source release of BOD on

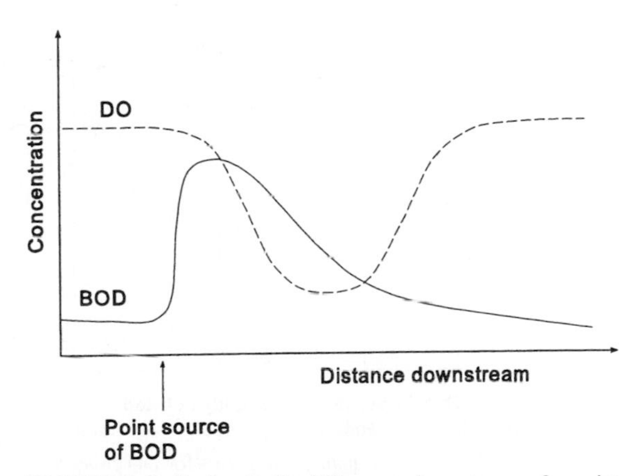

FIGURE 9.6 Sag in the DO curve downstream of a point source of BOD.

downstream DO. The increase in BOD causes DO to decrease. DO recovers downstream as the BOD breaks down and the stream is reaerated by gaseous exchange with the atmosphere and from photosynthesis by phytoplankton (algae) in the water. The graph of the DO deficit downstream from the point source is a classic DO "sag" curve.

In lakes the *epilimnion* (surface water) can experience a diurnal variation in DO caused by the metabolic processes of algae. Algae continuously consume DO for respiration and oxygenate the water by photosynthesis during the daylight hours. DO concentration decreases during the night when only respiration occurs and increases during the day. Another more significant variation in DO occurs in thermally stratified lakes. Deep lakes become thermally stratified during both the summer and the winter, though summer stratification is usually more pronounced. The warmer, less–dense surface water of the epilimnion in the summer does not mix with the colder, denser water of the *hypolimnion.* The epilimnion may be well oxygenated due to gaseous exchange and photosynthesis, but the deep water of the hypolimnion becomes depleted in oxygen during the summer and winter. In the fall when the lake becomes isothermal (uniform temperature throughout) the lake "overturns" and deep water comes to the surface and is oxygenated.

Figure 9.7 shows water quality categories based on the DO concentration at a temperature of 20°C. Water with less than 4 mg/liter is gravely polluted and can hardly sustain life beyond sludge worms and leeches. Dissolved oxygen concentrations above 8 mg/liter are considered good. Figure 9.8 shows the change in DO at selected locations around the country from the early–to–mid–1960s up through the late 1980s. At every location DO levels improved, but heavily–polluted water was still common around New York, Philadelphia, and Dallas/Fort Worth.

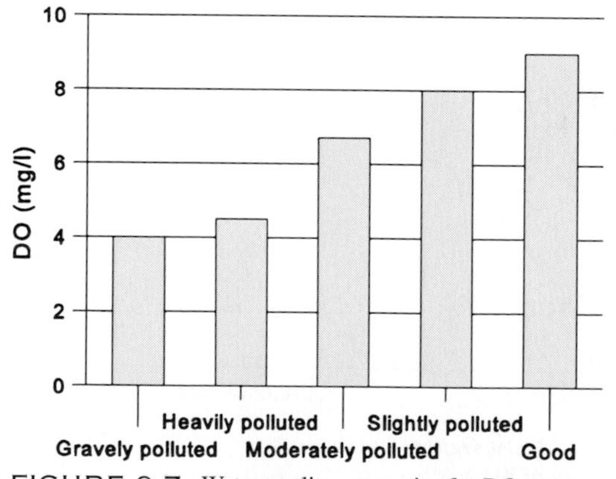

FIGURE 9.7 Water quality categories for DO concentration at 20°C (after Miller, 1997).

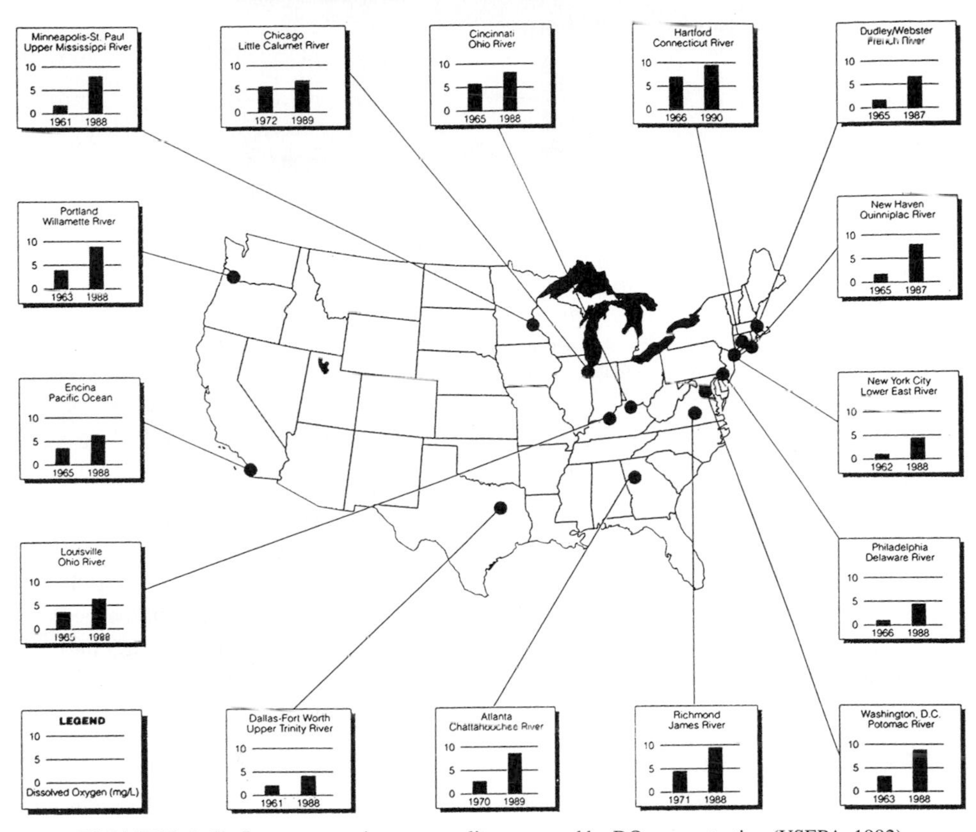

FIGURE 9.8 Improvement in water quality measured by DO concentration (USEPA, 1992).

SEDIMENT

Sediment enters streams and lakes through natural erosion and by accelerated erosion associated with agriculture, construction, logging, mining, and recreation. Of these land uses erosion from agriculture (both crop agriculture and grazing) contributes the most sediment to the Nation's waters (Figure 9.9 and Table 6.2). Over 1 billion tons of sediment enters the our waterways every year, with about one–third of the total washing into the Mississippi River system. Sediment clogs channels and harbors, and detrimentally interferes with the functioning of ecosystems. Port authorities in Oakland, California, estimated that between 1988 and 1991 siltation of the port cost $700 million in wages, taxes, and local sales, and 4000 jobs. Burgeoning international trade throughout the Pacific Rim has spurred the creation of larger, deeper, high–speed container ships. Oakland handles about

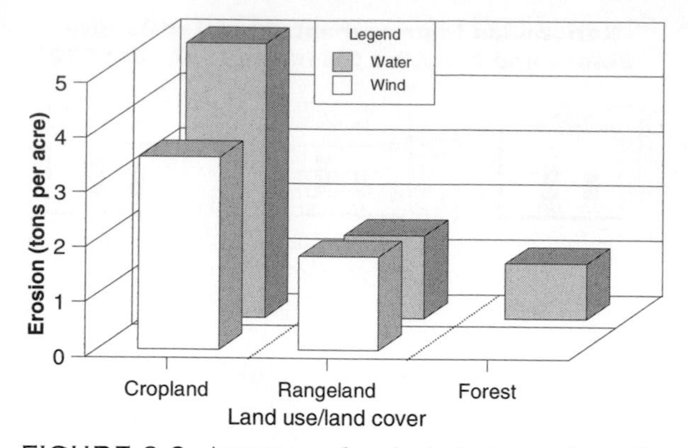

FIGURE 9.9 Average annual erosion by land use and agent for the United States (data from the SCS, no date).

90 percent of the region's container traffic but sedimentation is restricting the size of ships that can use the Oakland port. A significant portion of this sediment is still a legacy of the gold rush of the mid–19th century (Wood, 1991).

Sediment detrimentally affects ecosystems by clogging and scraping the gills of fish, reducing light penetration, which reduces photosynthesis by submerged aquatic vegetation, and covering the stream bed, interfering with spawning and feeding. Sediment also adds nutrients and other pollutants to the water because they are adsorbed on to the surfaces of the sediment particles.

Soil Conservation Measures

Over the years a number of effective measures have been developed to control soil erosion from farmland. Water is the primary agent of erosion, but in dry climates wind erosion becomes significant. Table 9.1 lists some methods for conserving soil. One of the main objectives is reducing the velocity of the water or wind. Low–velocity water (wind) is less able to erode material, and if the water (wind) is carrying material, when its velocity decreases, the sediment is deposited. Reducing the water's velocity also promotes infiltration, which reduces the amount of surface runoff. Many of the methods listed in Table 9.1 have additional environmental benefits. Crop rotation and strip cropping can improve soil fertility, and they help control pests naturally. This reduces the need for chemical fertilizers and pesticides.

In a recent analysis of the magnitude of human earth–moving activities Professor Hooke at the University of Minnesota concluded that humans are now one of the most important geomorphic agents in the United States. He estimated that housing construction, road construction, and mineral production displace about 7.6 billion tons of sediment annually. Put in perspective, this would fill the Grand Canyon in just 400 years—about 10,000 times faster than it took nature to carve

TABLE 9.1 Methods of Soil Conservation on Farmland

Method	Effects
Terracing	Shortens slope length, lowers slope angle, lowers water velocity, promotes infiltration
Strip cropping and crop rotation	Roughens surface, increases flow resistance, lowers water velocity, promotes infiltration; beneficial effects on nutrient and pest management
Contour plowing	Increases flow resistance, lowers water velocity, promotes infiltration
Minimum/no–till cultivation	Protects soil from rain impact, roughens the surface, increases flow resistance, lowers water velocity, promotes infiltration
Grassed waterways	Lowers water velocity, reduces gully erosion
Windbreaks	Lowers wind velocity, promotes deposition

it. Not all of the displaced material enters our waterways, but a significant amount eventually does.

In response to economic hardships in the agricultural sector in the early 1980s, Congress established the Conservation Reserve Program (CRP) in 1985. The CRP had a number of goals, one of which was to remove highly erodible farmland from agricultural use. Land accepted into the program was to be withdrawn from crop production for 10 years and managed using an approved soil conservation practice. By 1994, nearly 36.5 million acres were enrolled in the program nationwide. Most of this acreage was west of the Mississippi River (Osborne, *et al.,* 1992). The program has been generally regarded as a success in conserving soil. The Soil Conservation Service estimated that before the CRP the average erosion from CRP–eligible land was 20.6 tons per acre per year. After the CRP the average erosion was only 1.6 tons per acre per year. In 1994 it was estimated that over 693 million tons of soil were saved every year by the CRP (McKay, 1996)

NUTRIENTS

Aquatic ecosystems are often limited in the availability of two nutrients —nitrogen and phosphorous. In freshwater ecosystems phosphorous tends to be more important as a limiting factor, while nitrogen is more important in the ocean. Limited nutrient availability prevents the excessive growth of algae. Nutrient enrichment of a water body is called *eutrophication.* Like erosion eutrophication is a natural process but it is accelerated by human activities. Human–accelerated eutrophication is called *cultural eutrophication.* Figure 9.10 shows some of the potential sources of nutrients leading to eutrophication. Lakes are more susceptible to eutrophication than are rivers because lakes act as sinks for pollutants, and the

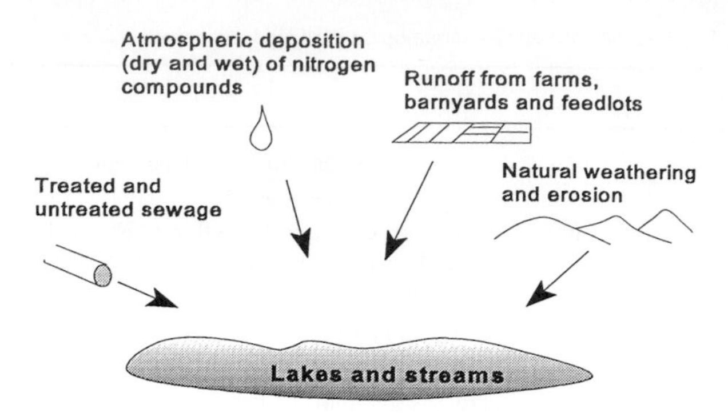

FIGURE 9.10 Sources of nutrients causing eutrophication.

residence time for water in lakes is much longer than that in rivers (Table 1.1), so pollutants are not flushed through the system as rapidly. Fertilization of the water increases primary productivity and the standing biomass. This can be detrimental to the existing ecosystem because algae on the surface reduces light penetration and inhibits photosynthesis by submerged aquatic vegetation. Also, algal blooms can deplete DO when the algae decompose.

Outbreaks of Pfiesteria, a particularly vicious and toxic organism, have occurred in estuarine waterways from North Carolina up to Maryland and Delaware. Pfiesteria was only identified as a new a organism in 1992. Pfiesteria attacks and kills fish in a matter of minutes. The organism also affects humans with symptoms ranging from skin rashes, to respiratory problems, to memory loss. Scientists do not know what causes the one–celled organism to change from a relatively benign spore on the bottom of brackish coastal streams into a potent killer. One thing that

NUTRIENT MANAGEMENT IN PENNSYLVANIA

The Commonwealth of Pennsylvania passed legislation in 1993 aimed at controlling nutrient water pollution from farms. The act states that "concentrated animal operations" are required to develop and maintain a nutrient management plan. Concentrated animal operations have an animal density exceeding 2 animal units per acre on an annual basis. One animal unit is 1000 pounds of live weight. One 1500-pound dairy cow is equivalent to 1.5 animal units, as are 500 chickens weighing 3 pounds each. Which acres will be counted in calculating animal density is unclear but presumably it will be the acres available for manure application. The driving force behind the new law is nutrient pollution and eutrophication of the Chesapeake Bay. In 1997 the EPA announced plans for national regulations to control manure use on farms.

TABLE 9.2 Select Characteristics for Comparing Oligotrophic and Eutrophic Lakes[a]

Oligotrophic	Eutrophic
Steep banks	Shallow, broad shore (littoral) zone
Blue to green water, clear	Green/yellow to brownish water, limited transparency
Water low in plant nutrients and Ca^{2+}	Water rich in plant nutrients and Ca^{2+}
Benthic sediment low in organic matter	Benthic sediment rich in organic matter
Oxygen abundant at all times at all levels	Oxygen depleted in summer hypolimnion
Few plants in the littoral zone	Abundant littoral zone vegetation
Low phytoplankton (algae) production	Abundant phytoplankton mass
Profundal benthos quantitatively poor	Profundal benthos biomass great

[a] Source: Bolsenga and Herdendorf, 1993.

is fairly well established is that the vast majority of outbreaks have occurred in waterways that are heavily polluted with excess nutrients. In March 1997 the EPA announced plans to regulate manure use on all farms. Undoubtedly, one of the events which precipitated the EPA's decision was the Pfiesteria outbreak.

Lakes are classified along a continuum according to their trophic status. *Oligotrophic* lakes are at one extreme and are characterized by low nutrient content, low primary productivity, and low standing biomass (Table 9.2). Oligotrophic lakes typically have clear, cold water and sandy bottoms. Lake trout and smallmouth bass are some fish species found in healthy oligotrophic lakes. At the other end of the spectrum are *eutrophic* lakes. Eutrophic lakes are nutrient enriched and support excessive algal growth. They typically have murky, turbid, warm water and muddy bottoms. Fish species associated with eutrophic lakes are bullhead, carp, catfish, and sunfish, which are often less–desirable species from a recreational fishing point of view. Between the two extremes are *mesotrophic* lakes. Large lakes, like Lake Erie, can have different trophic states throughout the lake (Box 9.1).

ORGANIC CHEMICALS

Organic molecules are defined by a molecular structure consisting of covalently bonded carbon atoms. Once upon a time only nature produced such molecules. Today synthetic organic molecules are routinely made in the laboratory. Pollutants in this category include solvents, PCBs, dioxin, food additives, petroleum and its derivative products, e.g., gasoline, diesel oil and jet fuel, and chemical pesticides. Many organic chemicals are either hazardous or toxic, but the truth is that most chemicals have never been tested to determine their potential carcinogenic (cancer causing), mutagenic (genetic mutations), or teratogenic (birth defects) effects. There are currently over 4 million distinct chemicals, with

BOX 9.1 EUTROPHICATION OF LAKE ERIE

Lake Erie is the shallowest of the five Great Lakes with a mean depth of 18.9 m (62 ft). The western basin is very shallow averaging only 7.4 m, the central basin averages 18.5 m and while the eastern basin is the deepest at 24.4 m. The lake's drainage basin is 18,800 km^2 (22,703 mi^2) and in the last 200 years has been transformed from wilderness into urban, industrial, and agricultural land uses. Since the turn of the century eutrophication has aged Lake Erie perhaps 15,000 years (Bolsenga and Herdendorf, 1993). Cultural eutrophication is the primary water quality problem. Phosphorous enters the lake from municipal sewage, industrial wastes, detergents, and runoff from farms including fertilizers and manure. By 1970 parts of the lake were nearly dead biologically because DO levels were so low. Programs to control phosphorous have improved DO in the last 15 years. The figure below shows the general trophic status of the lake today.

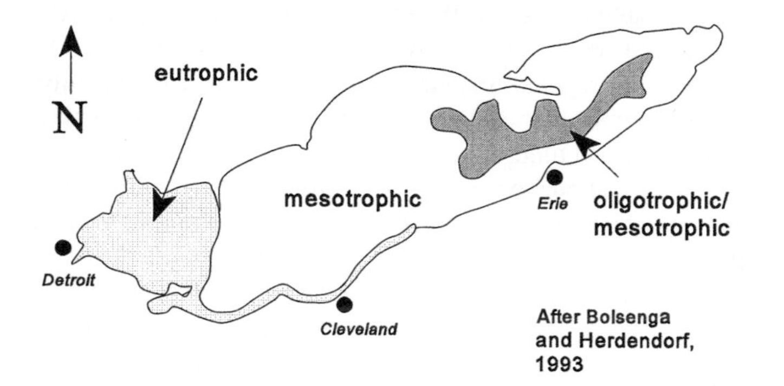

Water clarity has also improved dramatically since the early 1990s, but not because of any water pollution control program. The increased clarity has been due to the zebra mussel (*Dreissena polymorpha*), an alien species that has invaded Lake Erie. The zebra mussel is a very efficient filter feeder and is responsible for the increased clarity. Unfortunately, the zebra mussel has many negative economic and environmental impacts, such as encrusting water supply intake pipes and taking over the niche of native mussel species.

---

thousands of new ones created every year. Some 33,000 chemicals are commonly used in industrial processes as pharmaceuticals, food additives, and pesticides (Botkin and Keller, 1995).

Petroleum products can injure and kill fish by direct contact with their gills and scales. Some petroleum products are directly poisonous, and most cause suffocation, particularly of young fish and eggs that inhabit shallow waters where lighter–than–water oils tend to concentrate.

Chemical pesticides are some of the most notorious organic chemical pollutants. Pesticides include insecticides, herbicides, rodenticides, and fungicides. The use of chemical pesticides increased dramatically after World War II. They became an integral component of industrial agriculture in developed countries, and they were widely used to control infectious disease vectors in less–developed countries, like the *anopheles* mosquito that vectors malaria. In the late 1960s the agricultural use of chemical pesticides in less–developed countries started increasing as part of the "Green Revolution." The Green Revolution was an agricultural package that included new hybrid seeds, chemical fertilizers, chemical pesticides, and irrigation water.

Table 9.3 lists a few insecticides and herbicides and their environmental persistence. The chlorinated hydrocarbons tend to be the most persistent (nonbiodegradable) and can remain in the environment for years. When persistence is combined with high environmental mobility (carried by wind and water), chlorinated hydrocarbons can spread throughout the biosphere. This is why DDT is found in the fatty tissue of penguins in Antarctica even though DDT was never sprayed anywhere near Antarctica. The solubility of chlorinated hydrocarbons in fatty tissue along with persistence creates yet another problem called *biomagnification*. Biomagnification is when a substance at low concentrations in the soil or water is amplified through food chains, resulting in high concentrations in organisms at higher trophic levels. Only the chlorinated hydrocarbons in Table 9.3 biomagnify in the environment. In the United States some 25,000 different pesticide products are manufactured, exceeding 1 billion kilograms (2.2 billion pounds) every year (Miller, 1997).

TABLE 9.3 Some Insecticides and Herbicides and Their Environmental Persistence[a]

Pesticide type and examples	Environmental persistence
Insecticides	
Chlorinated hydrocarbons DDT, aldrin, dieldrin, heptachlor, toxaphene, lindane, mirex	High (2–15 years)
Organophosphates Malathion, parathion, methamidophos, methyl parathion, diazinon	Low to moderate (1–12 weeks) but some can last years
Carbamates Carbaryl (Seven), maneb, priopoxor, mexicabate, aldicarb	Usually low (days to weeks)
Herbicides	
Contact Atrazine, paraquat, simazine	Low (days to weeks)
Systemic 2,4–D, 2,4,5–T, silvex, daminozide (alar)	Mostly low (days to weeks)

[a]Source: Miller, 1997.

INORGANIC CHEMICALS

Metals such as mercury, lead, cadmium, arsenic, nickel, gold, silver, and copper are toxic or can become toxic after undergoing chemical changes in water. Copper is a trace element necessary for many forms of life, but copper sulfate is toxic to aquatic organisms. Like pesticides, some heavy metals biomagnify in living tissue. The concentration of heavy metals in the human body is called the *body burden.* Lead is a metal of special concern in drinking water because many older homes have lead–soldered pipes, and lead dissolves from the solder into the water. Lead is a known toxin that can cause brain damage, and developmental retardation in children. The average human body burden for lead in the 1970s was 150 mg in a 70 kg person, which was five times the body burden for the second–highest metal, cadmium (Lippman and Schlesinger, 1979). Most of the lead released into the environment in the last 50 years came from burning leaded gasoline. The replacement of leaded gasoline with unleaded gasoline in the 1970s has dramatically dropped lead emissions and consequently the concentration of lead in the environment. The U.S. Geological Survey has been monitoring the level of lead in rivers, lakes, and reservoirs across the country. The most recent study found lead levels in lakes and reservoirs down as much as 70 percent since the ban on leaded gasoline. Overall the lead concentrations in sediments from urban and suburban lakes and reservoirs are about double the baseline values prior to the 1950s and 1960s (Callender and van Metre, 1997).

In high concentrations iron can harm fish by clogging their gills. In drinking water iron imparts a disagreeable taste to the water, and may stain laundry and fixtures.

Dissolved solids are another group of inorganic chemicals. Calcium and bicarbonate ions enter water through chemical weathering of calcium–based rocks such as limestone and dolomite. As Equations (9.3) and (9.4) demonstrate, carbon dioxide dissolves in water, creating a weak carbonic acid. Carbonic acid reacts with calcium carbonate ($CaCO_3$) in the rocks, and the calcium (Ca^{2+}) and hydrogen carbonate (bicarbonate) (HCO_3^-) ions go into solution:

$$CO_2 + H_2O \Rightarrow H_2CO_3 \tag{9.3}$$

$$H_2CO_3 + CaCO_3 \Rightarrow Ca^{2+} + 2HCO_3^- . \tag{9.4}$$

Calcium and magnesium ions cause "hard" water. The term hard water came from the fact that the ions interfere with soaping agents, and it was hard to wash clothes in the water because the soap was less effective. Ion–exchange water softeners replace calcium ions with sodium (salt) ions.

Salt is itself a major pollutant in irrigated areas (Chapter 7). Other sources of salt pollution include oil and gas wells that allow deep brine to reach the surface, salt stored in piles and spread on roads to melt ice during the winter, and as a residual from the process of desalination.

Acids enter the water from mine drainage and by wet and dry deposition from the atmosphere. Acid mine drainage is a major problem in the abandoned coal fields of Pennsylvania and West Virginia. Acid mine drainage is considered the most serious water pollutant in Pennsylvania. The main culprit is iron sulfide (pyrite) minerals which react to produce sulfuric acid. Silt is another pollutant associated with mine drainage.

Sulfur dioxide (SO_2) is produced by burning sulfur–rich coal. Sulfur dioxide reacts in the atmosphere to form sulfuric acid (H_2SO_4). Sulfuric acid falls as acid rain, and is also dry deposited on land and water. Sulfuric acid from burning coal is the primary acid of concern in the eastern states. In the western states nitric acid (HNO_3) is a major concern. Nitric acid comes from chemical reactions involving the nitrogen oxides produced from burning petroleum–based fuels. There is no doubt that the acidification of lakes and streams disrupts ecosystems and eliminates species. Oligotrophic lakes are more susceptible to acidification because they lack bases, e.g., calcium ions, which buffer or neutralize acid. Acidification of the soil can accelerate nutrient leaching and mobilizes aluminum ions which are toxic to plants and fish.

The impact of acid rain on forests is less clear. Widespread forest damage in the late 1970s and early 1980s in northern Europe was attributed to acid rain. More recent evidence indicates that the decline in European forests was due more to drought and pests than acid rain. European forests have actually grown 20–30 percent in the last decade. Results from research in the United States are similarly inconclusive about the effects of acid rain on forests. The 10–year National Acid Precipitation Assessment Program (NAPAP) failed to detect any widespread damage to forests in the United States attributable directly to acid rain. New research on biogeochemical cycles suggests that acid rain is a more complex phenomenon than previously realized. We have reduced sulfur and nitrogen emissions through air pollution regulations. However, at the same time particulate matter (dust) in the atmosphere has been decreasing. Particulates are decreasing for a number of reasons, including air quality regulations. Particulate matter is composed of chemical bases such as calcium carbonate and magnesium carbonate. These bases act to neutralize acid rain in the air and on land. It may be that the reductions in atmospheric bases have been large enough to counteract the expected environmental benefits from reducing acid rain (Hedin and Likens, 1996).

HEAT

Heat can be harmful to ecosystems because aquatic organisms have specific tolerance ranges for water temperature. An increase from normal temperature of only 1.5 C° for lake fish and 3 C° for river fish is sufficient to cause migration away from the warmed area (Botkin and Keller, 1995). Excess heat interferes with fish metabolism and affects feeding, growth, and reproduction. Heat also reduces the solubility of DO in water (Figure 9.5), while at the same time speeding up biochemical reactions. Both of these intensify the DO deficit from oxygen–demanding wastes.

The most important sources of thermal pollution are discharges from industrial processes, especially cooling water from thermal electric power plants. In Chapter 4 we saw that cooling water for thermal power plants is the single largest withdrawal of water. The EPA has required the use of cooling towers and holding ponds to lower water temperature before discharge back into the receiving waters. Cooling towers increase evaporation and so some western states have objected to their use. This is a good example of a trade–off between water quality and water quantity concerns.

Urbanization and channel modification also elevate stream temperature. Runoff from urban land surfaces in the summer is warmer than that from natural surfaces. Straightening channels and removing streamside vegetation eliminates deep pools and reduces shade, thereby increasing water temperature.

RADIOACTIVE MATERIAL

Radioactive materials spontaneously decay emitting alpha or beta particles, or gamma radiation. All of these are considered *ionizing radiation* because they can dislodge electrons from atoms creating ions which may react with and damage living tissue. Damage to DNA may lead to cancer. Like pesticides and heavy metals, radioactive isotopes may biomagnify. Some sources of radioactivity in water are natural such as radon–222. Radon–222 is a colorless, odorless gas expelled naturally from rock and soil. The primary concern with radon–222 is as an indoor air pollutant seeping into homes through the basement and foundation. But water provides another pathway for radon to enter the home. Radon–222 dissolves in groundwater and escapes into the air when the water is used. In an enclosed space like a shower stall the concentration of radon may reach unsafe levels.

Nuclear power plants are potential sources of radioactive waste pollution of water. In the United States normal operation of commercial power plants has caused few instances of serious water pollution by radioactive isotopes. This has definitely not been true of the normal operation of government facilities that manufactured and processed radioactive material for atomic weapons. Intentional and accidental releases of radioactive material at weapons facilities have resulted in some of the most severe pollution anywhere on Earth. The 1450 km^2 (600 mi^2) Hanford, Washington, weapons complex is one of the most extreme examples in the United States. The Hanford site has widespread radioactive contamination of the soil, groundwater, and surface waters (see Box 9.2). In addition to Hanford, radioactive contamination exists at nuclear weapon facilities in Colorado, Idaho, Kentucky, Ohio, South Carolina, and Texas.

One of the most pressing issues facing the nuclear power industry and the federal government is the long–term storage of radioactive wastes. *High–level wastes* contain high levels of radioactivity and isotopes having half–lives of tens of thousands of years. Constructing a facility to safely contain high–level wastes for thousands of years is unprecedented in human history. As yet there is no

BOX 9.2 RADIOACTIVE POLLUTION AT HANFORD, WASHINGTON

The legacy of pollution from the Cold War at Hanford is staggering. In addition to dozens of sites where soil, groundwater, and surface-water pollution has already occurred. Hanford has 177 huge underground storage tanks containing a witches brew of highly radioactive waste, dozens of tons of plutonium, five gigantic and profoundly contaminated buildings, and 21,000 tons of spent nuclear fuel.

The ultimate cost to clean up the Hanford site is really unknown because there is no consensus on how clean is clean enough. For example, millions of dollars could be spent pumping and treating contaminated groundwater, and it would never be completely clean. A working estimate is that over the next 75 years it will cost at least $50 billion dollars to clean up the Hanford site. And Hanford represents only about one-fifth of the Department of Energy's (DOE) overall program to close down nuclear weapons facilities. Total cleanup costs range from $230 billion to half a trillion dollars. The cost of the cleanup at all DOE sites represents the largest single financial liability of the federal government with the exception of the federal deficit. This will be the largest public works project in our history.

Massive uncertainty is not limited to the cleanup costs. There is no consensus even on what should be cleaned up, or how the cleanup process should proceed. Overlapping regulations have complicated and frustrated the effort. The Tri-Party Agreement (TPA) between the state of Washington, the EPA, and the DOE governs the overall cleanup effort, but it complicates the process by placing Hanford under several environmental statutes. The two most important are RCRA and CERCLA (Superfund). Some nuclear reactors will be deactivated and decontaminated under RCRA guidelines, while others will follow different guidelines stipulated in CERCLA. As a result, the same type of cleanup will be done two different ways, and the wastes disposed of in different facilities.

One of the most daunting problems is simply where to store all the waste. The TPA requires much of it to be vitrified (encased in glass) and stored in another state. Vitrifying the high-level waste in the 177 storage tanks would create between 20,000 and 60,000 vitrified logs. The only high-level storage facility under investigation is at Yucca Mountain, Nevada. And that facility, if it ever opens, would have a capacity of only 6000 such logs, and it would have to accept waste from dozens of other facilities.

(Zorpette, 1996)

accepted plan for storing high–level wastes. The most promising site, from the federal government's perspective, is under Yucca Mountain in Nevada; however, the state of Nevada does not share the federal government's view. Evaluation of the Yucca Mountain site is years behind schedule and fraught with controversy.

The federal government would like to store *medium–level* waste at an underground facility called the Waste Isolation Pilot Plant (WIPP) near Carlsbad, New Mexico. Medium–level wastes do not contain as much radioactivity as high–level wastes. The transuranic wastes would be entombed within vaults carved into natural salt structures thousands of feet underground. These salt structures are geologically quite old and therefore thought to be tectonically stable.

The least dangerous waste category is *low–level wastes* produced by laboratories, medical facilities, and universities. Low–level wastes are typically buried in landfills. For all categories of radioactive waste, one of the most important considerations in evaluating its storage underground is whether the material could ever come in contact with groundwater and migrate to the surface.

In the former Soviet Union radioactive contamination of ground and surface water is widespread. The Soviets routinely dumped millions of cubic meters of highly radioactive wastewater into rivers, lakes, and wells in and around their nuclear weapons complex east of the Ural Mountains. They also dumped tons of radioactive material from submarines and icebreakers on and around the Kola Peninsula in the northwest. And who can forget Chernobyl, the world's worst nuclear power plant accident, which contaminated soil and water over thousands of square miles?

WATER QUALITY MANAGEMENT

The Clean Water Act (CWA) (1972) heralded the real beginning of the federal government's effort to reduce pollution and improve the quality of the Nation's waters. The goal of the CWA was to make the Nation's waters fishable and swimmable by 1983. The main thrust of the 1972 act was to control *point source* pollution.

POINT SOURCE POLLUTION

Point sources are discrete, identifiable sources such as tanks, pipes, or ditches, and are primarily associated with industries and municipal sewage plants. Section 208 of the CWA recognized that "areawide" or nonpoint sources of pollution were a problem, but the act did very little beyond acknowledging their existence. Pursuant to the CWA new municipal wastewater treatment plants were constructed and existing plants were upgraded. The municipal treatment plants were largely paid for with federal dollars through a massive construction–grant revolving fund.

The states, with approval from the EPA, designate beneficial uses of their waters, the water quality standards necessary to support those uses, and the amount and types of pollutants that may be discharged consistent with the designated uses. (These designated beneficial uses are unrelated to the beneficial uses in western water law.) The CWA says wastes cannot be discharged into surface waters without a National Pollution Discharge Elimination System (NPDES) permit. The NPDES permit specifies the types and amounts of pollutants that may be discharged, although a study in 1993 found that 7000 major industries found it cheaper to pay repeated fines for violating their permits than eliminate such pollution (Miller, 1997). Water quality is evaluated through a monitoring program run by the states, territories, Indian tribes, District of Columbia, and interstate water commissions. Water quality is assessed biennially and the data are submitted to the EPA in 305(b) reports. The EPA combines the individual 305(b) reports and produces a national water quality inventory report for Congress.

The focus on point sources, and our cultural proclivity to avoid regulations that require changing our behavior, favored a technology–based "cleanup approach" to water pollution control. Today industries treat their waste or send it to a municipal treatment plant. Certain industries are required to pretreat their waste to remove hazardous or toxic chemicals before sending it to the municipal facility. Controls on industrial pollution in 22 industries have reduced releases of toxic organic pollutants 99 percent, or nearly 660,000 pounds per day, since 1972. Releases of heavy metals have been reduced 98 percent, or 1.6 million pounds per day. If the EPA's estimates are correct, industrial pollution controls now prevent the release of nearly 1 billion pounds of toxic pollutants each year (Alder, 1993).

The crown jewel of the cleanup approach has been the municipal wastewater treatment plant. The proportion of the U.S. population served by municipal wastewater treatment plants increased from 47 percent in 1970 to 74 percent in 1985 (WRI, 1992).

Municipal Waste Treatment

Municipal wastewater treatment today is a multistage process (Figure 9.11). *Primary treatment* is mechanical treatment of wastewater using screens, comminutors (cutting blades), and settling basins to remove large floating debris and suspended material. After the primary settling stage the wastewater goes on to *secondary treatment*. Secondary treatment is mainly biological treatment to remove BOD. The wastewater enters an aeration tank where bacteria decompose the BOD. An alternative to an aeration tank is a *trickling filter*. Trickling filters are solid media having a large surface area, like a bed of rocks or honeycombed plastic. Bacteria growing on the media consume BOD as the wastewater slowly trickles over the surface. The objective of secondary treatment is to consume the BOD in the treatment plant rather than in the stream, lake, or ocean. Following the aeration tank is another period of settling where the leftover sludge and other suspended solids settle out. An *activated sludge* system returns some of this sludge back to the aeration tank, in effect reseeding the tank with bacteria. The

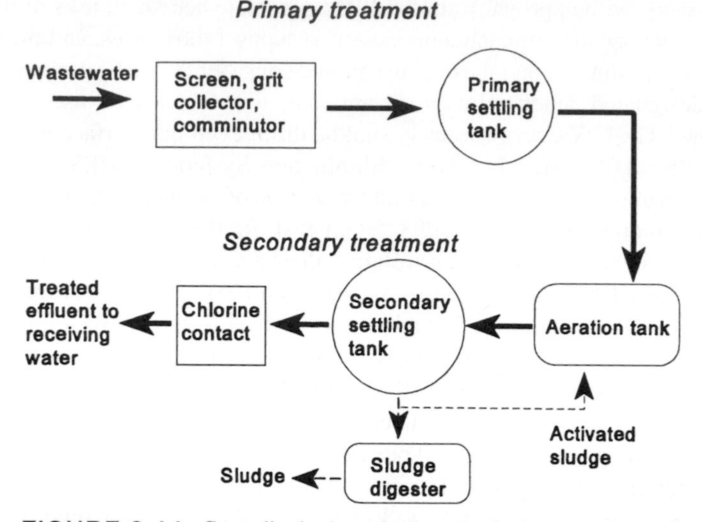

FIGURE 9.11 Generalized schematic diagram of primary and secondary wastewater treatment.

remaining sludge is sent to a digester where it is broken down further. The sludge is then dewatered and dried. From here it might be sent to a landfill, incinerated, or used as fertilizer. The use of sludge for fertilizer is often controversial. The most serious concerns are over pathogens and the potential accumulation of heavy metals in the soil, because heavy metals are not removed by primary or secondary treatment. Where sludge is used on farms it is usually restricted to animal feed crops. The last step in the treatment process is chlorination of the effluent to kill pathogens. The treated wastewater is then released. Most municipal wastewater systems consist solely of primary and secondary treatment.

Tertiary treatment is less common. This is chemical treatment to remove organic or inorganic chemical pollutants, like phosphorous and nitrogen compounds. Tertiary treatment involves adding chemicals to the wastewater, causing pollutants to flocculate and precipitate out of the water.

The economics of wastewater treatment follow the curves in Figure 5.9. Primary treatment is the least expensive and removes perhaps two–thirds of the suspended solids and one–third of the BOD. The marginal costs for secondary treatment are much higher. Primary and secondary treatment together remove up to 90 percent of the suspended solids and BOD. Tertiary treatment is more expensive still and is economically justified only in extreme circumstances. But regardless of the economics, tertiary treatment may be required by government regulation.

We have made significant progress in controlling point source pollution, and yet our waters are still impaired. The EPA estimates that the unmet needs for municipal waste treatment total \$110 billion (1990 dollars). Combined sewer overflows alone still discharge 3 to 11 billion pounds of organic material into our waters each year (Alder, 1993). As disturbing as these statistics sound, the most serious water quality problems today come not from point sources, but from nonpoint sources (NPS).

NONPOINT SOURCE POLLUTION

NPS are now the largest cause of water quality impairment in the United States. Nonpoint pollutants comes from atmospheric deposition, contaminated sediments, and land uses that generate polluted runoff like agriculture, urban land uses, logging, construction sites, and on–lot sewage disposal systems. NPS pollution also encompasses activities that either change the natural flow regime of a river or wetland, or result in habitat disturbance. NPS pollution is the reason that 36 percent of the surveyed rivers miles, 37 percent of the surveyed lake acreage, and 37 percent of the surveyed estuarine acreage are not clean enough to support their designated uses (USEPA, 1996b). Based on 1994 data from 305(b) reports, NPS pollution is the number–one source of pollution to rivers, lakes, and estuaries. Runoff from agriculture is most important for rivers and lakes, while urban runoff is the most important for estuaries (Table 9.4). Agricultural NPS pollution was responsible for 60 percent of the degraded river miles surveyed by the states and other jurisdictions. Municipal point sources were the second–most important cause of impairment to rivers, lakes, and estuaries. According to the EPA the most common NPS pollutants are sediment and nutrients washing off farms, animal feeding operations, construction sites, logging operations, and other disturbed areas. Other common NPS pollutants include bacteria, pesticides, salts, oil, grease, toxic chemicals, and heavy metals (USEPA, 1996c). NPS pollution is widespread because it can occur any time we disturb the land surface.

Section 319 of the amendments to the Clean Water Act in 1987 established the Nonpoint Source Management Program. In addition, the 1990 Coastal Zone

TABLE 9.4 Top Five Sources of Water Quality Impairment[a]

Rank	Rivers	Lakes	Estuaries
1	Agricultural runoff	Agricultural runoff	Urban runoff/storm sewers
2	Municipal point source	Municipal point source	Municipal point source
3	Hydrologic/habitat modification	Urban runoff/storm sewers	Agricultural runoff
4	Urban runoff/storm sewers	Unspecified nonpoint sources	Industrial point sources
5	Resource extraction	Hydrologic/habitat modification	Petroleum activities

[a]Source: U.S. EPA, 1996b.

SNY MAGILL CREEK, IOWA—A NATIONAL MONITORING PROGRAM SITE

The Sny Magill Creek is located in northeastern Iowa. The creek is widely used for recreational trout fishing. Land use in the 22,780-acre watershed includes agriculture (row crops and pasture), forest, and forested pasture. There are some 140 dairy, beef, and swine producers in the watershed. The average farm is size 275 acres.

Sediment is the primary NPS pollutant and is harming the trout fishery. The goal is to reduce sediment loading to the creek by 50 percent. BMPs include sediment control basins and stream bank stabilization. Nutrients and pesticides are also a concern, so BMPs to reduce fertilizer and pesticide use are planned, along with animal waste management.

(Source: EPA, Section 319 National Monitoring Program: An Overview, March, 1995)

Management Act Reauthorization Amendments established the Coastal Nonpoint Pollution Program. Section 319 provides federal grants for NPS pollution control, and as of 1995 the EPA had awarded states, territories, and tribes $370 million.

In 1991 the EPA established a National Monitoring Program to scientifically evaluate the effectiveness of various NPS pollution control strategies (USEPA, 1996d). As of 1997 the EPA had approved 11 NPS monitoring projects in 11 states. The most commonly cited pollutants at the 11 projects were nutrients, sediment, and bacteria. The results from these projects, and other projects added to the program in the future, will be assembled into a national NPS pollution database available to users in other states and watersheds.

Control strategies for NPS pollution will almost always involve some change in land–use or resource management strategy. Land–use management practices to control NPS pollution are called Best Management Practices or BMPs. The soil conservation methods in Table 9.1 are examples of BMPs for agriculture. Other agricultural BMPs include stream fencing and reestablishing riparian buffer zones. Fencing streams excludes animals from the water, reducing direct nutrient and bacterial input and streambank trampling and erosion. Riparian buffers are believed to intercept surface and subsurface pollutants before they reach the water. The effectiveness of riparian buffers and other BMPs are not well established, which is precisely the reason for the National Monitoring Program.

WATER QUALITY MONITORING

We monitor water quality for different purposes, including regulatory compliance, for public safety, and to assess current conditions and trends. These different purposes may have different requirements in terms of what is monitored, the frequency of sampling, and the level of sensitivity required. Nonstandard procedures for water quality monitoring and reporting have been a problem. The inability to directly combine and compare different data makes it difficult to answer the most basic questions that concern the public. For example, are the Nation's waters cleaner and biologically healthier today than they were 20 or 30 years ago? Have we gotten our money's worth from the billions of dollars spent on water pollution control? Ward (1996) argues that data–collection programs have not provided answers to these types of basic questions. He maintains we have produced a lot of water quality *data,* but very little *information* that is useful to the public. With decreasing program budgets, water quality monitoring and data collection should be driven first and foremost by the informational needs of the public. After all, public funds pay for most of these programs.

The most extensive water quality monitoring program to date has been in support of the 305(b) reports required by the CWA. Even here the states and other jurisdictions do not use identical monitoring methods and criteria. They have favored a more flexible approach that accommodates some of the natural geographic variability. There is a fundamental trade–off between flexibility and data comparability. Without known and consistent survey methods, the EPA (or any user) must exercise care in comparing data or judging the accuracy of data from different reports. In 1992 an Intergovernmental Task Force on Monitoring (ITFM) was convened to create a strategy for improving water quality monitoring nationwide. A permanent successor to the ITFM, the National Monitoring Council, will produce guidelines and conduct support activities to improve the protocols for national water quality monitoring.

The National Water Quality Assessment Program

A more recent water quality monitoring program is the U.S. Geological Survey's National Water Quality Assessment Program (NWQAP) which began in 1991. The two goals of the NWQAP are to (1) describe the water quality status and trends in the Nation's surface and groundwaters, and (2) provide a scientifically sound explanation of the natural and human factors affecting water quality in different parts of the country. The NWQAP investigations are being conducted on 59 study units. These study units are representative river basins and aquifers throughout the country. More than two–thirds of the Nation's water use occurs within the study units, more than two–thirds of the people served by public water systems live within their boundaries, and they encompass about one–half of the Nation's land area (U.S. Geological Survey, 1996). Study unit investigations all have a similar program design and use standardized methods for data collection and analysis. Standardization allows the results to be directly compared and synthesized into larger–scale regional and national water quality assessments. These regional and national assessments are referred to as National Syntheses (see Box 9.3) and focus on national priority issues such as nonpoint source pollution, sedimentation, and acid rain (U.S. Geological Survey, 1996).

BOX 9.3 NATIONAL SYNTHESES ON PESTICIDES AND NITRATES

A review of existing information found that pesticides in the atmosphere were detected in nearly every sample analyzed. The prevalence of pesticides can be explained by their degree of use and their environmental persistence.

Based on an analysis of nitrate in streams at about 150 sites in 10 states in the Midwest, the occurrence of nitrate was statistically related to the amount of precipitation, streamflow, acreage in the basin planted with corn or soybeans, cattle density, and human population density. A large percentage of point source loads occur near cities, while nonpoint sources vary widely and depend upon precipitation and runoff characteristics. No single nonpoint source for nitrate dominates everywhere.

(Source: U.S. Geological Survey, 1996)

THE QUALITY OF OUR NATION'S WATER

The most recent national assessment of water quality is the *National Water Quality Inventory: 1994 Report to Congress* (USEPA, 1996b). This report is based on data from the 61 individual 305(b) reports submitted to the EPA in 1994. The following discussion comes from Section 1 of the report, "National Summary of Water Quality Conditions."

As mentioned above, under the CWA the states and other jurisdictions designate, with the EPA's approval, the beneficial uses of their waters. They measure the attainment of the CWA's fishable and swimmable goals by how well their waters support their designated beneficial uses. The EPA recognizes six general beneficial uses: *aquatic life support, fish consumption, shellfish harvesting, drinking water supply, primary recreation contact,* and *secondary recreation contact.* In addition to these general beneficial uses the states, tribes, etc., can set their own individual beneficial uses such as groundwater recharge or the preservation of cultural/religious significance of the water. Water bodies can have more than one designated beneficial use.

Monitoring for the 305(b) reports provides the data to determine whether or not, and to what degree, water bodies support their designated use(s). There are five categories of use support (Table 9.5). The two levels "fully supporting" and "threatened" are the waters that do meet their designated uses. The three levels "partially supporting," "not supporting," and "not attainable" together compose the waters that do not meet their designated uses and are therefore degraded or impaired. Table 9.6 shows the proportion of the Nation's waters that were actually surveyed for the 1994 inventory. Since the surveys were, by in large, not based on random sampling strategies the results apply only to the surveyed waters, and it is not appropriate in most cases to generalize the results to the entire Nation.

TABLE 9.5 Levels of Use Support for Assessing Water Quality[a]

Use support level	Water quality	Definition
Fully supporting	Good	Water quality meets designated use criteria
Threatened	Good	Water quality supports beneficial uses now but may not in the future unless action is taken
Partially supporting	Fair (impaired)	Water quality fails to meet designated use criteria at times
Not supporting	Poor (impaired)	Water quality frequently fails to meet designated use criteria
Not attainable	Poor	The state, tribe, or other jurisdiction has performed a use–attainability analysis and demonstrated that use support is not attainable due to one of six biological, chemical, physical, or socioeconomic conditions

[a] Source: U.S. EPA, 1996b.

RIVERS AND STREAMS

The total surveyed length (Table 9.6) of over 3.5 million miles represents both perennial and nonperennial streams and rivers. Nonperennial streams are dry during at least some part of the year. The lower 48 states have about 1.3 million miles of perennial rivers and streams. The surveyed length (615,806 miles) represents 17 percent of the total, and 48 percent of the perennial channels. Sixty–four percent of the surveyed stream miles met their designated beneficial uses (Figure 9.12). The states and tribes reported that bacteria and sediment were the most prevalent pollutants in the surveyed rivers, followed by nutrients and oxygen–

TABLE 9.6 Total and Actual Waters Surveyed for the 1994 National Inventory[a]

Water body	Total length or area	Surveyed length or area	Surveyed length or area as a percentage of total
Rivers and streams	3,548,738 miles[b]	615,806 miles	17%
Lakes, ponds, and reservoirs	40,826,064 acres	17,134,153 acres	42%
Estuaries	34,388 square miles	26,946 square miles	78%
Ocean shore waters	36,000 miles	5,208 miles	9%
Great Lakes shoreline	5,559 miles	5,224 miles	94%

[a] Source: U.S. EPA, 1996b.
[b] Includes perennial and nonperennial watercourses.

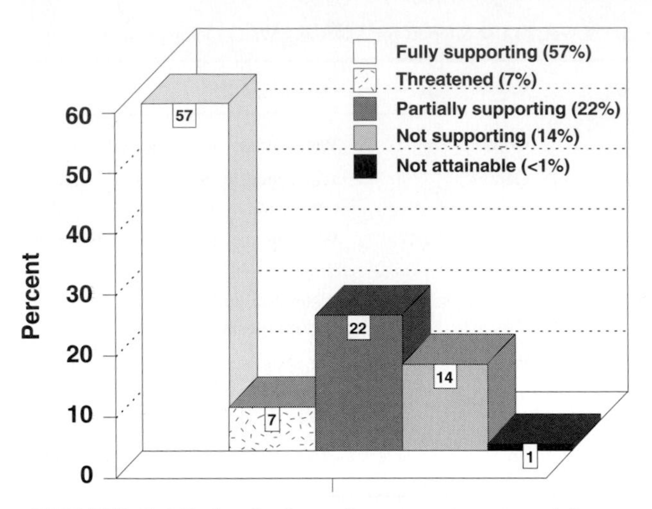

FIGURE 9.12 Levels of overall use support—surveyed rivers.

demanding wastes. Agriculture (crop production, grazing, and feedlots) was responsible for 60 percent of the impaired river miles, while municipal points sources impaired 17 percent, and hydrologic/habitat modification impaired another 17 percent.

LAKES, PONDS, AND RESERVOIRS

Forty–two percent of the total lake area of the Nation was surveyed. Sixty–three percent of the survey acreage met its designated beneficial uses, while 37 percent was impaired (Figure 9.13). Of the states and territories that reported types

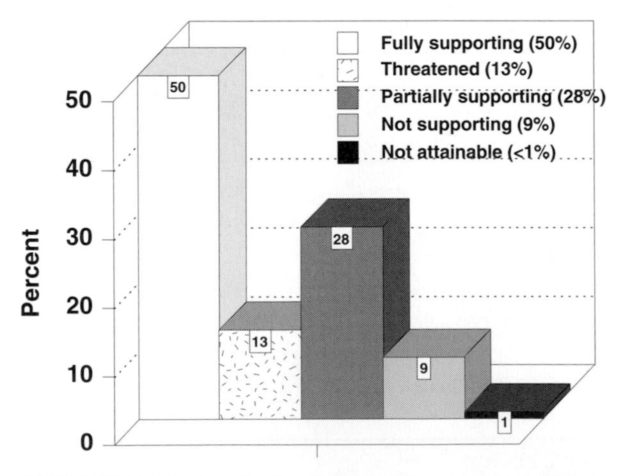

FIGURE 9.13 Levels of overall use support—surveyed lakes.

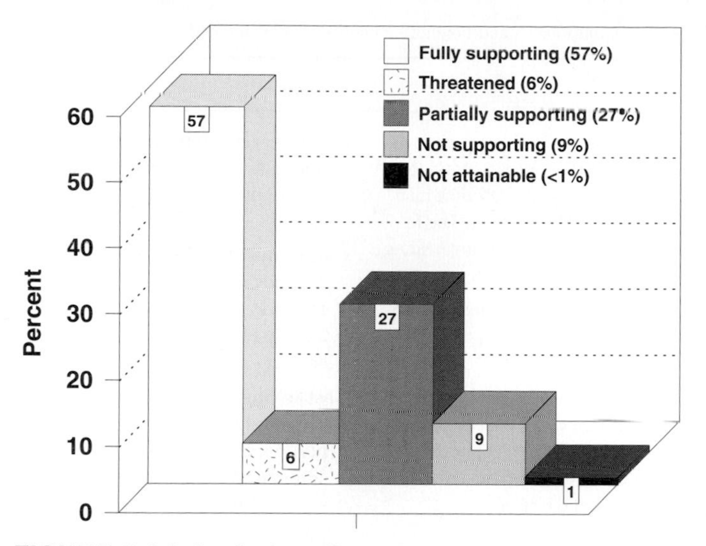

FIGURE 9.14 Levels of overall use support—surveyed estuaries.

of pollutants, nutrients were the most common pollutant affecting lakes. Siltation and oxygen–demanding wastes were also common lake pollutants. Forty–one states assessed the trophic status of their lakes. These states reported that 18 percent were oligotrophic, 32 percent were mesotrophic, and 36 percent were eutrophic. Again, agriculture was the leading source of lake water quality impairment and was responsible for causing half of the impaired lake acres.

ESTUARIES

More than three–quarters of the Nation's estuaries were surveyed. Sixty–three percent of the estuaries supported their designated uses; 37 percent did not (Figure 9.14). The most common pollutants were nutrients and bacteria. The number one source was urban runoff and storm sewer effluent, but municipal point sources, agriculture, and industrial points sources were important sources as well.

THE WATERSHED APPROACH TO WATER QUALITY

Since 1991 the EPA has supported the development and use of a *watershed approach* for improving water quality and the biological integrity of our waters. The dominance of NPS pollution, the necessity of local control over decisions, and the need for more creative and comprehensive approaches to persistent and emerging problems are driving the watershed approach. In Chapter 2 we saw that the basin–wide management concept evolved in the first decade of the 20th century. What is new today is the reemergence of watershed management driven by

TABLE 9.7 Components and Potential Benefits of a Watershed Approach to Water Quality[a]

Component	Benefits
A geographic **focus on watersheds**—nature's boundaries for water resources—provides a more logical basis for water–resource management than political boundaries. The focus is not just the water, but the surrounding land from which the water drains.	An **improved environment** due to greater attention focused on the resource and achievement of ecological goals rather than administrative requirements.
The continuous **improvement in the environment** and in water management programs based on sound science.	**Community building** through cooperation and collaboration as stakeholders gain a sense of common purpose. Stakeholder involvement helps create public support, lasting solutions, and economic stability.
The **involvement of all stakeholders** and affected parties in setting goals for the watershed. Stakeholders can take a comprehensive look at ecosystem issues and define goals to address local concerns.	**Cost savings** by streamlining and coordinating programs for monitoring, permitting and reporting, and predictability for the regulated community. The comprehensive, long–term nature of watershed planning provides the regulated community with a clearer picture of environmental policies.

[a] Source: U.S. EPA, 1996e.

water quality rather than by water quantity issues. According to the EPA the watershed approach has the components and benefits listed in Table 9.7.

The EPA is backing its evolution to a watershed approach by changing the way it administers financial assistance from its various water programs. The EPA has changed the NPS pollution grant program under Section 319 to provide states more flexibility to focus on high–priority problems on a watershed basis. The state revolving loan fund under the CWA (Section 604(b)(3)), which traditionally financed the construction of municipal wastewater treatment plants, has been used to finance NPS projects, habitat restoration, and urban stormwater projects. States are also using the fund to pay for assessments of environmental conditions throughout watersheds and for developing watershed protection plans. EPA is continuing to develop ways that the fund can be used in support of watershed protection activities. The agency has plans to begin offering Performance Partnership Grants. In essence Performance Partnership Grants will allow states and tribes to use money allocated through categorical grant programs to address high–priority and multimedia pollution problems within watersheds. Other proposals to encourage the watershed approach include longer review cycles (10 years instead of 5) for NPDES permits for states that use watershed management; longer cycles (5 years instead of 3) between state review and issuance of updated water quality standards; longer cycles (5 years instead of 2) between 305b assessment reports; and exemption from certain requirements under the Safe Drinking Water Act (USEPA, 1996e).

Two areas where the watershed approach is compelling are the management of wetlands and the use of market–based effluent trading. When wetlands mitigation programs are developed within the context of an entire watershed they can help communities balance the need for new development with the need to protect and maintain wetland habitat. Addressing wetlands management at the watershed level will likely increase the chances of successful approval of a compensatory wetlands mitigation program.

Effluent trading is a market–based approach to pollution control (Chapter 5). By creating tradable pollution credits industries are encouraged to take advantage of economies of scale and other cost–effective ways to control pollution. In theory, effluent trading is a more economically efficient means of achieving environmental quality goals than are uniform government regulations. Effluent trading can benefit from a watershed approach because such a program requires a comprehensive understanding of all pollution sources within a watershed (USEPA, 1996e). Point source effluent trading seems feasible, but trading in nonpoint source pollution is less certain.

ECOSYSTEMS AND WATER QUALITY

We have to do a better job at maintaining suitable habitat for both terrestrial and aquatic species. The human population is increasing rapidly, though the rates of increase in many countries are slowing. People need homes, roads to travel on, and stores in which to buy goods and services. In pursuit of our day–to–day existence we massively alter the natural landscape, and in the process alter or destroy natural habitats and entire ecosystems. Our enormous consumption of natural resources further disrupts habitats, while the waste products of resource use (actually wasted resources) often end up polluting our air, water, and soil. A pervasive consequence of all of our activities is the accelerated extinction of other species. It is generally recognized that we are now living in a period of human–caused mass extinction. The primary causes of this mass extinction are habitat alteration/destruction, hunting and fishing, introduction of alien species, pollution, predator and pest control activities, and the collection of exotic species for pets and decorative plants.

Estimates of human–induced extinction rates worldwide are rough, but range from 1000 to 10,000 times greater than the natural background rates (Lawton and May, 1995). How many species are there on Earth? No one knows. A low estimate is 5 million; a high estimate is 30 million. Only approximately 1.75 million species have been identified worldwide. The biota of the United States is relatively well known compared to the rest of the world. Something like 100,000 species native to the United States have been identified, though the actual total number will undoubtedly be much higher (Eisner *et al.,* 1995).

The Nature Conservancy recently undertook an assessment of the status of plants and animals in the United States (Stein and Flack, 1997a). The researchers reviewed the status of 20,439 species, about one–third of the known native species. They developed and ranked individual species using a seven–category scale

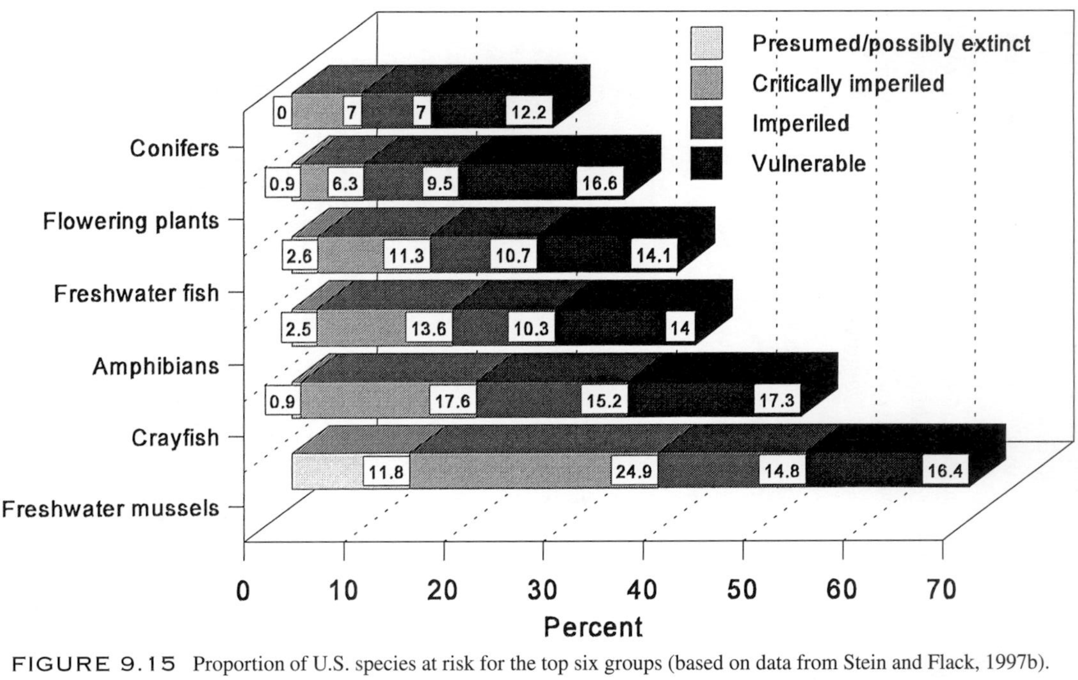

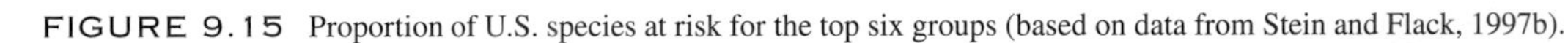

FIGURE 9.15 Proportion of U.S. species at risk for the top six groups (based on data from Stein and Flack, 1997b).

ranging from presumed extinct to secure. The researchers found that, "the most striking pattern evident in these assessments is the dire condition of those species that depend upon freshwater aquatic systems or wetlands for all or part of their life cycle" (Stein and Flack, 1997b, p. 10). Of the 13 total species groups, the top 4 in terms of risk were freshwater mussels, crayfish, amphibians, and freshwater fish (Figure 9.15). Freshwater mussels have been hit harder than any other species group. Nearly 12 percent of native freshwater mussel species are presumed or possibly extinct, while almost another 25 percent are critically imperiled. Clearly our approaches to water quality management in the past have not worked to protect aquatic organisms. The 1972 CWA's overtly anthropocentric goal of fishable and swimmable waters has not sustained aquatic ecosystems. Hopefully the trend toward watershed–based water quality management and the increasing interest in the natural values provided by water resources will improve the habitats for aquatic species. There is some disagreement, however, as to whether watersheds are the most appropriate spatial unit for water–related ecosystem management (see Omernik and Bailey, 1997).

The watershed approach to water quality can be integrated with new and emerging approaches to floodplain management, the last topic of this book. These new and evolving approaches to floodplain management have many of the same objectives—ecosystem and habitat preservation, water quality enhancement, and greater involvement of local stakeholders in decision making.

10

Floods and Droughts

The Human–Ecological Model of Hazards

Floods

The Physical System

Streamflow Analysis

The Human System

A Brief Review of Flood Management in the United States

Typology of Human Adjustments to Reduce Flood Losses

The Evolving Approach to Flood Mitigation

Drought

Drought Definition

Drought Analysis

Human Adjustment to Drought in the United States

In this chapter we examine two water–related natural hazards—floods and droughts. Socioeconomic systems adapt to the normal availability of water in a region. That is to say we *expect* a certain amount of water to be available to satisfy our needs for water supply, hydropower generation, recreational enjoyment, transportation, and ecosystem maintenance. This expectation is based on our experience with the natural cycle of water renewal accompanying the change of seasons. Experience also shows, sometimes painfully, that the hydrologic cycle is naturally variable. Too much water causes flooding, while too little for too long leads to desiccating droughts.

THE HUMAN–ECOLOGICAL MODEL OF HAZARDS

Once upon a time we viewed hazards as totally natural, that is to say as acts of God. This is no longer the case. People choose to live in floodplains, and they choose to live in areas that are more or less prone to drought. (From a political economy viewpoint a great debate can be engaged over just how much choice poor people have over their location.) It may sound trite but if people were not living in floodplains there would be no flood hazard threatening society. There would still be floods, of course, but their impacts on society would be minimal. This view of natural hazards as an interaction between a natural system (hydrologic cycle) and an established human socioeconomic system is the "human–ecological model" of natural hazards. The human–ecological model as a paradigm for geographical inquiry was advocated eloquently and early on by Harlan H. Barrows in his presidential address to the Association of American Geographers (Barrows, 1923). As a student of Barrows, Gilbert F. White adopted this paradigm to guide his research into human adjustments to the flood hazard in the United States (White, 1945). The human–ecological model is one way to view the interaction of extreme geophysical phenomena and human societies; other models have been proposed. Figure 10.1 summarizes the main components and processes in this systems–based, human–ecological model. Characteristics of the physical system that define the geophysical event include its frequency of occurrence, magnitude, duration, speed

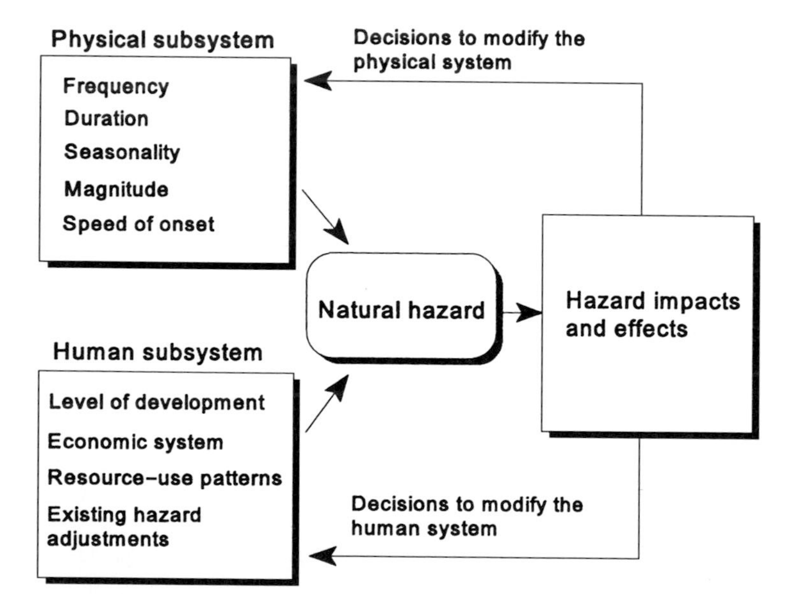

FIGURE 10.1 Simplified diagram of the human-ecological model of natural hazards.

of onset, and seasonality. Characteristics of the human system that define society's vulnerability include the level of economic development, the type of economic system, resource–use patterns (especially land use), and the types of hazard adjustments adopted by the society. It is the interaction of these two subsystems that defines the character and extent of the natural hazard. When an extreme geophysical event occurs, the societal impacts depend upon characteristics of both the physical and the human system. A relatively small flood may have little or no impact on a town protected by a levee, while another town similarly situated near the river, but lacking such protection, may suffer damage. The natural event triggers a systematic cascade of impacts, decision making, and planning, which may lead to the adoption of new hazard adjustments. Adjustments may focus on modifying the human system in such a way as to make it less susceptible to future floods. This might involve new building codes, land–use regulations, zoning ordinances, or even the relocation of existing structures off the floodplain. Alternatively the decision may be to modify the physical system by building a levee or flood control dam. The process of individual and societal adjustment to the flood hazard is a fine example of water–resource planning and decision making as discussed in Chapter 6. This chapter begins by examining floods. The second part of the chapter is an introductory overview of the drought hazard. For both hazards we discuss characteristics of the physical system, including methods used to analyze the phenomena, and then society's adjustments to these hazards.

FLOODS

THE PHYSICAL SYSTEM

Floods along rivers are termed *riverine floods,* while *coastal floods* occur adjacent to lakes and the ocean due mainly to wind–augmented wave action. Riverine floods are generated by many different physical mechanisms. The simplest flood–generating mechanism is when high–intensity precipitation exceeds the soil's infiltration capacity and generates hortonian surface runoff (Chapter 1). But lower–intensity, long–duration precipitation can slowly saturate the ground and generate floods by saturation overland flow and higher than normal groundwater levels. Floods in the Upper Mississippi and Missouri basins in the summer of 1993 were caused by a combination of intense rainfall from storm systems that persistently followed the same track month after month. High groundwater and sewer backups, not riverine flooding, caused damage to nearly 50 percent of the approximately 100,000 homes that suffered losses (Galloway, 1995). Other generating mechanisms for riverine flooding include ice jams, rapid snowmelt, and dam failure. Near record flooding along the Susquehanna River in Pennsylvania in February 1996 resulted from rapid melting of nearly 2 feet of snow. The rapid melting was caused by abnormally warm temperatures and rain falling directly on the snowpack.

Coastal lake flooding results from high lake levels and waves pushed on shore by strong winds. Shoreline erosion is a concomitant problem along shorelines composed of erodible sediments. Adjacent to oceans, tropical storms and hurricanes cause flooding through a combination of storm surge and heavy rainfall. Storm surge is the elevated sea surface pushed inland by strong winds. When combined with high tides, storm surges have been known to exceeded 25 feet. Storm surge exacerbates riverine flooding near the coast by damming the rain–swollen rivers and forcing them over their banks.

Riverine floodplains can be defined in different ways. A geomorphological definition uses topography and surficial material to delineate the floodplain. A *geomorphological floodplain* is the level land composed of alluvium adjacent to and actively worked by the river. This definition emphasizes the fluvial processes of sediment erosion, transport, and deposition. A *regulatory floodplain* is the area inundated by a flood having a certain probability of occurrence. The 100–year regulatory floodplain is the area inundated by the 100–year flood discharge (Figure 10.2). The 100–year floodplain is subdivided into the *floodway* and the *flood fringe*. The floodway is the area that can convey the 100–year flood with no more than a 1 foot rise above the unconstricted (undeveloped) floodplain. For communities participating in the National Flood Insurance Program (NFIP) new development is generally not allowed in the floodway. The flood fringe is adjacent to the floodway. Development is allowed in the flood fringe, but must not cause the 100–year flood to rise more than 1 foot above its naturally occurring elevation. Regulation of land use in the floodplain is a key element of the NFIP. In desert environments delineation of the regulatory floodplain along normally dry arroyos is more difficult. In these environments bank erosion may move the channel tens of feet laterally during a single flood. Many areas consider erosion and sediment deposition a more significant hazard than the water itself. The Federal Emergency Management Agency (FEMA) now prefers the term Special Flood Hazard Area (SPHA) rather than the 100–year floodplain. While the two terms define the same area, the change in terminology was apparently motivated by the desire to minimize the public's misperception of the flood risk (see Chapter 6).

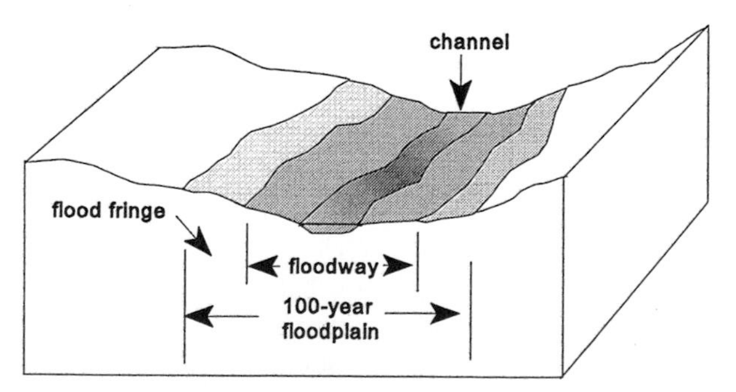

FIGURE 10.2 Cross-sectional diagram of the components of the 100-year floodplain.

STREAMFLOW ANALYSIS

Hydrologists use a variety of methods for analyzing streamflows and floods. The more advanced methods use computer models or scale models. The methods discussed here are three of the more basic and include the *rational method,* a *unit hydrograph,* and *flood frequency analysis.* Each method provides different information about floods, and each requires different types of data.

Rational Method

The rational method calculates peak discharge and should be used only on small (less than 300 acres) watersheds. The best results are for urban watersheds having a high percentage of impermeable surface, because the rational method assumes storm runoff is generated solely by hortonian overland flow. The equation for the rational method is

$$Q_{pk} = kCIA, \tag{10.1}$$

where Q_{pk} is peak discharge, C is the rational runoff coefficient (Table 10.1), I is the rainfall intensity, A is the basin area, and, k is a unit conversion coefficient.

TABLE 10.1 Values of the Rational Runoff Coefficient C for Storms with Return Periods of 5–10 years[a]

Land use	C	Land use	C
Urban areas		Rural areas	
Business		Sandy and gravelly soils	
Downtown areas	0.70–0.95	Pasture	0.35
Neighborhood areas	0.50–0.70	Woodland	0.30
Residential		Cultivated	0.40
Single family	0.30–0.50	Heavy clay soils or shallow soils over bedrock	
Multiunits, detached	0.40–0.60	Pasture	0.45
Multiunits, attached	0.60–0.75	Woodland	0.40
Apartments	0.50–0.70	Cultivated	0.50
Industrial	0.50–0.90		
Parks	0.10–0.25		
Unimproved areas	0.10–0.30		
Streets			
Asphalt	0.70–0.95		
Concrete	0.80–0.95		
Lawns, sand soil			
Flat, 2%	0.05–0.10		
Steep, 7%	0.15–0.20		
Lawns, heavy soil			
Flat, 2%	0.13–0.17		
Steep, 7%	0.25–0.35		

[a] Source: Viessman and Welty, 1985; Dunne and Leopold, 1978.

The conversion coefficient k depends upon the system of units used. For customary English units $k = 1.008$ and is usually assumed to equal 1.0. Peak discharge Q_{pk} is calculated in cubic feet per second when rainfall intensity I is in inches per hour and A is in acres. For I in millimeters per hour and A in square kilometers, $k = 0.278$, and Q_{pk} is in cubic meters per second. Some values for the rational runoff coefficient C are given in Table 10.1. The C coefficient is simply the fraction of precipitation that runs off. The C values in Table 10.1 are for rainfall of moderate intensity. For more intense, less frequent storms, the greater depth of water on the surface drowns out surface irregularities and increases runoff. It has been suggested that the values for C be increased for higher–intensity storms.

The rational method assumes rainfall is uniform in both space and time over the basin. For small basins this assumption is acceptable. An important requirement of the rational method is that the duration of the storm equal or exceed the *time of concentration* for the basin. The time of concentration is the time it takes water to travel from the most distant part of the basin to the outlet. This requirement ensures that the entire basin is generating runoff, thus producing the greatest peak discharge. An advantage of the rational method is that it calculates peak discharge using measurable basin properties and can be used on basins without streamflow records (see Example 10.1). The next two methods require historical records.

EXAMPLE 10.1

Calculate the peak discharge using the rational method and the following data.

Basin area = 200 acres
Total rainfall = 0.6 inches
Storm duration = 30 minutes (0.5 hours)
Land use: single–family residences cover 60 percent of the basin and commercial land covers the remaining 40 percent.

The rational method requires rainfall intensity in inches/hour, so the intensity is

$$I = (0.6 \text{ in}/0.5 \text{ hr}) = 1.2 \text{ in/hr}$$

From Table 10.1 the C value for single–family residential zones ranges from 0.30 to 0.50. Field inspection and experience are used to select the most representative value. For this example we use a median value of $C = 0.40$. Commercial land use is not listed in Table 10.1. Again, using professional judgement and field observation the hydrologist selects a representative value. Here we will use a value of $C = 0.80$ for commercial land. The final runoff coefficient is an area–weighted average of the two values. Each coefficient is weighted in proportion to the area of the basin it represents. The weighted value is $C = 0.6(0.40) + 0.4(0.80) = 0.56$.

Peak discharge is $Q_{pk} = 0.56(1.2)200 = 134$ cfs.

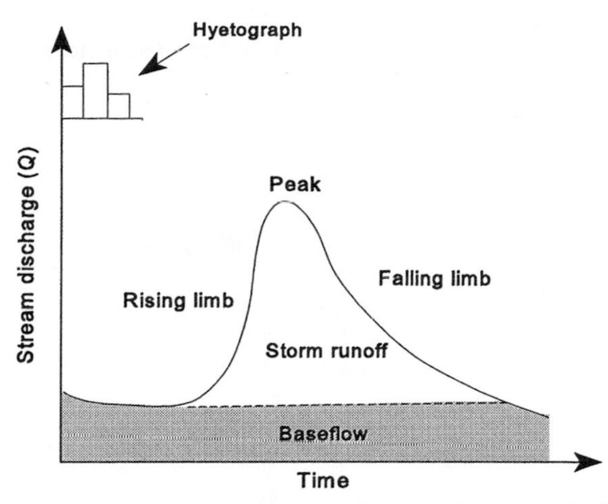

FIGURE 10.3 A total stream hydrograph composed of baseflow and storm runoff.

Unit Hydrograph

A stream *hydrograph* is the graph of streamflow versus time. Figure 10.3 is the hydrograph of a simple (single peak) runoff event. The figure also includes the hyetograph of rainfall which generated the flood wave. The hydrograph has a rising limb, a peak (Q_{pk}), and a falling or recession limb. Complex storms with multiple bursts of rainfall produce complex hydrographs with multiple peaks. The total stream hydrograph is composed of both groundwater–derived baseflow and storm runoff.

The shape of a hydrograph is controlled by both transient and permanent factors. Transient factors are related mainly to the storm and include the amount of rainfall, the storm's duration, and the speed and path of the storm over the basin. Transient factors change from storm to storm. Permanent factors influencing the hydrograph's shape include the basin area, basin shape, the density and pattern of the stream channel network, and the steepness of the slopes within the basin and of the channel itself. Unlike transient factors the permanent factors remain constant from storm to storm. The unit hydrograph technique assumes that for the same basin and the same duration storm, the differences between hydrographs are due solely to the amount of rainfall.

Developing a unit hydrograph for a drainage basin requires having paired records of precipitation and streamflow. I will not describe how to create a unit hydrograph; instead the focus here is on using a unit hydrograph once it has been developed. *A unit hydrograph is the hydrograph from a unit duration storm that generates a unit amount of runoff.* The definition says the duration of the unit hydrograph is the duration of the generating storm. A unit hydrograph can be developed for any duration storm. For example, a 2–hour unit hydrograph is the unit hydrograph for a 2–hour storm. A 3–hour unit hydrograph is the unit

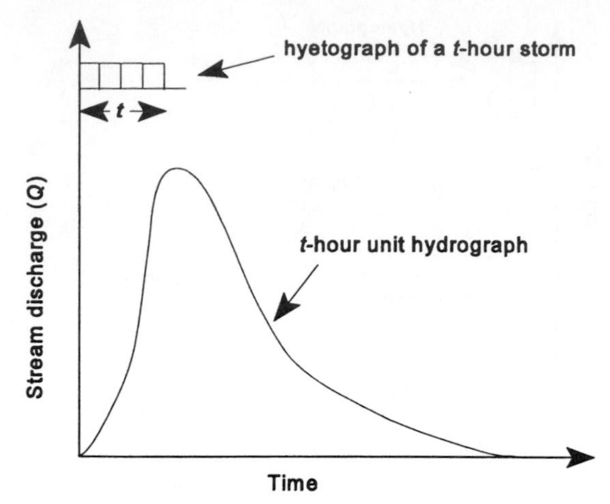

FIGURE 10.4 Hypothetical t-hour unit hydrograph for a basin.

hydrograph for a storm lasting 3 hours, and so on. You cannot use a 2–hour unit hydrograph with a 3–hour storm. Generally speaking then, the t–hour unit hydrograph is the unit hydrograph for a t–hour duration storm.

The second part of the definition deals with the runoff. The unit amount of runoff is 1 *inch* for English units and typically 1 *centimeter* for metric units. So, for example, the 2–hour unit hydrograph is the hydrograph of 1 inch of runoff from a 2–hour storm. Figure 10.4 is a hypothetical t–hour unit hydrograph. Notice that the unit hydrograph applies only to the storm runoff; no baseflow is included. Using the t–hour unit hydrograph to calculate streamflow from a t–hour storm is straightforward. Since the unit hydrograph gives the storm–generated streamflow from 1 inch (centimeter) of runoff, you just multiple the unit hydrograph ordinates by the depth of runoff from a particular t–hour storm. Example 10.2 demonstrates the use of a unit hydrograph. The unit hydrograph is a very popular method for analyzing streamflows. Compared to the rational method the unit hydrograph is much more informative since it calculates the entire hydrograph, not just the peak flow. The unit hydrograph is also more flexible as to basin size and has been used on basins as large as 2000 square miles. While unit hydrographs are derived using streamflow records, there are procedures for creating *synthetic unit hydrographs* using physiographic basin data. Synthetic unit hydrographs are explained in most standard hydrology textbooks.

EXAMPLE 10.2

Given in Table 10.2 and plotted as Figure 10.5 are the ordinate values of a hypothetical triangular–shaped 1.5–hour unit hydrograph.

TABLE 10.2 Hypothetical 1.5–Hour Unit Hydrograph

Time (hours)	Unit hydrograph (cfs)
0	0
1	15
2	30
3	20
4	10
5	0

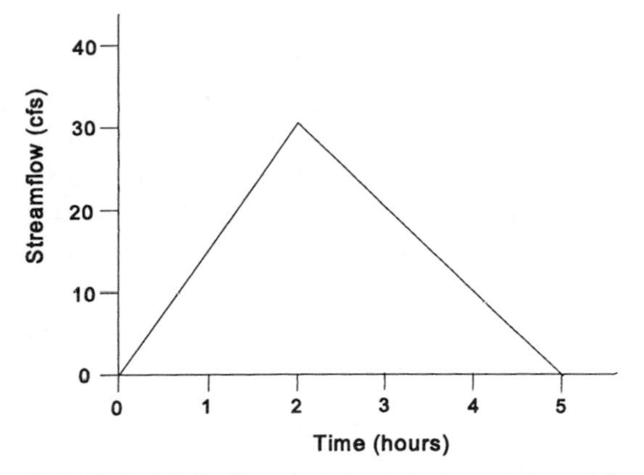

FIGURE 10.5 Hypothetical unit hydrograph for a 1.5-hour duration storm.

Calculate the total stream hydrograph using the following data:

storm duration = 1.5 hrs
precipitation = 3.0 in
infiltration = 0.5 in
baseflow = 5 cfs (constant).

The storm runoff (precipitation − infiltration) is 2.5 inches. Multiplying the storm runoff by the unit hydrograph ordinates in Table 10.2 (or Figure 10.5) gives the storm runoff hydrograph (see Table 10.3). Finally, adding 5 cfs of baseflow to the storm runoff hydrograph produces the total stream hydrograph.

TABLE 10.3 Calculation of the Total Streamflow Hydrograph Using the Unit Hydrograph

Time (hours)	Unit hydrograph (cfs)		Storm runoff (inches)		Storm runoff hydrograph (cfs)		Baseflow (cfs)		Total stream hydrograph (cfs)
0	0.0	×	2.5	=	0.0	+	5.0	=	5.0
1	15.0	×	2.5	=	37.5	+	5.0	=	42.5
2	30.0	×	2.5	=	75.0	+	5.0	=	80.0
3	20.0	×	2.5	=	50.0	+	5.0	=	55.0
4	10.0	×	2.5	=	25.0	+	5.0	=	30.0
5	0.0	×	2.5	=	0.0	+	5.0	=	5.0

Flood Frequency Analysis

Flood frequency analysis requires data on historical peak discharges and produces an estimate of their exceedence probability. The basic data for flood frequency analysis is the *annual maximum series.* For each year of record the single largest (maximum) instantaneous streamflow is collected. For n years of record there are n annual maximums, assuming of course there are no missing values. (For desert arroyos there can be many years when the maximum is zero.) Flood frequency analysis assumes (and requires) that the annual maximum values be independent of each other. Independent means that the maximum flow in one year does not influence the maximum flow in any other year. This was *not* the case in Chapter 6 where we used the Markov model to generate time–dependent annual average streamflows.

The basic flood frequency analysis ranks the annual maximum series from largest to smallest, and then assigns an estimate of the exceedence probability (p) to each ranked value. This is exactly the procedure we used for the frequency analysis of reservoir storage in Example 6.2. Weibull's plotting position equation (Equation (6.3)) can again be used to estimate the exceedence probability. A more sophisticated flood frequency analysis would fit the data to some theoretical probability distribution. In flood frequency analysis it is common to report the *return period* or *recurrence interval* (T) for a peak discharge as well as its exceedence probability. The return period is the average number of years between floods of this magnitude. Return period is just the reciprocal of the exceedence probability:

$$T = 1/p. \tag{10.2}$$

Thus, the 100–year flood has a 1 percent chance of being equaled or exceeded each and every year (see Example 10.3).

EXAMPLE 10.3

The data for this flood frequency analysis are the annual maximum discharges for the Conestoga River at Lancaster, Pennsylvania from 1929 to 1993 ($n = 65$ years). The data were ranked and assigned exceedence probabilities and return periods. Table 10.4 gives a partial listing of the ranked data. The first value in the table (50,300 cfs) is the maximum flood of record and was caused by Hurricane Agnes in 1972. You may notice that this flood was not ranked. This discharge was deemed to be an "outlier" and excluded from the analysis. The reasoning is that the return period for this event is much longer than would be assigned by the procedure used here. Figure 10.6 is the plot of discharge Q versus return

TABLE 10.4 Annual Maximum Discharges for the Conestoga River at Lancaster, Pennsylvania (1929–1993)[a]

Rank (m)	Q (cfs)	$p = m/(n+1)$	$T = 1/p$ (years)
	50300		
1	25300	0.015	66.00
2	22800	0.030	33.00
3	20500	0.045	22.00
4	17300	0.061	16.50
5	16300	0.076	13.20
6	15000	0.091	11.00
7	13600	0.106	9.43
8	13000	0.121	8.25
9	12600	0.136	7.33
10	12600	0.152	6.60
11	12400	0.167	6.00
12	11200	0.182	5.50
13	11200	0.197	5.08
14	11000	0.212	4.71
15	10800	0.227	4.40
.	.	.	.
.	.	.	.
.	.	.	.
55	4220	0.883	1.20
56	4040	0.848	1.18
57	3860	0.864	1.16
58	3530	0.879	1.14
59	3260	0.894	1.12
60	3240	0.909	1.10
61	3200	0.924	1.08
62	2900	0.939	1.06
63	2680	0.955	1.05
64	2130	0.970	1.03
65	1790	0.985	1.02

[a] Source: USGS, 1998.

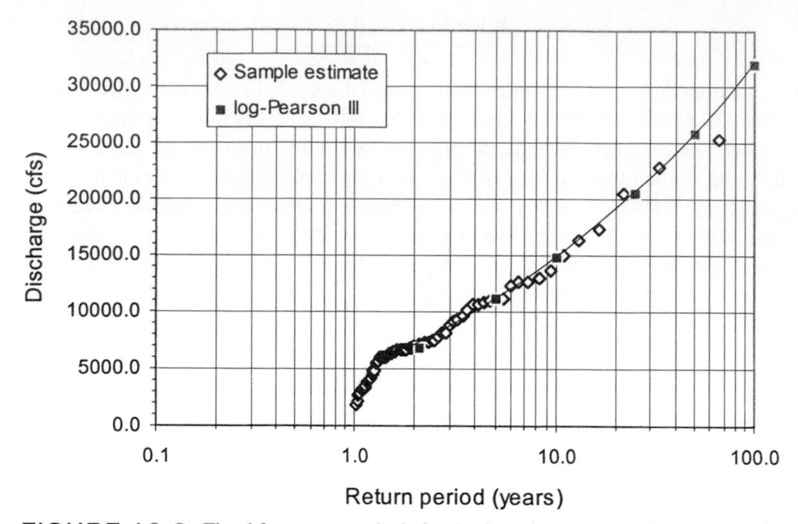

FIGURE 10.6 Flood frequency analysis for the Conestoga River at Lancaster, Pennsylvania for the period 1929 to 1993. The diamonds are discharge versus estimated return period. The squares and line are the plot of the log-Pearson III theoretical probability distribution fitted to the data.

period *T* on semilog graph paper. The line on the graph is the plot of the log–Pearson III probability distribution fitted to the data. The log–Pearson III distribution is the theoretical probability distribution for flood frequency analysis preferred by federal government agencies.

What value does the log–Pearson III distribution calculate for the 100–year flood discharge?

What would the return period be for the Hurricane Agnes flood discharge (50,300 cfs)?

THE HUMAN SYSTEM

People live and work near rivers to take advantage of the benefits provided by the water and the floodplain. Access to water for water supply, waste disposal, and water–based transportation, level land and good soils for building sites and agriculture, and the recreational and aesthetic pleasure of being near the water are some of the many benefits. We tend to forget that the floodplain belongs to the river, and the river must occasionally use it to store excess water. When the river wants to use its floodplain, we may or may not be able to prevent it, and there may or may not be damages to society depending upon the characteristics of the human system (Figure 10.1). If the benefits of using the floodplain exceed the costs of periodic flood damage, then using the floodplain is a rational use of resources. A major problem confounding the assessment of benefits and costs has been governmental policies that distort the true cost of living on floodplains. When the government offers disaster assistance and/or low–interest loans following a flood, then taxpayers are paying some of the cost of floodplain occupation—taxpayers are subsidizing the people living on the floodplain. This "distributive" economic policy lowers the direct costs of flooding to the floodplain residents by distributing the cost among taxpayers at large. It may sound incredibly hard–hearted but if the government would steadfastly refuse to help flood victims, they (victims) would have a more accurate picture of the true cost of living in harm's way. Of course, this viewpoint is not very popular and it seems cold and uncaring, especially following a disaster. So instead of making those directly at risk shoulder the full burden of that risk, the government comes to the rescue and helps the residents recover and rebuild. Unfortunately this has resulted in a continuing cycle of flood damage, rebuilding, flood damage etc.

One of the objectives of the National Flood Insurance Program was to have those directly at risk bear more of the cost of that risk by purchasing flood insurance. Private insurance companies do not offer flood insurance because they know they will lose money. The NFIP offers federal (taxpayer–subsidized) flood insurance to communities that agree to adopt floodplain land–use regulations. Over the years the amount of the subsidy has decreased and federal flood insurance today is fairly expensive. Following a flood disaster insurance payments are used to make repairs rather than government loans and grants. According to NFIP policy, those without flood insurance are simply out of luck when they are flooded and suffer damage. That is the theory; reality is quite different. The problem has been the government's own lack of resolve in sticking to its policy. In the wake of a disaster pleas for help have proven too hard to resist, and the government comes to the rescue whether or not people are insured. Of course doing this undermines the very purpose of the NFIP. Why buy insurance if you know you will get "free" federal assistance? The 1993 flood along the Upper Mississippi may have marked the beginning of real change in federal policy, and there now seems to be a genuine desire to break the cycle of repeated flooding and rebuilding. In the wake of the 1993 flood a record number of people asked for federal assistance, not to rebuild their homes, but to permanently relocate out of the floodplain. By 1995 some 8000 structures and 20,000 people had been relocated. Another change following from the 1993 flood was the National Flood Insurance Reform Act (NFIRA) (1994). The NFIRA strengthened the NFIP by establishing a grant program for flood mitigation planning. The NFIRA also codified the Community Rating System (CRS). The CRS is an incentive program whereby communities that exceed the minimum requirements of the NFIP receive reductions in the insurance premiums for their residents. Approximately 940 communities were participating in the CRS in 1997, and the flood insurance policies in these communities represent over 60 percent of all NFIP flood insurance policies in place (FEMA, 1997). Overall, individual purchase of flood insurance has been disappointing. Penetration of flood insurance into the target market (floodplain occupants) is only 20–30 percent of those at risk (Galloway, 1995).

A BRIEF REVIEW OF FLOOD MANAGEMENT IN THE UNITED STATES

In the 1800s and early 1900s dealing with floods was the responsibility of private individuals, local governments, and quasi–governmental levee and conservancy districts. The federal government slowly became involved and it was not until 1917 that floods were considered a water–resource problem worthy of full federal involvement from coast to coast. The natural–resource philosophies of exploitation and wise use that dominated at the turn of the century, combined with a "man over nature" ethic, defined our approach to floods—that approach was *flood control.* We were going to control rivers—tame a rude nature—using human ingenuity and the power of new technology. In the fight against nature our primary weapons were structural devices such as levees, diversion channels, and dams. While federal legislation at the time recognized watershed treatment to reduce runoff, and even nonstructural measures, the emphasis was clearly on flood control through single–purpose structural solutions. The primary federal agency in the fight against floods was the Corps of Engineers, but the Tennessee Valley Authority, the Bureau of Reclamation, the U.S. Forest Service, and the Soil Conservation Service all began including flood control as a component of their water projects. In addition to flood control were federal programs for disaster relief. The first federal disaster act was passed in 1950, and the Small Business Administration began disaster relief programs in the 1950s as well (FEMA, 1992)

By the 1960s the federal government had spent $7 billion trying to control floods, and while acknowledging the effectiveness of these structures, the disconcerting reality was that the annual toll from flooding continued to rise. Structural protection had encouraged further encroachment onto the Nation's floodplains and increased the total property at risk. This so–called flood control paradox forced a critical examination and rethinking of the structural approach. In 1966 a Congressional Task Force on Flood Policy issued its report *A Unified Program for Managing Flood Losses* (U.S. Congress, 1966). The report made recommendations for legislation, specific studies, and new programs for collecting and disseminating flood–related information (FEMA, 1992). Ten years later the Water Resources Council updated and revised the unified program in its report, *A Unified Program for Floodplain Management* (WRC, 1976). The name change was subtle but it signaled a significant change in the program's emphasis by recognizing flooding as more than just a problem of flood losses—flooding was seen as a much broader natural–resource management issue. The report highlighted the lack of intergovernmental coordination as the most serious impediment to improved floodplain management. Further revisions to the unified program came in 1979 and 1986. Periodic revisions were necessary because of the program's effectiveness, which is to say the adoption and implementation of recommendations quickly made the existing program obsolete. Acceptance and adoption of comprehensive flood hazard mitigation was especially impressive by state and local entities. Reflecting the increasing concern with the environment, revisions to the program in 1979 explicitly identified the importance of preserving and restoring the "natural values" of floodplains.

Throughout the decade of the 1980s and continuing into the 1990s comprehensive floodplain management continued to evolve. Instead of primarily a builder of large flood control projects, the federal government's role evolved into one of a provider of technical information and assistance to local and state governments. This increasingly decentralized approach has fostered integrated structural and nonstructural approaches tailored to local circumstances in thousands of communities nationwide. In established urban areas structural approaches are the most efficacious, but in newly developing areas nonstructural approaches centered on land–use regulation offer the best adjustment over the long term.

Following the 1993 flood a federal interagency committee was formed to review the Nation's floodplain management activities (Galloway, 1995). The committee reviewed the events that combined to create the 1993 flood, and the Nation's approach to floodplain management in general. Overall, the committee saw a need to improve intergovernmental coordination, increase state and local government's stake in floodplain management, improve and streamline the NFIP, and update national goals and objectives to more accurately reflect environmental and social values. A few of the committee's recommendations have been implemented, such as changes to the NFIP, but others are more problematic. One interesting recommendation was to reactivate the Water Resources Council and the river basin commissions that were eliminated by President Reagan in 1981. History continues to repeat itself.

TYPOLOGY OF HUMAN ADJUSTMENTS TO REDUCE FLOOD LOSSES

Researchers have identified the four general classes of adjustments to the flood hazard shown in Box 10.1. This adjustment typology is generic and is applicable to any hazard, not just floods.

Modify the Natural Event

The flood event can be modified at different points in the hydrologic system. *Structural protection* such as dams, levees, and channel modifications are used to affect streamflow. Drawbacks of structural measures include the facts that they encourage a false sense of security which can lead to unwise floodplain encroachment, they usually involve lengthy and complex planning and design, they can result in major losses when a flood exceeds the structure's design level (as happened with levees in 1993 along the Missouri and Mississippi), and they often cause detrimental impacts to the natural environment.

Watershed treatment aims to modify the flood–generation process earlier in the hydrologic cycle by reducing runoff from the land surface. Contour plowing, terracing, and strip cropping all slow surface runoff and promote infiltration. However, transforming natural land covers (forests and grasslands) into farmed

BOX 10.1 TYPOLOGY OF HUMAN ADJUSTMENT TO FLOODS

Modify the event
- Structural protection
  - Dams
  - Levees
  - Diversion channels
- Watershed treatment
  - Wetlands preservation and restoration
  - Contour plowing, terracing, strip cropping
  - On-site retention
  - Revegetation
- Weather modification

Modify the damage susceptibility
- Land use regulation
  - Safe land-use practices
  - Safe construction practices
- Emergency measures
  - Warning systems
  - Flood fighting
  - Flood preparedness plans

Modify the loss burden
- Flood insurance
- Disaster relief
- Low-interest loans
- Tax write-offs

Do nothing

fields can reduce a soil's infiltration capacity by compaction, breakdown of soil structure, and the reduction of organic matter.

Some of the beneficial functions of wetlands are floodwater storage and attenuation of downstream flood peaks. A report by the Illinois State Water Survey (Demissie and Kahn, 1993) estimated that for every 1 percent increase in wetland area within the basins they studied, peak flows from the basins are reduced an average of 3.9 percent. Whether or not more wetlands could have helped moderate the 1993 flood is debatable, but one can imagine that if the more than 19 million acres of wetlands that have been eliminated in the last 200 years from the Missouri and Mississippi basins north of Saint Louis were still in place, there might have been some effect on the flood. Illinois, Iowa, and Missouri, the states that suffered the most in damage the 1993 flood, have lost 85 percent or more of their original

BOX 10.2 WETLANDS ACREAGE CONTINUES TO DECREASE

In September 1997 the U.S. Fish and Wildlife Service released the first sample-based survey of the Nation's wetlands since 1990. The study showed a net loss rate of 117,000 acres per year between 1985 and 1995. The greatest losses were to freshwater forested wetlands of the South. The good news was that the rate of loss was nearly 60 percent slower than loss rates in the 1970s and 1980s, and 74 percent slower than loss rates from the 1950s to 1970s. But clearly the "no net loss" policy first articulated by President George Bush is not yet being met.

Agricultural activities were responsible for 79 percent of the total loss from 1985 to 1995. Urban and other development accounted for the remaining 21 percent.

Between the 1780s and the 1980s the country lost 54 percent of its original wetland area, or 221 million acres. This loss amounts to 60 acres per hour for 200 years.

(Source: U.S. Fish and Wildlife Service, News Release, September 1997)

wetlands (Faber, 1993). California has lost over 90 percent of its wetlands, and overall the lower 48 states have lost 54 percent. A recent assessment of the status of wetlands shows that total wetland acreage continues to decrease across the country, though the rate of decrease has slowed (see Box 10.2).

As a flood adjustment, *Weather modification* remains more theoretical than real. While there were experiments with seeding weather disturbances, i.e., hurricanes, a few decades ago, there are no plans—research or otherwise—to try modifying weather systems with the explicit goal of reducing their magnitude and frequency.

Modify the Damage Susceptibility

Adjustments aimed at the human system to reduce damage potential are at the core of the nonstructural approach to flood hazard mitigation. By restricting development in the floodplain we recognize the river's priority for using the floodplain to store floodwater. Land–use regulations are also used to preserve and restore natural floodplain values. Undeveloped floodplain provides habitat for wildlife, nutrient recycling, water purification, and recreational and scientific benefits. Safe land–use practices are land uses that can withstand flooding with little or no damage. Safe land uses include open space, parks, and recreational areas like hiking and biking trails. Safe construction practices include floodproofing and elevating structures above the flood level.

Emergency measures can reduce damages but their effectiveness depends upon the amount of lead time available, and whether individuals and the community have developed preparedness plans. For large riverine floods that take days to weeks to crest there may be ample time to move personal items out of the house or from the first floor to the second floor. There may also be time to organize flood–fighting efforts, e.g., sand bags and pumping equipment. Sufficient time is not always available, as when the levees failed along the Mississippi River in 1993, turning a normally slow process into one of rapid inundation. *Flash floods* give little time for emergency response. Flash floods are characteristic of steep mountain valleys and desert arroyos. Intense thunderstorms can send a wall of water down a mountain canyon with little or no warning. The Big Thompson Canyon flood in Colorado in June of 1976 killed 139 people. The wall of water came so quickly there was very little time to react. Some people tried to outrun the water in their cars, and most of these people died. The most effective response in a flash flood is to climb vertically up the canyon wall. Catastrophic dam failure is another cause of flash flooding.

Modify the Loss Burden

Leading the list of adjustments here is flood insurance. When (not if) people get flooded, their flood insurance policy pays for the property damage. The NFIP offers federal flood insurance to communities that agree to regulate floodplain land use. When the NFIP was initiated in 1968 few communities joined the program voluntarily. Record flooding in Pennsylvania in June 1972 from Hurricane Agnes was followed just one month later by a devastating flash flood in Rapid City, South Dakota. These back–to–back disasters prompted Congress to change the NFIP in 1973. Amendments to the program tied future federal disaster relief, as well as Small Business Administration and federally backed home mortgage loans, to community participation in the NFIP. The number of communities participating in the NFIP increased markedly in the following years.

Other measures to reduce the burden of flooding include disaster relief, low–interest loans, and tax write–offs. In theory these measures are now tied to participation in the NFIP, but as mentioned above, the government has not always followed its own policy and has offered assistance to insured and uninsured alike.

The last adjustment is to do nothing and just suffer the loss. This approach is less common in developed countries; however, worldwide this is probably the most common adjustment of all.

THE EVOLVING APPROACH TO FLOOD MITIGATION

Flood hazard mitigation now emphasizes a mix of structural and nonstructural adjustments, but there are still many problems. One ongoing problem is getting accurate and reliable information on areas at risk. Despite expenditures of over $870 million only a little more than one–half of the Nation's 178 million floodplain acres have been mapped. Unmapped areas are generally not subject to regulations restricting floodplain development. Even in mapped areas the maps are too general and do not provide reliable information on potential floodwater elevations. Still other areas have special risks, like the many western states that suffer high erosion hazard, areas subject to high groundwater, or rivers that are prone to ice–jam flooding. By and large these special risks are not incorporated into standard flood hazard maps. Interestingly, 31 percent of flood insurance claims have been paid for damages to property located outside the 100–year regulatory floodplain (Kusler and Larson, 1993).

Another continuing problem is mitigating future losses. In the 1980s, 30,000 structures (2 percent of all NFIP–insured structures) filed multiple (two or more) insurance claims, accounting for about 30 percent of the total claims (Kusler and Larson, 1993). Until the flood of 1993, there had been little progress in breaking the cycle of repetitive losses. That disaster and the major flood disasters that followed in California (winter of 1993–1994), Pennsylvania (winter of 1996), California again (winter of 1996–1997), and in North Dakota and Minnesota (spring of 1997), may provide the needed incentive and the resolve to get serious about mitigating future losses.

Multiobjective River Corridor Management

The typology of human adjustments to floods outlined above is limited in its emphasis on reducing flood losses to humans and society. A more comprehensive approach to floods and floodplain management is beginning to emerge. It is a view that sees floods and floodplain land use as part of a larger, more comprehensive water–resource system—one having multiple objectives. In addition to flood–loss reduction, floodplains should be managed to enhance water quality, to provide wetland and riparian habitats, to maintain ecosystem integrity, and to provide aesthetic and recreational opportunities. This more comprehensive view of floodplain management is variously called "multiobjective river corridor management," "greenways," or "environmental corridor management" (Kusler and Larson, 1993). This multiobjective approach is, as you might have guessed, founded on

TWO EXAMPLES OF MULTIOBJECTIVE RIVER CORRIDOR MANAGEMENT

South Platte River: The city of Littleton, Colorado, south of Denver, established a 625-acre floodplain park along the South Platte River. The 2.5-mile linear park incorporates old gravel pits reclaimed to create natural areas for fish, wildlife, and recreation. The fish habitat in the South Platte was restored, and a nature center and recreation trail were built.

Mingo Creek: Mingo Creek runs through the city of Tulsa, Oklahoma. The city developed a greenway plan for Mingo Creek including parks and trails linking together existing flood control structures. The plan also includes restoration of riparian vegetation and other recreational facilities.

(Kusler and Larson, 1993)

basin–wide planning. Here is another example of where a watershed approach is being offered as the most effective approach. Given the site–specific nature of the management objectives, the multiobjective watershed approach should be a local, community–based program, with federal and state government providing technical and funding assistance.

DROUGHT

The same human–ecological model (Figure 10.1) can be used to study the phenomenon of drought. The physical characteristics of drought are nearly the exact opposite of those of floods in every way. Whereas floods are well defined in time and space, droughts are much more difficult to delineate. Just defining what a drought is has proven a challenge. Droughts affect large areas over long periods of time; floods generally affect smaller areas over much shorter periods. Recovery and reconstruction following a flood, however, can last many years. As with all natural hazards, a drought's impact is a function of the characteristics of the human subsystem including the level of development, the types of economic activities, resource–use patterns, and existing drought adjustments.

DROUGHT DEFINITION

There are many underlying problems to developing a suitable definition of drought. It may take weeks or months of persistent dryness before a drought is finally acknowledged, and it may take a similar period of sustained wetness to bring the drought to an end. Just exactly when a drought begins and ends may not be clear. A drought may not affect all components of the hydrologic system simultaneously. It is possible for soil moisture to be abnormally low, while streamflow and lake levels are near normal. A farmer may be affected by the low soil moisture levels, while customers of an urban water supply system see no immediate problem. Another problem is that drought is not an absolute condition but rather a relative lack of moisture. A particular level of water deficit must be evaluated in terms of the normal expectation for water at a given time and place. And an increase in demand can lead to drought conditions just as surely as a decrease in the water supply.

Dracup *et al.* (1980) suggest that the first step in defining drought be the clear identification of the subject of primary interest. The term *agricultural drought* refers to drought affecting agriculture and focuses on soil moisture availability. *Hydrologic drought* focuses on water supply and might use an index of streamflow and/or reservoir storage to define drought. *Meteorological drought* refers to a period of below normal precipitation and above normal temperature. Some human activities defy such straightforward identification of the hydrologic component of primary interest. Irrigation agriculture is a good example. Being agriculture, soil moisture is obviously important, but because it is irrigated, streamflow, reservoir storage, or groundwater levels are also of primary importance. Human manipulation of the hydrologic cycle further complicates the picture by shifting drought impacts in time and space. Irrigation agriculture along the east slope of the Rocky Mountains in Colorado depends upon snowfall from the previous winter. A drought during the winter can cause water shortages and economic hardship six months to a year later, even if weather patterns have returned to normal. Much of northern California and the Pacific Northwest suffered a six–year drought from 1987 to 1993. Southern California suffered as well because of its hydrological dependence on water from the north.

Most drought definitions incorporate the idea that drought is an extended period of time during which water availability is significantly below normal. The time period may be weeks, months, or even years depending upon the situation. What is considered significantly below normal also depends upon the situation because some activities are more sensitive to moisture deficits than others. Even the same activity may be more sensitive to drought at different times. A corn crop is extremely sensitive to soil moisture deficits at certain times in its growing cycle. Inadequate moisture for only a short period near the corn's maturation phase can reduce yields dramatically compared to the same level of moisture stress earlier or later in the season.

More comprehensive definitions of drought expand the concept of departure from normal and recognize that human societies are adapted to the average, or normally expected, moisture availability at that location. As Hounam *et al.* (1975, p. 12) put it,

> . . . a pastoralist, raising fat lambs on improved pastures with a uniformly distributed rainfall averaging, say, 1000 mm a year, might be troubled by the relative "dryness" in a year producing only 750 mm, irrespective of its temporal distribution. To another pastoralist in semi–arid country normally receiving 300 mm a year, this total 750 mm would represent a record wet year, bringing with it the troubles associated with excessive moisture. . . .The agriculturalist or pastoralist, especially in the drier regions, has assessed the nature of the local rainfall and, through years of long and sometimes bitter experience, has learnt to adapt his operation to rainfall characteristics of the area.

Human manipulation of the hydrologic cycle through interbasin transfers, water storage, or groundwater use may complicate the situation, but the fundamental concept remains valid.

One last comment on drought definition is to recognize that drought is a *stochastic* hydrologic phenomenon. Stochastic means the hydrologic process is random, but not purely random. The sequence of hydrologic values are dependent in time (serially correlated). Once a drought becomes established, it tends to persist. We learned about hydrologic persistence earlier when simulating synthetic annual average streamflows using the Markov model.

There are different ways to quantitatively define and analyze drought. The first is by using a *runs* approach. A runs approach defines and describes drought using properties of a stochastic time series. Drought *indexes* are a second way to define and describe drought. Some indexes are extremely simple; some are quite complex. An example of a simple index is to define a threshold value for a hydrologic variable and some duration of time when water availability is below that threshold.

TABLE 10.5 Advantages and Limitations of Different Methods for Analyzing Drought[a]

Approach	Capable of defining drought characteristics (occurrence, length, etc.)	Capable of describing stochastic properties	Capable of describing frequency properties (e.g., return period)
Runs	yes	yes	no
Simple index	yes	no	yes, when combined with frequency analysis
PDSI	yes	yes	no
Frequency analysis	no	no	yes

[a] Source: Thompson, 1998.

An example would be, say, less than 80 percent of normal precipitation in a two–month period. The Palmer Drought Severity Index (PDSI) is a more complex index based on a water balance equation. A third way to study drought is by a *frequency analysis* similar to that used for floods. Frequency analysis does not define droughts, it only assigns a return period to drought events that are defined by some other method. Drought frequency analysis can be applied to low streamflows exactly as was done for floods. The basic data for drought frequency analysis are the *annual minimum series.* In most cases, however, it is not the single lowest instantaneous streamflow that interests the planner, but rather the lowest flow over some period of time, such as the driest five–day period. Table 10.5 lists some advantages and limitations of the different methods for analyzing drought. Runs and the Palmer Drought Severity Index are discussed below.

DROUGHT ANALYSIS

Runs Analysis of Drought

The following discussion of runs is taken largely from Thompson (1998). The theory of runs is founded on the concept of two processes crossing one another. By process we mean a stationary, stochastic time series for a hydrologic variable. A monthly precipitation time series or a series of daily streamflows are stochastic hydrologic processes. Recall from our discussion of synthetic streamflow modeling that "stationary" means that the parameters of the process (mean and variance) are not changing with time. When applied to drought, runs theory describes how a hydrologic process crosses above and below some critical threshold value. This threshold value is called a *truncation level.*

Figure 10.7 is an illustrative example of a time series for a hydrologic variable x. The increments for time (t) could be weeks, months, or even years. The truncation level x_0 is the value for which negative departures of the hydrologic

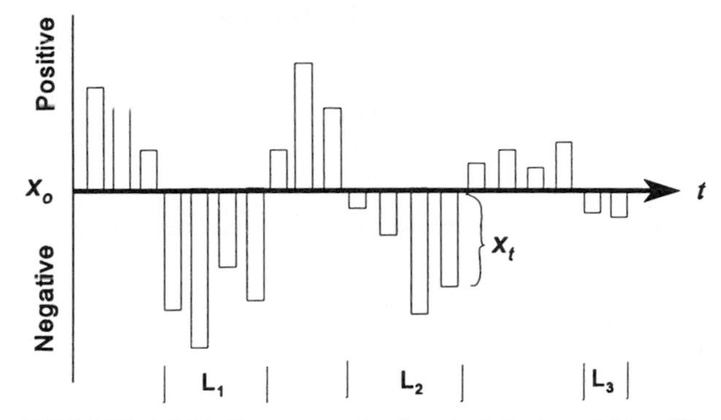

FIGURE 10.7 Runs properties for a hydrologic variable x. The truncation value x_0 is used to define negative departures (droughts). L_i is the negative run length and defines the duration of event i. (Source: Thompson, 1998)

variable are defined. Only negative departures are used to analyze drought, but analogous properties exist for positive departures. A negative departure $(x - x_0) < 0$ represents a drought. The truncation level can be set to any value, for example, at the mean. Setting x_0 slightly below the mean recognizes that small departures from the mean are still within the range of what would be considered normal. A downcrossing occurs when at time $t-1$, $(x - x_0) > 0$, and at time t, $(x - x_0) < 0$. This marks the beginning of a drought as defined by that truncation level. For any drought event i the negative run length L_i is defined as a consecutive sequence of negative deviations preceded and succeeded by a positive deviation. Run length defines the *duration* of a drought event. In Figure 10.7 there are three drought events. The first and second events are each four time periods long, while the third drought lasts two periods. The sum of the negative deviations in a run length i is called the run sum S_i and measures drought *magnitude*. Magnitude measures the cumulative water deficiency for a given drought event. The ratio (S_i / L_i) measures the average drought *intensity*. And lastly, drought *severity* at time t is given by the magnitude of an individual negative deviation x_t . A runs–based definition is applicable to any hydrologic subsystem, e.g., precipitation, soil moisture, or streamflow, and quantifies drought characteristics such as duration, magnitude, intensity, and severity (see Example 10.4).

EXAMPLE 10.4

Assume that the stochastic time series in Figure 10.7 represents deviations of monthly precipitation from the long–term mean monthly precipitation. Negative deviations are thus months when the precipitation fell below the long–term mean for that month. The first drought event begins in month 4 and ends in month 7. Assume the values for the negative departures for the four drought months are as follows:

Month	*Deviation (inches)*
4	−1.56
5	−2.05
6	−1.13
7	−1.33

The runs approach calculates the following drought characteristics:

duration, $L_1 = 4$ months
magnitude, $S_1 = (-1.56 - 2.05 - 1.13 - 1.33) = -6.07$ in
average intensity, $I_1 = -6.07/4 = -1.52$ in
severity in month 5 = −2.05 in.

Palmer Drought Severity Index (PDSI)

The most widely used drought index in the United States is the PDSI (Palmer, 1965). The PDSI usually ranges between +6.0 (extremely wet) and −6.0 (extreme drought) (Table 10.6). The PDSI is calculated monthly for climate divisions throughout the country by the National Climatic Data Center (NCDC) in Asheville, North Carolina. The PDSI is based on a weighted water balance equation. Human alterations of the natural water balance, e.g., reservoir storage, are not considered. The PDSI is a meteorological drought index. This means that the first month the weather begins to change from dry (wet) to near normal or wet (dry) conditions, the drought (wet spell) ends despite the fact that other components of the hydrological system such as soil moisture, rivers, or lakes may be below (above) their normal state (Karl, 1983). This is why the NCDC now calculates a

TABLE 10.6 Categories for Wet Spells and Droughts Using the PDSI[a]

PDSI	Classification
≥ 4.00	Extremely wet
3.00 to 3.99	Very wet
2.00 to 2.99	Moderately wet
1.00 to 1.99	Slightly wet
0.50 to 0.99	Incipient wet spell
0.49 to −0.49	Near normal
−0.50 to −0.99	Incipient drought
−1.00 to −1.99	Mild drought
−2.00 to −2.99	Moderate drought
−3.00 to −3.99	Severe drought
≤−4.00	Extreme drought

[a] Source: Palmer, 1965.

modified version of the index called the Palmer Hydrological Drought Index (PHDI). The PHDI is less sensitive to variations in weather and is thought to more accurately reflect hydrological conditions. During a sustained drought or wet spell the two indices are nearly identical in value.

The popularity of the PDSI (or PHDI) is due to its ability to capture the persistent nature of droughts while providing a single quantitative index of environmental conditions. Also, the PDSI is a standardized index, which means it is directly comparable between different climatic regions, and at different times of the year. A PDSI value of -1.5 in Arizona in September is directly comparable to one of -1.5 in Florida in June. The PDSI reflects departures from climatic conditions that are normal or expected for that location at that particular time of the year.

The calculation of the PDSI uses the water balance

$$P = ET + R + RO + L, \tag{10.3}$$

where P is precipitation, ET is actual evapotranspiration, R is soil water recharge, RO is runoff, and L is the water loss from the soil. The change in soil moisture storage from one month to the next is calculated as $\Delta S = (R - L)$. The relationship between actual and potential evapotranspiration was described in Chapter 1. What Palmer did, and this really is the essence of his innovative approach, was to extend the actual–to–potential relationship to the remaining three variables on the right side of Equation (10.3) by defining new potential values. He thus defined potential recharge (PR), potential runoff (PRO), and potential soil water loss (PL). For example, PR is defined as the unused water storage capacity of the soil,

$$PR = (AWC - S) \tag{10.4}$$

where AWC is the total available water capacity, and S is the actual amount of water stored in the soil at the beginning of the month. His next step was to quantify these actual–to–potential relationships as parameters using long–term averages. For each month the normal relationship between the actual and the potential value are parameterized as

$$\alpha = (\overline{ET}/\overline{PE}) \tag{10.5}$$

$$\beta = (\overline{R}/\overline{PR}) \tag{10.6}$$

$$\delta = (\overline{L}/\overline{PL}) \tag{10.7}$$

$$\gamma = (\overline{RO}/\overline{PRO}) \tag{10.8}$$

The overbar quantities are the monthly means calculated for the period of record. For example, if the average ET for June at some location is 3.5 inches, and the average PE for June is 6 inches, then the normal relationship in June between the two is $\overline{ET}/\overline{PE} = 3.5/6.0 = 0.583$. (This type of calculation is done for every month and for each of the four parameters, yielding 48 parameter values for each location.) Now if a particular June were warmer than normal and had, say, $PE =$ 7.0 inches, then ET should be adjusted by α so that ET bears its normal relation to the climatic demand for moisture (Palmer, 1965, p. 12). In this example the adjustment gives $ET = 0.583(7.0) = 4.08$ inches. Palmer called this adjusted value *Climatically Appropriate For Existing Conditions* (CAFEC), and designated it with a circumflex symbol, $\hat{ET}$. Palmer defined a new water balance equation, identical in form to Equation (10.3), composed of CAFEC components:

$$\hat{P} = \hat{ET} + \hat{R} + \hat{RO} + \hat{L}. \tag{10.9}$$

The CAFEC quantities on the right side are defined by

$$\hat{ET} = \alpha PE \tag{10.10}$$

$$\hat{R} = \beta PR \tag{10.11}$$

$$\hat{RO} = \gamma PRO \tag{10.12}$$

$$\hat{L} = \delta PL \tag{10.13}$$

Equation (10.9) calculates a CAFEC precipitation value $\hat{P}$ that is appropriate for the existing climatic conditions. The difference between the observed precipitation P and the CAFEC precipitation $\hat{P}$ is the moisture departure from normal:

$$d = P - \hat{P} \tag{10.14}$$

The moisture departure d is the heart of Palmer's index. The d's are then standardized to be directly comparable between different places and times of the year. Palmer conceived of meteorological drought as a departure of precipitation from what the established economy in a region has come to expect as normal. According to Palmer (1965),

> A drought period may now be defined as an interval of time, generally on the order of months to years in duration, during which the actual moisture supply at the given place rather consistently falls short of the climatically expected or climatically appropriate moisture supply. . . .the problem here is to develop a method for computing the amount of precipitation that should have occurred in a given area during a given period of time in order for the "weather" during the period to have been normal—normal in the sense that the moisture supply during the period satisfied the average or climatically expected percentage of the absolute moisture requirements during the period. In other words, the question is how much precipitation should have occurred during a given period to have kept the water resources of the area commensurate with their established use? (pp. 3–4)

Table 10.7 and Figure 10.8 show the PDSI values for a 12–year period encompassing the Dust Bowl drought in northwestern Oklahoma (OK climate division 4). From 1929 to 1933 environmental conditions oscillated between mild wetness and mild drought. The year 1933 began with near normal conditions that turned into incipient drought by the spring. Conditions deteriorated significantly and the drought became severe (PDSI $= -3.0$) by June. For the next 58 months, from June 1933 to April 1938, drought conditions ranged between moderate and

TABLE 10.7 PDSI Values for Northwestern Oklahoma (Climate Division 4)[a]

Year	J	F	M	A	M	J	J	A	S	O	N	D
1929	0.88	0.99	1.44	0.97	1.87	1.61	1.97	-0.72	0.10	0.20	0.66	−0.36
1930	−0.17	−0.67	−1.17	−1.34	−1.49	−1.36	−1.64	−2.10	−2.51	0.75	0.87	0.99
1931	0.87	0.94	1.74	2.18	0.01	−0.50	−0.65	−0.88	−1.64	−1.58	0.79	0.75
1932	1.41	0.02	−0.33	−0.14	−0.40	1.47	1.53	1.55	−0.55	−0.73	−1.28	1.44
1933	−0.42	−0.34	−0.50	−0.86	−1.65	−3.00	−3.57	−2.97	−2.79	−2.82	−3.02	−2.84
1934	−2.58	−2.71	−2.65	−2.74	−3.20	−3.87	−5.12	−4.79	−4.07	−4.19	−4.23	−4.16
1935	−4.20	−4.08	−3.74	−4.06	−3.21	−2.99	−3.36	−3.54	−3.45	−3.54	−2.90	−2.56
1936	−2.49	−2.65	−3.19	−3.68	−3.20	−3.53	−4.66	−5.60	−3.14	−2.80	−2.84	−2.58
1937	−2.33	−2.47	−2.18	−2.46	−2.23	−2.13	−2.81	−2.80	−2.71	−2.68	−2.70	−2.77
1938	−3.03	−2.46	−2.03	−1.72	−1.43	−1.00	−1.04	−1.67	−1.88	−2.59	−2.19	−2.42
1939	−1.05	−1.26	−1.13	−1.35	−1.79	−1.08	−1.44	−1.55	−2.71	−3.11	−3.16	−3.11
1940	−2.90	−2.47	−3.05	−2.47	−2.85	−3.45	−3.62	−3.37	−3.20	−3.51	0.98	0.90

[a] Data from the U.S. Department of Commerce, 1990.

extreme, with the years 1934 and 1936 having the two most extreme months. The drought ended abruptly in November of 1940 when conditions turned wetter than normal. Figures 10.9 and 10.10 each show the PDHI for 2 months in the spring and summer of 1936, one of the worst years of the Dust Bowl period.

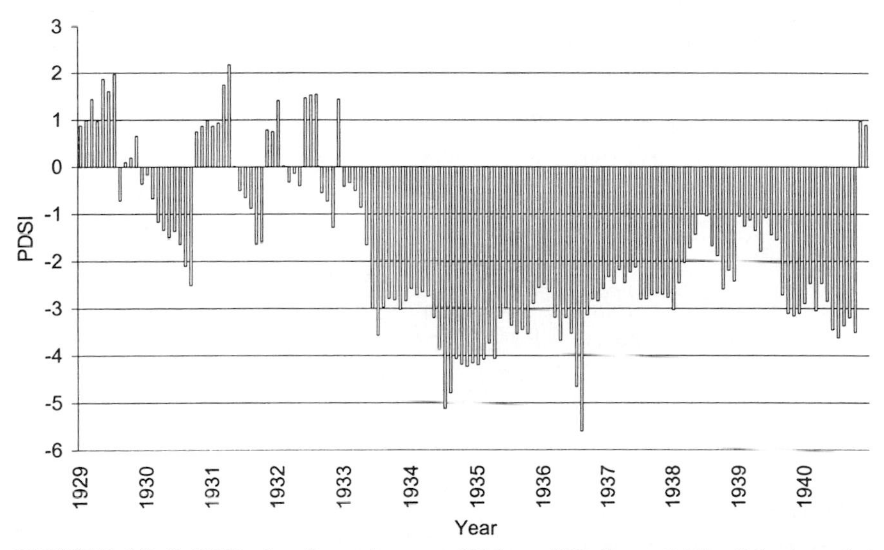

FIGURE 10.8 PDSI values for northwestern Oklahoma (OK climate division 4) for the period 1929 to 1940. Actual index values are given in Table 10.8. (Source: Thompson, 1998)

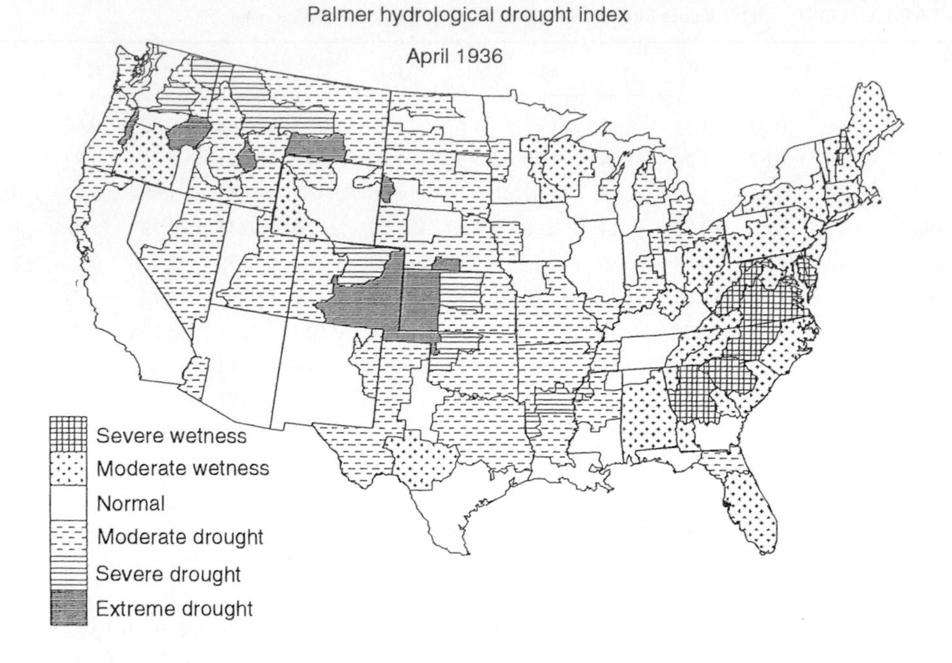

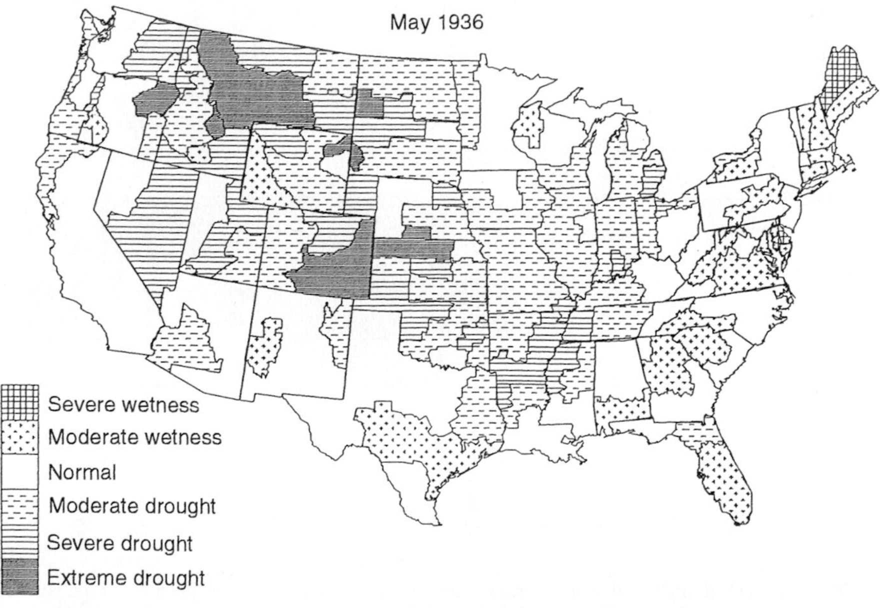

FIGURE 10.9 PHDI for April (top) and May (bottom) of 1936. (Source: U.S. Department of Commerce, 1990)

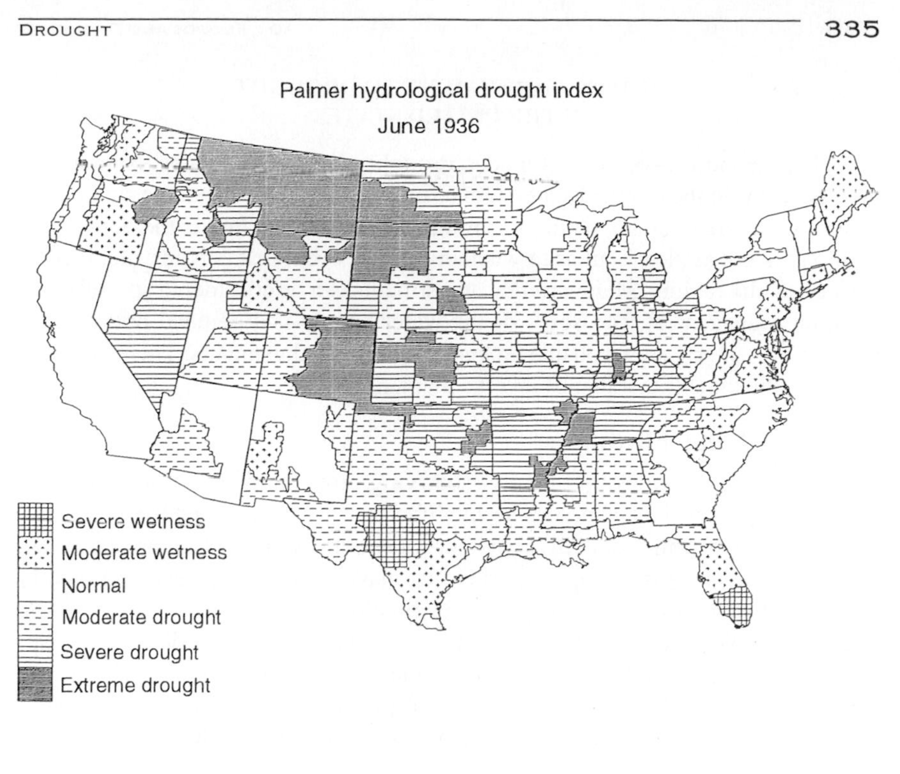

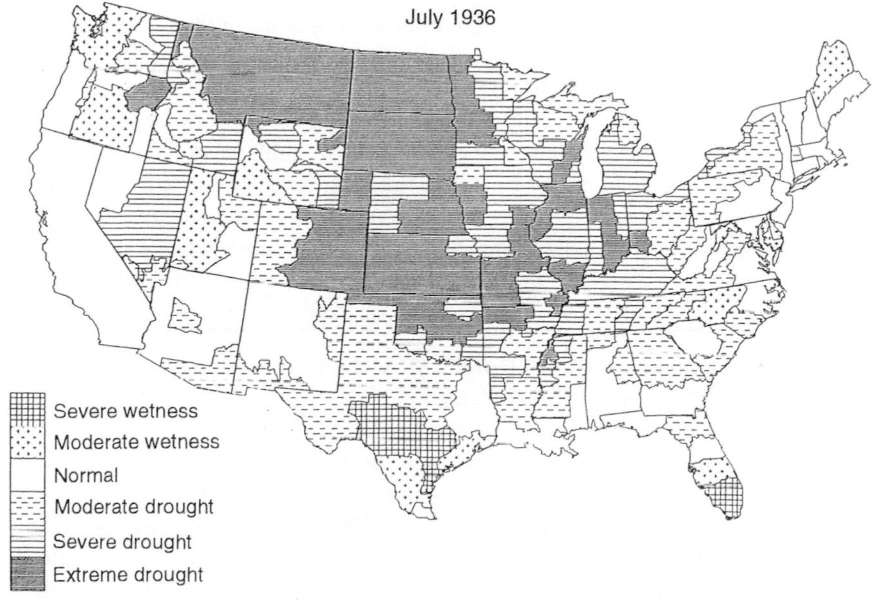

FIGURE 10.10 PHDI for June (top) and July (bottom) of 1936. (Source: U.S. Department of Commerce, 1990).

HUMAN ADJUSTMENT TO DROUGHT IN THE UNITED STATES

Using the human–ecological model, drought adjustments can be examined as modifications to the physical subsystem or modifications to the human subsystem. Due to the more complex nature of droughts compared to floods the adjustment typology is not as clear–cut as to whether it is the event that is being modified or society's vulnerability. Many drought adjustments can be considered special cases of water management, and some of the water supply and demand options presented in Chapter 7 are reconsidered here as drought adjustments. There are other drought adjustments that do not involve managing water.

Modify the Natural Event

There are few options for directly modifying droughts. We do not possess the understanding nor the technology to modify large–scale atmospheric processes. Whether we ever will is doubtful; whether we should even try is a more important question. The clearest example of where we have tried to intervene and modify drought is by cloud seeding. The problem with cloud seeding is that, by definition, there are precious few clouds to seed during a drought. As mentioned in Chapter 7, cloud seeding under ideal conditions may increase precipitation 10 percent. The Bureau of Reclamation, which was the primary federal agency for cloud–seeding research, announced in 1994 that it was no longer in the cloud–seeding business.

Also included under this adjustment category are cultural/religious practices beseeching divine intervention to end a drought. The Hopi of the Southwest have ancient ceremonies to call forth rain from the sky. In May of 1996, after 11 months of drought, residents of a small Texas community gathered together and prayed for rain. The event was captured by the national news networks, but it did not produce any rain. Texas officials were calling this drought the worst natural disaster in the state's history—even worse than the Dust Bowl of the 1930s. Damages to agriculture and related industries through May 1996 in Texas alone were estimated to exceed $2.5 billion. Damages in Oklahoma were estimated at over a billion dollars.

POTENTIAL IMPACTS FROM A SEVERE SUSTAINED DROUGHT IN THE COLORADO RIVER BASIN

Reconstructed streamflows from tree-ring records suggest that a severe drought lasting several decades occurred in the Colorado River basin in the late 1500s. What would be the impacts of a similar event happening today? This was the basic research question posed by an interdisciplinary team studying the potential impacts of a severe sustained drought (SSD) in the Colorado River basin (Young, 1995). They used river flows as the basic indicator (definition) of drought. The study was divided into two phases. Phase 1 was a hydrologic analysis aimed at defining severe drought scenarios. The "worst case" scenario used for the study was a 38-year hypothetical drought created by rearranging the sequence of streamflows from the most severe drought that occurred in the tree-ring reconstruction of streamflows at Lees Ferry (Meko *et al.,* 1995). The reconstructed streamflows were rearranged so that the driest years (lowest streamflows) came at the end of the drought. The annual flow volume for this SSD scenario was 10.95 million acre-feet, or 81 percent of the long-term mean annual flow volume.

Phase 2 of the study was to assess the societal and environmental impacts of the SSD scenario. Societal and environmental impacts depend upon both the structural and the institutional development of the basin. Structural development refers to the reservoirs and distribution facilities that store and move water within and outside the basin. Federally sponsored development of the Colorado River has created a tremendous amount of reservoir storage capacity. The combined capacity of the two largest reservoirs—Lake Mead and Lake Powell—can store more than 4 years of the river's average native flow. Reservoir storge provides substantial protection from drought, but is it sufficient to mitigate the effects of a worst case scenario? The other major consideration in assessing impacts is institutional—the "Law of the River"—which is the complex of laws, regulations, and operating procedures that govern how, when, where, and by whom Colorado River water is used. Impacts were assessed using computer simulations and a gaming experiment that allowed the Law of the River to be changed.

The results of the study indicated that Lake Powell and other major Upper Basin reservoirs would be emptied by the worst case SSD scenario, and Lake Meade very nearly emptied. One reason for this is that the Upper Basin bears the greatest drought risk because it must deliver an absolute quantity of water to the Lower Basin (Table 3.2). Overall, however, basin-wide water shortages were less than 25 percent of normal, which is not too extreme. Consumptive uses in the Lower Basin were largely unaffected, while water deliveries for consumptive uses in the Upper Basin were reduced by about one-half. The greatest impacts overall were to instream uses like hydropower generation, recreation, ecosystem maintenance, and salinity control. The single largest economic loss was electric power generation, with an average loss of $600 million per year. The estimated present value of the total quantifiable costs, excluding salinity, from the SSD scenario was $5 billion of which 45 percent was to consumptive uses. This underestimates the real total cost because salinity and intangible environmental impacts are not included. Some environmental impacts, like the loss of wetlands, are reversible, but some, like species extinction, are not (Lord *et al.,* 1995).

Water storage in reservoirs might also be classified as an adjustment that modifies the drought event, since releasing stored water modifies the river's natural (low) flow regime. Interbasin water transfers, desalination, and groundwater pumping might be included here as well, although they could also be considered adjustments aimed at limiting damage susceptibility.

Modify Damage Susceptibility

Adjustments that modify damage susceptibility can be considered by water–use sector. Urban/domestic water users can reduce drought susceptibility by reducing demand through water conservation. Low–flow water fixtures and drought–resistant landscaping permanently alter the demand for water. Emergency measures might be employed too. Drought emergency measures include restricting or banning water uses like car washing, lawn watering, and the filling of swimming pools. Under more extreme conditions indoor water use may even be curtailed. Many California cities allocated households an average monthly quota of water during the 1987–1993 drought. Meeting the quota required using less water inside as well as outside the home. Restaurants often stop serving water to patrons unless they specifically request it. Urban water supply systems might also employ emergency measures to increase water supply. Drilling a new well, temporary pipelines to new sources, desalination, and the temporary or permanent reallocation of water from agriculture to cities are possibilities.

Many farmers in the eastern United States have installed irrigation systems specifically as a drought adjustment. This strategy works during mild droughts when lack of rainfall causes low soil moisture but surface or groundwater supplies are still available. During extreme drought these sources too may dry up. In the West, farmers might try increasing their water–use efficiency to mitigate a drought's impact. Improving irrigation efficiency raises an interesting theoretical question, depending upon whether the efficiency improvement is temporary or permanent. Are higher–efficiency water–use systems more vulnerable to drought? If farmers are normally somewhat inefficient in their use of water, that is, they have more water allocated to them than they need under normal conditions, then during drought when their allocation is reduced, they use this normally surplus water to get through the drought. In doing so they increase their efficiency. In this way a little excess "slop" in the system may act to reduce farmers' drought vulnerability. On the other hand if farmers permanently improved their water–use efficiency, and the conserved water is reallocated for use by cities, then when a drought occurs the farmers may not be able to increase their efficiency much beyond this already high–efficiency level. Farmers might be more vulnerable to drought by operating at a normally higher level of efficiency, because they no longer have access to that surplus water in times of drought. Urban areas too may now be more vulnerable because they no longer have a potential water supply source within the agricultural sector to tap during an emergency, because that surplus water is already committed for use by city residents. Increasing water–use efficiency is widely advocated, but it means operating closer to the edge, with less of a margin between supply and demand, and it may make us more vulnerable to drought. This idea is offered only as a possibility, not as fact.

Farmers can also modify damage susceptibility by planting different types of crops or drought–tolerant plant varieties, cultivating in ways that conserve soil moisture, purchasing supplemental feed for livestock, and using or not using certain areas of land.

Regardless of the water–use sector, having a drought response plan in place can help minimize drought impacts. A drought response plan tells officials what to do, when to do it, and where the necessary resources are located to get the job done. Quite a few states have developed drought emergency plans in the last decade (Wilhite, 1997).

Modify the Loss Burden

Adjustments that modify the loss burden are aimed almost exclusively at the agricultural sector. Federal crop insurance is available to compensate farmers for crop losses resulting from natural hazards. Loans and grants are also available. Not all of the adjustments in this category involve the government. During the 1988 Midwest drought, farmers in the Southeast donated and shipped truckloads of hay to their fellow farmers to the north. A few years later Midwestern farmers reciprocated and shipped hay to drought–stricken farmers in the South. Another adjustment to modify the loss burden is to sell off part or all of the herd to minimize economic losses. Over time the evolution of agricultural drought adjustments in the United States has shifted the loss burden from the individual to the Nation as a whole. The Dust Bowl drought of the 1930s had devastating impacts on individuals and families. Twenty some years later the drought of the mid–1950s, which was hydrologically more severe than that of the 1930s, resulted in fewer local and regional impacts (foreclosed farms, forced migrations, etc.) because government policies and programs distributed the impacts nationally among taxpayers at large (Warrick, 1975).

APPENDIX 1

UNIT CONVERSIONS AND EQUIVALENTS

Length

multiply	by	to get
feet	30.48	centimeters
feet	0.3048	meters
miles	1.6093	kilometers
miles	63360	inches
miles	5280	feet
kilometers	0.6214	miles
kilometers	1000	meters
meters	100	centimeters
meters	39.37	inches

Area

multiply	by	to get
acres	43560	square feet
acres	4047	square meters
acres	0.4047	hectares
square miles	640	acres
square miles	2.590	square kilometers

Temperature

$°C = 5/9(°F - 32)$

$°F = 9/5(°C) + 32$

$°K = °C + 273$

Energy and power

multiply	by	to get
Joules	0.2389	calories

1 watt = 1 joule per second
1 watt per square meter = 0.001433 calories per square centimeter per minute
1 langley = 1 calorie per square centimeter per minute

Volume

multiply	by	to get
cubic feet	7.481	gallons
cubic feet	0.02832	cubic meters
acre-feet	43560	cubic feet
acre-feet	1233.62	cubic meters
cubic meters	264.2	gallons

Pressure

multiply	by	to get
feet of water	0.02950	atmospheres
feet of water	62.43	pounds per sq. foot
bars	2.089×10^{-3}	pounds per sq. foot
bars	0.987	atmospheres
bars	1000	millibars
millibars	100	pascals

Velocity

multiply	by	to get
meters per second	3.6	kilometers per hour
meters per second	2.237	miles per hour
feet per second	0.6818	miles per hour
feet per second	0.3048	meters per second

Discharge

multiply	by	to get
cubic meters per second	35.31	cubic feet per second

Important Equivalents

1 cfs for 1 day = 1.98 acre-feet
1 cfs = 2447 cubic meters per day
1 gallon per minute = 0.002228 cfs
1 inch of runoff per hour from 1 acre = 1.008 cfs
1 inch of runoff per hour from 1 square mile = 645.3 cfs
1 inch of runoff from 1 square mile = 2,323,200 cubic feet = 53.33 acre-feet

Important Information

Acceleration of gravity = 9.81 m s^{-2}
Standard sea level pressure = 1013.25 mb = 14.7 lbs in^{-2} = 29.92 inches of mercury

Properties of water at 15°C
- Specific gravity = 0.99913
- Density = 0.99910 g cm^{-3} = 62.36 lbs ft^{-3}
- Heat of vaporization = 588.9 cal g^{-1}
- Vapor pressure = 17.04 mb

APPENDIX 2

Sources of Water–Related Information

Data	Internet/WWW address
Precipitation data	
National Weather Service	
Climatological Data (state)	
Local Climatological Data	
National Climatic Data Center, Asheville, North Carolina	*http://www.ncdc.noaa.gov*
USDA Agricultural experiment stations	*http://hydrolab.arsusda.gov/arswater.html*
Water data	
U.S. Geological Survey (home page)	*http://water.usgs.gov*
Water Supply Papers	
Water Resources Data (state)	
Hydroclimatic Data Network	
Water Data Storage and Retrieval (WATSTOR)	
Water–use data	*http://water.usgs.gov/public/wateruse/*
U.S. Environmental Protection Agency	*http://www.epa.gov/waterhome/programs.html*
Storage and Retrieval System (STORET)	
Water Body System (WBS)	
Residential water–use data	*http://www.waterwiser.org/wateruse/main.html*
Water-related journals	
Water Resources Research	
Journal of the American Water Resources Association (formerly the *Water Resources Bulletin*)	
Journal of Hydrology	
Bulletin of the American Meteorological Society	
Monthly Weather Review	

APPENDIX 3

Discount and Investment Factors, $i = 0.040$

n	F/P	P/F	F/A	A/F	P/A	A/P	P/G
1	1.0400	0.9615	1.0000	1.0000	0.9615	1.0400	0.9615
2	1.0816	0.9246	2.0400	0.4902	1.8861	0.5302	2.8107
3	1.1249	0.8890	3.1216	0.3203	2.7751	0.3603	5.4776
4	1.1699	0.8548	4.2465	0.2355	3.6299	0.2755	8.8969
5	1.2167	0.8219	5.4163	0.1846	4.4518	0.2246	13.0065
6	1.2653	0.7903	6.6330	0.1508	5.2421	0.1908	17.7484
7	1.3159	0.7599	7.8983	0.1266	6.0021	0.1666	23.0678
8	1.3686	0.7307	9.2142	0.1085	6.7327	0.1485	28.9133
9	1.4233	0.7026	10.5828	0.0945	7.4353	0.1345	35.2366
10	1.4802	0.6756	12.0061	0.0833	8.1109	0.1233	41.9922
11	1.5395	0.6496	13.4864	0.0741	8.7605	0.1141	49.1376
12	1.6010	0.6246	15.0258	0.0666	9.3851	0.1066	56.6328
13	1.6651	0.6006	16.6268	0.0601	9.9856	0.1001	64.4403
14	1.7317	0.5775	18.2919	0.0547	10.5631	0.0947	72.5249
15	1.8009	0.5553	20.0236	0.0499	11.1184	0.0899	80.8539
16	1.8730	0.5339	21.8245	0.0458	11.6523	0.0858	89.3964
17	1.9479	0.5134	23.6975	0.0422	12.1657	0.0822	98.1238
18	2.0258	0.4936	25.6454	0.0390	12.6593	0.0790	107.0091
19	2.1068	0.4746	27.6712	0.0361	13.1339	0.0761	116.0273
20	2.1911	0.4564	29.7781	0.0336	13.5903	0.0736	125.1550
21	2.2788	0.4388	31.9692	0.0313	14.0292	0.0713	134.3705
22	2.3699	0.4220	34.2480	0.0292	14.4511	0.0692	143.6535
23	2.4647	0.4057	36.6179	0.0273	14.8568	0.0673	152.9852
24	2.5633	0.3901	39.0826	0.0256	15.2470	0.0656	162.3482
25	2.6658	0.3751	41.6459	0.0240	15.6221	0.0640	171.7261
26	2.7725	0.3607	44.3117	0.0226	15.9828	0.0626	181.1040
27	2.8834	0.3468	47.0842	0.0212	16.3296	0.0612	190.4680
28	2.9987	0.3335	49.9676	0.0200	16.6631	0.0600	199.8054
29	3.1187	0.3207	52.9663	0.0189	16.9837	0.0589	209.1043
30	3.2434	0.3083	56.0849	0.0178	17.2920	0.0578	218.3539
40	4.8010	0.2083	95.0255	0.0105	19.7928	0.0505	306.3231
50	7.1067	0.1407	152.6671	0.0066	21.4822	0.0466	382.6460
100	50.5049	0.0198	1237.6237	0.0008	24.5050	0.0408	587.6299

Discount and Investment Factors, $i = 0.045$

n	F/P	P/F	F/A	A/F	P/A	A/P	P/G
1	1.0450	0.9569	1.0000	1.0000	0.9569	1.0450	0.9569
2	1.0920	0.9157	2.0450	0.4890	1.8727	0.5340	2.7884
3	1.1412	0.8763	3.1370	0.3188	2.7490	0.3638	5.4173
4	1.1925	0.8386	4.2782	0.2337	3.5875	0.2787	8.7715
5	1.2462	0.8025	5.4707	0.1828	4.3900	0.2278	12.7838
6	1.3023	0.7679	6.7169	0.1489	5.1579	0.1939	17.3912
7	1.3609	0.7348	8.0192	0.1247	5.8927	0.1697	22.5350
8	1.4221	0.7032	9.3800	0.1066	6.5959	0.1516	28.1604
9	1.4861	0.6729	10.8021	0.0926	7.2688	0.1376	34.2166
10	1.5530	0.6439	12.2882	0.0814	7.9127	0.1264	40.6559
11	1.6229	0.6162	13.8412	0.0722	8.5289	0.1172	47.4340
12	1.6959	0.5897	15.4640	0.0647	9.1186	0.1097	54.5100
13	1.7722	0.5643	17.1599	0.0583	9.6829	0.1033	61.8455
14	1.8519	0.5400	18.9321	0.0528	10.2228	0.0978	69.4052
15	1.9353	0.5167	20.7841	0.0481	10.7395	0.0931	77.1560
16	2.0224	0.4945	22.7193	0.0440	11.2340	0.0890	85.0675
17	2.1134	0.4732	24.7417	0.0404	11.7072	0.0854	93.1115
18	2.2085	0.4528	26.8551	0.0372	12.1600	0.0822	101.2619
19	2.3079	0.4333	29.0636	0.0344	12.5933	0.0794	109.4946
20	2.4117	0.4146	31.3714	0.0319	13.0079	0.0769	117.7875
21	2.5202	0.3958	33.7831	0.0296	13.4047	0.0746	126.1200
22	2.6337	0.3797	36.3034	0.0275	13.7844	0.0725	134.4734
23	2.7522	0.3634	38.9370	0.0257	14.1478	0.0707	142.8305
24	2.8760	0.3477	41.6892	0.0240	14.4955	0.0690	151.1754
25	3.0054	0.3327	44.5652	0.0224	14.8282	0.0674	159.4936
26	3.1407	0.3184	47.5706	0.0210	15.1466	0.0660	167.7721
27	3.2820	0.3047	50.7113	0.0197	15.4513	0.0647	175.9988
28	3.4297	0.2916	53.9933	0.0185	15.7429	0.0635	184.1627
29	3.5840	0.2790	57.4230	0.0174	16.0219	0.0624	192.2542
30	3.7453	0.2670	61.0071	0.0164	16.2889	0.0614	200.2642
40	5.8164	0.1719	107.0303	0.0093	18.4016	0.0543	274.5002
50	9.0326	0.1107	178.5030	0.0056	19.7620	0.0506	335.9070
100	81.5885	0.0123	1790.8560	0.0006	21.9499	0.0456	482.4874

Discount and Investment Factors, $i = 0.050$

n	F/P	P/F	F/A	A/F	P/A	A/P	P/G
1	1.0500	0.9524	1.0000	1.0000	0.9524	1.0500	0.9524
2	1.1025	0.9070	2.0500	0.4878	1.8594	0.5378	2.7664
3	1.1576	0.8638	3.1525	0.3172	2.7232	0.3672	5.3580
4	1.2155	0.8227	4.3101	0.2320	3.5460	0.2820	8.6488
5	1.2763	0.7835	5.5256	0.1810	4.3295	0.2310	12.5664
6	1.3401	0.7462	6.8019	0.1470	5.0757	0.1970	17.0437
7	1.4071	0.7107	8.1420	0.1228	5.7864	0.1728	22.0185
8	1.4775	0.6768	9.5491	0.1047	6.4632	0.1547	27.4332
9	1.5513	0.6446	11.0266	0.0907	7.1078	0.1407	33.2347
10	1.6289	0.6139	12.5779	0.0795	7.7217	0.1295	39.3738
11	1.7103	0.5847	14.2068	0.0704	8.3064	0.1204	45.8053
12	1.7959	0.5568	15.9171	0.0628	8.8633	0.1128	52.4873
13	1.8856	0.5303	17.7130	0.0565	9.3936	0.1065	59.3815
14	1.9799	0.5051	19.5986	0.0510	9.8986	0.1010	66.4524
15	2.0789	0.4810	21.5786	0.0463	10.3797	0.0963	73.6677
16	2.1829	0.4581	23.6575	0.0423	10.8378	0.0923	80.9975
17	2.2920	0.4363	25.8404	0.0387	11.2741	0.0887	88.4145
18	2.4066	0.4155	28.1324	0.0355	11.6896	0.0855	95.8939
19	2.5270	0.3957	30.5390	0.0327	12.0853	0.0827	103.4128
20	2.6533	0.3769	33.0660	0.0302	12.4622	0.0802	110.9506
21	2.7860	0.3589	35.7193	0.0280	12.8212	0.0780	118.4884
22	2.9253	0.3418	38.5052	0.0260	13.1630	0.0760	126.0091
23	3.0715	0.3256	41.4305	0.0241	13.4886	0.0741	133.4973
24	3.2251	0.3101	44.5020	0.0225	13.7986	0.0725	140.9389
25	3.3864	0.2953	47.7271	0.0210	14.0939	0.0710	148.3215
26	3.5557	0.2812	51.1135	0.0196	14.3752	0.0696	155.6337
27	3.7335	0.2678	54.6691	0.0183	14.6430	0.0683	162.8656
28	3.9201	0.2551	58.4026	0.0171	14.8981	0.0671	170.0082
29	4.1161	0.2429	62.3227	0.0160	15.1411	0.0660	177.0537
30	4.3219	0.2314	66.4388	0.0151	15.3725	0.0651	183.9950
40	7.0400	0.1420	120.7998	0.0083	17.1591	0.0583	246.7043
50	11.4674	0.0872	209.3480	0.0048	18.2559	0.0548	296.1707
100	131.5013	0.0076	2610.0252	0.0004	19.8479	0.0504	401.5971

Discount and Investment Factors, $i = 0.055$

n	F/P	P/F	F/A	A/F	P/A	A/P	P/G
1	1.0550	0.9479	1.0000	1.0000	0.9479	1.0550	0.9479
2	1.1130	0.8985	2.0550	0.4866	1.8463	0.5416	2.7448
3	1.1742	0.8516	3.1680	0.3157	2.6979	0.3707	5.2996
4	1.2388	0.8072	4.3423	0.2303	3.5052	0.2853	8.5285
5	1.3070	0.7651	5.5811	0.1792	4.2703	0.2342	12.3542
6	1.3788	0.7252	6.8881	0.1452	4.9955	0.2002	16.7056
7	1.4547	0.6874	8.2669	0.1210	5.6830	0.1760	21.5177
8	1.5347	0.6516	9.7216	0.1029	6.3346	0.1579	26.7305
9	1.6191	0.6176	11.2563	0.0888	6.9522	0.1438	32.2891
10	1.7081	0.5854	12.8754	0.0777	7.5376	0.1327	38.1434
11	1.8021	0.5549	14.5835	0.0686	8.0925	0.1236	44.2475
12	1.9012	0.5260	16.3856	0.0610	8.6185	0.1160	50.5592
13	2.0058	0.4986	18.2868	0.0547	9.1171	0.1097	57.0405
14	2.1161	0.4726	20.2926	0.0493	9.5896	0.1043	63.6565
15	2.2325	0.4479	22.4087	0.0446	10.0376	0.0996	70.3755
16	2.3553	0.4246	24.6411	0.0406	10.4622	0.0956	77.1688
17	2.4848	0.4024	26.9964	0.0370	10.8646	0.0920	84.0104
18	2.6215	0.3815	29.4812	0.0339	11.2461	0.0889	90.8768
19	2.7656	0.3616	32.1027	0.0312	11.6077	0.0862	97.7468
20	2.9178	0.3427	34.8683	0.0287	11.9504	0.0837	104.6014
21	3.0782	0.3249	37.7861	0.0265	12.2752	0.0815	111.4234
22	3.2475	0.3079	40.8643	0.0245	12.5832	0.0795	118.1978
23	3.4262	0.2919	44.1118	0.0227	12.8750	0.0777	124.9109
24	3.6146	0.2767	47.5380	0.0210	13.1517	0.0760	131.5506
25	3.8134	0.2622	51.1526	0.0195	13.4139	0.0745	138.1065
26	4.0231	0.2486	54.9660	0.0182	13.6625	0.0732	144.5691
27	4.2444	0.2356	58.9891	0.0170	13.8981	0.0720	150.9304
28	4.4778	0.2233	63.2335	0.0158	14.1214	0.0708	157.1834
29	4.7241	0.2117	67.7114	0.0148	14.3331	0.0698	163.3221
30	4.9840	0.2006	72.4355	0.0138	14.5337	0.0688	169.3415
40	8.5133	0.1175	136.6056	0.0073	16.0461	0.0623	222.3661
50	14.5420	0.0688	246.2175	0.0041	16.9315	0.0591	262.2623
100	211.4686	0.0047	3826.7025	0.0003	18.0958	0.0553	338.5132

Discount and Investment Factors, $i = 0.060$

n	F/P	P/F	F/A	A/F	P/A	A/P	P/G
1	1.0600	0.9434	1.0000	1.0000	0.9434	1.0600	0.9434
2	1.1236	0.8900	2.0600	0.4854	1.8334	0.5454	2.7234
3	1.1910	0.8396	3.1836	0.3141	2.6730	0.3741	5.2422
4	1.2625	0.7921	4.3746	0.2286	3.4651	0.2886	8.4106
5	1.3382	0.7473	5.6371	0.1774	4.2124	0.2374	12.1469
6	1.4185	0.7050	6.9753	0.1434	4.9173	0.2034	16.3767
7	1.5036	0.6651	8.3938	0.1191	5.5824	0.1791	21.0321
8	1.5938	0.6274	9.8975	0.1010	6.2098	0.1610	26.0514
9	1.6895	0.5919	11.4913	0.0870	6.8017	0.1470	31.3785
10	1.7908	0.5584	13.1808	0.0759	7.3601	0.1359	36.9624
11	1.8983	0.5268	14.9716	0.0668	7.8869	0.1268	42.7571
12	2.0122	0.4970	16.8699	0.0593	8.3838	0.1193	48.7207
13	2.1329	0.4688	18.8821	0.0530	8.8527	0.1130	54.8156
14	2.2609	0.4423	21.0151	0.0476	9.2950	0.1076	61.0078
15	2.3966	0.4173	23.2760	0.0430	9.7122	0.1030	67.2668
16	2.5404	0.3936	25.6725	0.0390	10.1059	0.0990	73.5651
17	2.6928	0.3714	28.2129	0.0354	10.4773	0.0954	79.8783
18	2.8543	0.3503	30.9057	0.0324	10.8276	0.0924	86.1845
19	3.0256	0.3305	33.7600	0.0296	11.1581	0.0896	92.4643
20	3.2071	0.3118	36.7856	0.0272	11.4699	0.0872	98.7004
21	3.3996	0.2942	39.9927	0.0250	11.7641	0.0850	104.8776
22	3.6035	0.2775	43.3923	0.0230	12.0416	0.0830	110.9827
23	3.8197	0.2618	46.9958	0.0213	12.3034	0.0813	117.0041
24	4.0489	0.2470	50.8156	0.0197	12.5504	0.0797	122.9316
25	4.2919	0.2330	54.8645	0.0182	12.7834	0.0782	128.7565
26	4.5494	0.2198	59.1564	0.0169	13.0032	0.0769	134.4716
27	4.8223	0.2074	63.7058	0.0157	13.2105	0.0757	140.0705
28	5.1117	0.1956	68.5281	0.0146	13.4062	0.0746	145.5482
29	5.4184	0.1846	73.6398	0.0136	13.5907	0.0736	150.9003
30	5.7435	0.1741	79.0582	0.0126	13.7648	0.0726	156.1236
40	10.2857	0.0972	154.7620	0.0065	15.0463	0.0665	201.0031
50	18.4202	0.0543	290.3359	0.0034	15.7619	0.0634	233.2192
100	339.3021	0.0029	5638.3681	0.0002	16.6175	0.0602	288.6646

Discount and Investment Factors, $i = 0.065$

n	F/P	P/F	F/A	A/F	P/A	A/P	P/G
1	1.0650	0.9390	1.0000	1.0000	0.9390	1.0650	0.9390
2	1.1342	0.8817	2.0650	0.4843	1.8206	0.5493	2.7023
3	1.2079	0.8278	3.1992	0.3126	2.6485	0.3776	5.1858
4	1.2865	0.7773	4.4072	0.2269	3.4258	0.2919	8.2951
5	1.3701	0.7299	5.6936	0.1756	4.1557	0.2406	11.9445
6	1.4591	0.6853	7.0637	0.1416	4.8410	0.2066	16.0565
7	1.5540	0.6435	8.5229	0.1173	5.4845	0.1823	20.5611
8	1.6550	0.6042	10.0769	0.0992	6.0888	0.1642	25.3949
9	1.7626	0.5674	11.7319	0.0852	6.6561	0.1502	30.5011
10	1.8771	0.5327	13.4944	0.0741	7.1888	0.1391	35.8284
11	1.9992	0.5002	15.3716	0.0651	7.6890	0.1301	41.3307
12	2.1291	0.4697	17.3707	0.0576	8.1587	0.1226	46.9669
13	2.2675	0.4410	19.4998	0.0513	8.5997	0.1163	52.7001
14	2.4149	0.4141	21.7673	0.0459	9.0138	0.1109	58.4975
15	2.5718	0.3888	24.1822	0.0414	9.4027	0.1064	64.3299
16	2.7390	0.3651	26.7540	0.0374	9.7678	0.1024	70.1714
17	2.9170	0.3428	29.4930	0.0339	10.1106	0.0989	75.9993
18	3.1067	0.3219	32.4101	0.0309	10.4325	0.0959	81.7933
19	3.3086	0.3022	35.5167	0.0282	10.7347	0.0932	87.5359
20	3.5236	0.2838	38.8253	0.0258	11.0185	0.0908	93.2118
21	3.7527	0.2665	42.3490	0.0236	11.2850	0.0886	98.8078
22	3.9966	0.2502	46.1016	0.0217	11.5352	0.0867	104.3125
23	4.2564	0.2349	50.0982	0.0200	11.7701	0.0850	109.7162
24	4.5331	0.2206	54.3546	0.0184	11.9907	0.0834	115.0106
25	4.8277	0.2071	58.8877	0.0170	12.1979	0.0820	120.1891
26	5.1415	0.1945	63.7154	0.0157	12.3924	0.0807	125.2459
27	5.4757	0.1826	68.8569	0.0145	12.5750	0.0795	130.1768
28	5.8316	0.1715	74.3326	0.0135	12.7465	0.0785	134.9782
29	6.2107	0.1610	80.1642	0.0125	12.9075	0.0775	139.6476
30	6.6144	0.1512	86.3749	0.0116	13.0587	0.0766	144.1832
40	12.4161	0.0805	175.6319	0.0057	14.1455	0.0707	182.2055
50	23.3067	0.0429	343.1797	0.0029	14.7245	0.0679	208.2509
100	543.2013	0.0018	8341.5580	0.0001	15.3563	0.0651	248.7747

Discount and Investment Factors, $i = 0.070$

n	F/P	P/F	F/A	A/F	P/A	A/P	P/G
1	1.0700	0.9346	1.0000	1.0000	0.9346	1.0700	0.9346
2	1.1449	0.8734	2.0700	0.4831	1.8080	0.5531	2.6815
3	1.2250	0.8163	3.2149	0.3111	2.6243	0.3811	5.1304
4	1.3108	0.7629	4.4399	0.2252	3.3872	0.2952	8.1819
5	1.4026	0.7130	5.7507	0.1739	4.1002	0.2439	11.7469
6	1.5007	0.6663	7.1533	0.1398	4.7665	0.2098	15.7449
7	1.6058	0.6227	8.6540	0.1156	5.3893	0.1856	20.1042
8	1.7182	0.5820	10.2598	0.0975	5.9713	0.1675	24.7602
9	1.8385	0.5439	11.9780	0.0835	6.5152	0.1535	29.6556
10	1.9672	0.5083	13.8164	0.0724	7.0236	0.1424	34.7391
11	2.1049	0.4751	15.7836	0.0634	7.4987	0.1334	39.9652
12	2.2522	0.4440	17.8885	0.0559	7.9427	0.1259	45.2933
13	2.4098	0.4150	20.1406	0.0497	8.3577	0.1197	50.6878
14	2.5785	0.3878	22.5505	0.0443	8.7455	0.1143	56.1173
15	2.7590	0.3624	25.1290	0.0398	9.1079	0.1098	61.5540
16	2.9522	0.3387	27.8881	0.0359	9.4466	0.1059	66.9737
17	3.1588	0.3166	30.8402	0.0324	9.7632	0.1024	72.3555
18	3.3799	0.2959	33.9990	0.0294	10.0591	0.0994	77.6810
19	3.6165	0.2765	37.3790	0.0268	10.3356	0.0968	82.9347
20	3.8697	0.2584	40.9955	0.0244	10.5940	0.0944	88.1031
21	4.1406	0.2415	44.8652	0.0223	10.8355	0.0923	93.1748
22	4.4304	0.2257	49.0057	0.0204	11.0612	0.0904	98.1405
23	4.7405	0.2109	53.4361	0.0187	11.2722	0.0887	102.9923
24	5.0724	0.1971	58.1767	0.0172	11.4693	0.0872	107.7238
25	5.4274	0.1842	63.2490	0.0158	11.6536	0.0858	112.3301
26	5.8074	0.1722	68.6765	0.0146	11.8258	0.0846	116.8071
27	6.2139	0.1609	74.4838	0.0134	11.9867	0.0834	121.1523
28	6.6488	0.1504	80.6977	0.0124	12.1371	0.0824	125.3635
29	7.1143	0.1406	87.3465	0.0114	12.2777	0.0814	129.4399
30	7.6123	0.1314	94.4608	0.0106	12.4090	0.0806	133.3809
40	14.9745	0.0668	199.6351	0.0050	13.3317	0.0750	165.6245
50	29.4570	0.0339	406.5289	0.0025	13.8007	0.0725	186.7059
100	867.7163	0.0012	12381.6618	0.0001	14.2693	0.0701	216.4693

Discount and Investment Factors, $i = 0.075$

n	F/P	P/F	F/A	A/F	P/A	A/P	P/G
1	1.0750	0.9302	1.0000	1.0000	0.9302	1.0750	0.9302
2	1.1556	0.8653	2.0750	0.4819	1.7956	0.5569	2.6609
3	1.2423	0.8050	3.2306	0.3095	2.6005	0.3845	5.0758
4	1.3355	0.7488	4.4729	0.2236	3.3493	0.2986	8.0710
5	1.4356	0.6966	5.8084	0.1722	4.0459	0.2472	11.5538
6	1.5433	0.6480	7.2440	0.1380	4.6938	0.2130	15.4415
7	1.6590	0.6028	8.7873	0.1138	5.2966	0.1888	19.6608
8	1.7835	0.5607	10.4464	0.0957	5.8573	0.1707	24.1464
9	1.9172	0.5216	12.2298	0.0818	6.3789	0.1568	28.8407
10	2.0610	0.4852	14.1471	0.0707	6.8641	0.1457	33.6926
11	2.2156	0.4513	16.2081	0.0617	7.3154	0.1367	38.6574
12	2.3818	0.4199	18.4237	0.0543	7.7353	0.1293	43.6957
13	2.5604	0.3906	20.8055	0.0481	8.1258	0.1231	48.7730
14	2.7524	0.3633	23.3659	0.0428	8.4892	0.1178	53.8594
15	2.9589	0.3380	26.1184	0.0383	8.8271	0.1133	58.9288
16	3.1808	0.3144	29.0772	0.0344	9.1415	0.1094	63.9590
17	3.4194	0.2925	32.2580	0.0310	9.4340	0.1060	68.9307
18	3.6758	0.2720	35.6774	0.0280	9.7060	0.1030	73.8276
19	3.9515	0.2531	39.3532	0.0254	9.9591	0.1004	78.6359
20	4.2479	0.2354	43.3047	0.0231	10.1945	0.0981	83.3442
21	4.5664	0.2190	47.5525	0.0210	10.4135	0.0960	87.9430
22	4.9089	0.2037	52.1190	0.0192	10.6172	0.0942	92.4246
23	5.2771	0.1895	57.0279	0.0175	10.8067	0.0925	96.7831
24	5.6729	0.1763	62.3050	0.0161	10.9830	0.0911	101.0137
25	6.0983	0.1640	67.9779	0.0147	11.1469	0.0897	105.1132
26	6.5557	0.1525	74.0762	0.0135	11.2995	0.0885	109.0792
27	7.0474	0.1419	80.6319	0.0124	11.4414	0.0874	112.9104
28	7.5759	0.1320	87.6793	0.0114	11.5734	0.0864	116.6063
29	8.1441	0.1228	95.2553	0.0105	11.6962	0.0855	120.1672
30	8.7550	0.1142	103.3994	0.0097	11.8104	0.0847	123.5938
40	18.0442	0.0554	227.2565	0.0044	12.5944	0.0794	150.9629
50	37.1897	0.0269	482.5299	0.0021	12.9748	0.0771	168.0462
100	1383.0772	0.0007	18427.6961	0.0001	13.3237	0.0751	190.0089

Discount and Investment Factors, $i = 0.080$

n	F/P	P/F	F/A	A/F	P/A	A/P	P/G
1	1.0800	0.9259	1.0000	1.0000	0.9259	1.0800	0.9259
2	1.1664	0.8573	2.0800	0.4808	1.7833	0.5608	2.6406
3	1.2597	0.7938	3.2464	0.3080	2.5771	0.3880	5.0221
4	1.3605	0.7350	4.5061	0.2219	3.3121	0.3019	7.9622
5	1.4693	0.6806	5.8666	0.1705	3.9927	0.2505	11.3651
6	1.5869	0.6302	7.3359	0.1363	4.6229	0.2163	15.1462
7	1.7138	0.5835	8.9228	0.1121	5.2064	0.1921	19.2306
8	1.8509	0.5403	10.6366	0.0940	5.7466	0.1740	23.5527
9	1.9990	0.5002	12.4876	0.0801	6.2469	0.1601	28.0550
10	2.1589	0.4632	14.4866	0.0690	6.7101	0.1490	32.6869
11	2.3316	0.4289	16.6455	0.0601	7.1390	0.1401	37.4046
12	2.5182	0.3971	18.9771	0.0527	7.5361	0.1327	42.1700
13	2.7196	0.3677	21.4953	0.0465	7.9038	0.1265	46.9501
14	2.9372	0.3405	24.2149	0.0413	8.2442	0.1213	51.7165
15	3.1722	0.3152	27.1521	0.0368	8.5595	0.1168	56.4451
16	3.4259	0.2919	30.3243	0.0330	8.8514	0.1130	61.1154
17	3.7000	0.2703	33.7502	0.0296	9.1216	0.1096	65.7100
18	3.9960	0.2502	37.4502	0.0267	9.3719	0.1067	70.2144
19	4.3157	0.2317	41.4463	0.0241	9.6036	0.1040	74.6170
20	4.6610	0.2145	45.7620	0.0219	9.8181	0.1019	78.9079
21	5.0338	0.1987	50.4229	0.0198	10.0168	0.0998	83.0797
22	5.4365	0.1839	55.4568	0.0180	10.2007	0.0980	87.1264
23	5.8715	0.1703	60.8933	0.0164	10.3711	0.0964	91.0437
24	6.3412	0.1577	66.7648	0.0150	10.5288	0.0950	94.8284
25	6.8485	0.1460	73.1059	0.0137	10.6748	0.0937	98.4789
26	7.3964	0.1352	79.9544	0.0125	10.8100	0.0925	101.9941
27	7.9881	0.1252	87.3508	0.0114	10.9352	0.0914	105.3742
28	8.6271	0.1159	95.3388	0.0105	11.0511	0.0905	108.6198
29	9.3173	0.1073	103.9659	0.0096	11.1584	0.0896	111.7323
30	10.0627	0.0994	113.2832	0.0088	11.2578	0.0888	114.7136
40	21.7245	0.0460	259.0565	0.0039	11.9246	0.0839	137.9668
50	46.9016	0.0213	573.7702	0.0017	12.2335	0.0817	151.8263
100	2199.7613	0.0005	27484.5157	0.0000	12.4943	0.0800	168.1050

Discount and Investment Factors, $i = 0.085$

n	F/P	P/F	F/A	A/F	P/A	A/P	P/G
1	1.0850	0.9217	1.0000	1.0000	0.9217	1.0850	0.9217
2	1.1772	0.8495	2.0850	0.4796	1.7711	0.5646	2.6206
3	1.2773	0.7829	3.2622	0.3065	2.5540	0.3915	4.9693
4	1.3859	0.7216	4.5395	0.2203	3.2756	0.3053	7.8556
5	1.5037	0.6650	5.9254	0.1688	3.9406	0.2538	11.1808
6	1.6315	0.6129	7.4290	0.1346	4.5536	0.2196	14.8585
7	1.7701	0.5649	9.0605	0.1104	5.1185	0.1954	18.8130
8	1.9206	0.5207	10.8306	0.0923	5.6392	0.1773	22.9783
9	2.0839	0.4799	12.7512	0.0784	6.1191	0.1634	27.2972
10	2.2610	0.4423	14.8351	0.0674	6.5613	0.1524	31.7201
11	2.4532	0.4076	17.0961	0.0585	6.9690	0.1435	36.2041
12	2.6617	0.3757	19.5492	0.0512	7.3447	0.1362	40.7125
13	2.8879	0.3463	22.2109	0.0450	7.6910	0.1300	45.2140
14	3.1334	0.3191	25.0989	0.0398	8.0101	0.1248	49.6820
15	3.3997	0.2941	28.2323	0.0354	8.3042	0.1204	54.0941
16	3.6887	0.2711	31.6320	0.0316	8.5753	0.1166	58.4316
17	4.0023	0.2499	35.3207	0.0283	8.8252	0.1133	62.6792
18	4.3425	0.2303	39.3230	0.0254	9.0555	0.1104	66.8244
19	4.7116	0.2122	43.6654	0.0229	9.2677	0.1079	70.8570
20	5.1120	0.1956	48.3770	0.0207	9.4633	0.1057	74.7693
21	5.5466	0.1803	53.4891	0.0187	9.6436	0.1037	78.5554
22	6.0180	0.1662	59.0356	0.0169	9.8098	0.1019	82.2111
23	6.5296	0.1531	65.0537	0.0154	9.9629	0.1004	85.7336
24	7.0846	0.1412	71.5832	0.0140	10.1041	0.0990	89.1212
25	7.6868	0.1301	78.6678	0.0127	10.2342	0.0977	92.3736
26	8.3401	0.1199	86.3546	0.0116	10.3541	0.0966	95.4910
27	9.0490	0.1105	94.6947	0.0106	10.4646	0.0956	98.4748
28	9.8182	0.1019	103.7437	0.0096	10.5665	0.0946	101.3266
29	10.6528	0.0939	113.5620	0.0088	10.6603	0.0938	104.0489
30	11.5583	0.0865	124.2147	0.0081	10.7468	0.0931	106.6444
40	26.1330	0.0383	295.6825	0.0034	11.3145	0.0884	126.4191
50	59.0863	0.0169	683.3684	0.0015	11.5656	0.0865	137.6759
100	3491.1927	0.0003	41061.0904	0.0000	11.7613	0.0850	149.7930

BIBLIOGRAPHY

Alder, R. W., 1993. Revitalizing the Clean Water Act. *Environment,* Vol. 35, No. 9, pp. 4–5 and 40.

American Society of Civil Engineers (ASCE), 1968. *Urban Water Resources Research,* Appendix G. p. G 34, Urban Hydrology Research Council, ASCE., New York.

American Water Resources Association (AWRA), 1994. Cut down that water! *HYDATA News and Views,* Vol. 14, No. 2.

Bajwa, R. S., W. M. Crosswhite, and J. E. Hostetler, 1987. *Agricultural Irrigation and Water Supply,* Agricultural Information Bulletin No. 532. U.S. Department of Agriculture, Economic Research Service, Washington, D.C.

Balling, R. C., Jr. 1993. *The Heated Debate,* Pacific Research Institute for Public Policy, San Francisco, California.

Barrows, H. H., 1923. Geography as human ecology. *Annals of the Association of American Geographers,* Vol. XIII, No. 1., pp. 1–14.

Beaumont, P., 1978. Man's impact on river systems: a world–wide view. *Area* 10, pp. 38–41.

Benarde, M. A., 1989. *Our Precious Habitat,* John Wiley & Sons, New York.

Black, P. E., 1987. *Conservation of Water and Related Land Resources,* Rowman & Littlefield, Totowa, New Jersey.

Blaney, H. F. 1955. Climate as an index of irrigation needs. *Water, The Yearbook of Agriculture 1955,* U.S. Department of Agriculture, USGPO, Washington, D.C.

Blaney, H. F., 1959. Monthly consumptive use requirements for irrigated crops. *Proceedings of the American Society of Civil Engineers, Journal of Irrigation and Drainage Division,* Vol. 84, No. IR1, pp. 1–12.

Blaney, H. F., H. R. Haise, and M. E. Jensen, 1960. *Monthly Consumptive Use by Irrigated Crops in the Western United States,* Provisional Supplement to SCS TP-96. U.S. Soil Conservation Service, USGPO, Washington, D.C.

Bolsenga, S. J. and C. E. Herdendorf, 1993. *Lake Erie and Lake St. Clair: Handbook,* Wayne State University Press, Detroit, Michigan.

Botkin, D. B. and E. A. Keller, 1995. *Environmental Science: Earth as a Living Planet,* John Wiley & Sons, New York.

Burby, R. J. and S. P. French, 1985. *Flood Plain Land Use Management, A National Assessment,* Studies in Water Policy and Management, No. 5. Westview Press, Boulder, Colorado.

Callender, E. and P. C. van Metre, 1997. Reservoir sediment cores show U.S. lead declines. *Environmental Science & Technology,* Vol. 31, No. 9, pp. 424A–248A.

Clawson, M., 1981. *New Deal Planning: The National Resources Planning Board.* Published for Resources for the Future, Johns Hopkins Press, Baltimore, Maryland.

Charney, J. G., 1975. Dynamics of deserts and droughts in the Sahel. *Quarterly Journal of the Royal Meteorological Society,* Vol. 101, pp. 193–202.

Chronicle of America, 1993. JL International Publishing, Liberty, Missouri.

Coggins, G. C. and C. F. Wilkinson, 1981. *Federal Public Land and Resources Law,* University Casebook Series, The Foundation Press, Mineola, New York.

Collier, M. P., R. H. Webb, and E. D. Andrews, 1997. Experimental flooding in the Grand Canyon. *Scientific American,* Vol. 276, No.1, January, pp. 82–89.

Cordell, H. K., J. C. Bergstrom, L. A. Hartman, and D. B. K. English, 1990. An Analysis of Outdoor Recreation and Wilderness Situation in the United States 1989–2040. A Technical Document Supporting the 1989 USDA Forest Service RPA Assessment, General Technical Report RM-189. Rocky Mountain Forest and Range Experiment Station, USFS, USDA, Fort Collins, CO.

Ciracy-Wantrup, S. V., 1961. Philosophy and objectives of watershed policy. In G. Toley and F. Riggs (eds.), *Economics of Watershed Planning,* Symposium on the economics of watershed planning sponsored by the Southeast Land Tenure Research Committee, The Farm Foundation, and the Tennessee Valley Authority. Iowa State University Press, Ames.

Dellapenna, J. W., 1994. The regulated riparian version of the ASCE Model Water Code: The third way to allocate water. *Water Resources Bulletin,* Vol. 30, No. 2, pp. 197–204.

Dellapenna, J. W., September, 1997. Personal communication.

Demissie, M. and A. Kahn, 1993. *Influence of Wetlands on Streamflow in Illinois,* Illinois Department of Conservation, Springfield.

Doorenbos, J. and W. O. Pruitt, 1977. *Crop Water Requirements,* FAO Irrigation and Drainage Paper 24 (revised). U.N. Food and Agricultural Organization (FAO), Rome.

Dracup, J. A., K. S. Lee, and E. G. Paulson, Jr. 1980. On the definition of droughts. *Water Resources Research,* Vol. 16, No. 2, pp. 297–302.

Dunne, T. and L. B. Leopold, 1978. *Water in Environmental Planning,* W. H. Freeman and Company, New York.

Dzurik, A. A., 1990. *Water Resources Planning,* Rowman & Littlefield Publishers, Savage, Maryland.

Eisner, T., J. Lubchenco, E. O. Wilson, D. S. Wilcove, and M. J. Bean, 1995. Building a scientifically sound policy for protecting endangered species. *Science,* Vol. 268, pp. 1231–1232.

Reviving the Everglades. *Environment,* 1995. Vol. 35, No. 7, p. 21.

Faber, S., 1993. The Mississippi flood. *Environment,* Vol. 35, No. 10, pp. 2–3.

Food and Agriculture Organization (FAO), 1990. *Food and Agriculture Production Yearbook, 1990.* FAO Statistics Series No. 99, Vol. 44. Statistics and Economics Department, FAO, United Nations, Rome.

Federal Inter-Agency River Basin Committee (FIARBC), Subcommittee on Benefits and Costs, 1950. *Proposed Practices for Economic Analysis of River Basin Projects,* Washington, D.C.

Federal Emergency Management Agency (FEMA), 1996. *Dam Mitigation,* on-line information, *http://www.fema.gov/home/mit/damsafe.htm*

Federal Emergency Management Agency (FEMA), 1997. *Flood Mitigation: Floodplain Managment Summary,* online information at *http://www.fema.gov/home/mit/fldmit.htm*

Federal Emergency Management Agency (FEMA), 1992. *Floodplain Management in the United States: An Assessment Report, Vol. 1, Summary.* Prepared by the Natural Hazards Research and Applications Information Center, University of Colorado, Boulder.

Foxworthy, B. L. and D. W. Moody, 1986. National perspectives on surface water resources. In *National Water Summary, 1985, Hydrologic Events and Surface Water Resources,* U.S. Geological Survey Water Supply Paper 2300. USGPO, Washington, D.C.

Gardner, J. S., 1977. *Physical Geography,* Harper's College Press, New York.

Gallowy, G. E., 1995. New directions in floodplain management. *Water Resources Bulletin,* Vol. 31, No. 3, pp. 351–357.

Gleick, P. H., 1993. *Water in Crisis,* Oxford University Press, New York.

Gomez-Ferrer, R., D. W. Hendricks, and C. D. Turner, 1983. Salt transport by the South Platte River in Northeast Colorado. *Water Resources Bulletin,* Vol. 19, No. 2, pp. 183–190.

Grund, F. J., 1837. *The American People in Their Moral, Social and Political Relations,* Longman, London, England.

Hall, D. W. and D. R. Risser, 1993. Effects of agricultural management on nitrogen fate and transport in Lancaster County, Pennsylvania. *Water Resources Bulletin,* Vol. 29, No. 1, pp. 55–76.

Haan C. T., 1977. *Statistical Methods in Hydrology,* Iowa State University Press, Ames.

Hammitt, W. E. and D. N. Cole, 1987. *Wildland Recreation: Ecology and Management,* John Wiley & Sons, New York.

Hansen, V. E., O. W. Israelsen, and G. E. Stringham, 1980. *Irrigation Principles and Practices,* John Wiley & Sons, New York.

Hedin, L. O. and G. E. Likens, 1996. Atmospheric dust and acid rain. *Scientific American,* Vol. 275, No. 6, December, pp. 88–92.

Henderson-Sellers, A. and K. McGuffie, 1987. *A Climate Primer,* John Wiley & Sons, New York.

High Plains Associates, 1982. *Six-State High Plains Ogallala Aquifer Regional Resources Study.* A Report to the U.S. Department of Commerce and the High Plains Study Council. Austin, Texas.

Holmes, B. H., 1979. *History of Federal Water Resource Programs and Policies, 1961–1970.* Miscellaneous Publication No. 1379. U.S., Department of Agriculture, USGPO, Washington, D.C.

Hounam, C. E., J. J. Burgos, M. S. Kalik, W. C. Palmer, and J. Rodda, 1975. *Drought and Agriculture,* Technical Note No. 138. World Meteorological Organization, Geneva.

Howe, C. H. and F. P. Linaweaver, 1967. The impact of price on residential water demand and its relation to system design and price structure. *Water Resources Research,* Vol. 13, No. 1, pp. 13–32.

Hundley, N., Jr. 1986. The West against itself: The Colorado River, an institutional history. In G. D. Weatherford and F. L. Brown (eds.), *New Courses for the Colorado River, Major Issues for the Next Century,* University of New Mexico Press, Albuquerque.

Inland Waterways Commission, 1908. *Report of the Inland Waterways Commission,* Senate Document No. 325, 60th Congress, 1st Session, IWC, USGPO, Washington, D.C.

Jackson, R. D. and S. B. Idso, 1975. Surface albedo and desertification. *Science,* Vol. 189, pp. 1012–1013.

Jackson, W. L., B. Shelby, A. Martinez, and B. P. van Haveren, 1989. An interdisciplinary process for protecting instream flows. *Journal of Soil and Water Conservation,* Vol. 44, No. 2, pp. 121–126.

James, L. D. and R. R. Lee, 1971. *Economics of Water Resources Planning,* McGraw-Hill, New York.

Jensen, M. E., 1984. Improving irrigation systems. In E. A. Engelbert and A. F. Scheuring (eds.), *Water Scarcity: Impacts on Western Agriculture,* University of California Press, Berkeley.

Job, C. A., 1996. Benefits and costs of wellhead protection. *Ground Water Monitoring and Remediation,* Spring, pp. 65–68.

Jones, P. D., D. E. Parker, T. J. Osborn, and K. R. Briffa, 1998. *Global and Hemispheric Temperature Anomalies—Land and Marine,* Carbon Dioxide Analysis Center, Oak Ridge National Laboratory, Oak Ridge, Tennessee. Data download from *http:/cdiac.ESD.ORNL.GOV/ftp/trends/temp/jonescru/global.dat*

Jordan, J. L., 1994. The effectiveness of pricing as a stand-alone water conservation program. *Water Resources Bulletin,* Vol. 30, No. 5, pp. 871–877.

Karl, T. R., 1983. Some spatial characteristics of drought duration in the United States. *Journal of Climate and Applied Meteorology,* Vol. 22, No. 8, pp. 1356–1366.

Karl, T. R., R. W. Knight, D. R. Easterling, and R. G. Quayle, 1995. Trends in the U.S. climate during the Twentieth Century. *Consequences,* Vol. 1, No. 1, pp. 2–12.

Karl, T. R., D. R. Easterling, R. W. Knight, and P. Y. Hughes, 1994a. U.S. National and regional temperature anomalies. In T. A. Boden, D. P. Kaiser, R. J. Sepanski, and F. W. Stoss (eds.), *Trends '93: A Compendium of Data on Global Change,* Carbon Dioxide Analysis Center, Oak Ridge National Laboratory, Oak Ridge, Tennessee.

Karl, T. R., D. R. Easterling, and P. Ya. Groisman, 1994b. United States historical climatology network—National and regional estimates of monthly and annual precipitation. In T. A. Boden, D. P. Kaiser, R. J. Sepanski, and F. W. Stoss (eds.), *Trends '93: A Compendium of Data on Global Change,* Carbon Dioxide Analysis Center, Oak Ridge National Laboratory, Oak Ridge, Tennessee.

Kates, R. W., 1962. *Hazard and Choice Perception in Flood Plain Management,* Department of Geography Research Paper 70. University of Chicago, Chicago, Illinois.

Kusler, J. and L. Larson, 1993. Beyond the ark: A new approach to U.S. floodplain management. *Environment,* Vol. 35, No. 5, pp. 7–11 and 31–34.

Lamb, J. C., III, 1985. *Water Quality and Its Control,* John Wiley & Sons, New York.

Lantis, D. W., R. Steiner, and A. E. Karinen, 1970. *California: Land of Contrast,* Wadsworth Publishing Company, Belmont, California.

Lawton, J. H. and R. M. May, 1995. *Extinction Rates,* Oxford University Press, Oxford, U.K.

Leopold, L. B. and T. Maddock, 1953. The hydraulic geometry of stream channels and some physiographic implications. *U.S. Geological Survey Professional Paper 252,* Reston, VA.

Leshy, J. D., 1983. Special water districts—The historical background. In J. N. Corbridge Jr. (ed.), *Special Water Districts: Challenges for the Future,* Proceedings of a workshop on special water districts held at the University of Colorado, September 12–13, 1983, Boulder.

Lettenmair, D. P, T. Y. Gan, and D. R. Dawdy, 1988. *Interpretation of Hydrologic Effects of Climate Change in the Sacramento–San Joaquin River Basin, California,* Water Resources Technological Report No. 110. University of Washington, Seattle.

Lindblom, C. E., 1964. The science of muddling through. In W. J. Gore and J. W. Dyson (eds.), *The Making of Decisions,* The Free Press, New York.

Lins, H. F. and P. J. Michaels, 1994. Increasing U.S. streamflow linked to greenhouse forcing. *EOS, Transactions,* American Geophysical Union, Vol. 75, No. 25, p.281 and pp. 284–285.

Linsley, R. K., M. A. Kohler, and J. L. H. Paulhus, 1982. *Hydrology for Engineers,* Fourth Edition, McGraw-Hill, New York.

Lippman, M. and R. B. Schlesinger, 1979. *Chemical Contamination in the Human Environment,* Oxford University Press, New York.

Lord, W. B., J. F. Booker, D. M. Getches, B. L. Harding, D. S. Kenney, and R. A. Young, 1995. Managing the Colorado River in a severe sustained drought: An evaluation of institutional options. *Water Resources Bulletin,* Vol. 31, No. 5, pp. 939–944.

Madariaga, B. and K. E. McConnell, 1987. Exploring existence value. *Water Resources Research,* Vol. 23, No. 5, pp. 936–942.

Marsh, G. P., 1864. *Man and Nature or Physical Geography as Modified by Human Action,* reprinted with D. Lowenthal (ed.), Harvard University Press, Cambridge, Massachusetts, 1965.

Marsh, W. M. and J. M Grossa, 1996. *Environmental Geography: Science, Land Use and Earth Systems,* John Wiley & Sons, New York.

Mather, J. R., 1984. *Water Resources: Distribution, Use, and Management,* Wiley Interscience, John Wiley & Sons, New York.

Matthews, O. P., 1994. Judicial resolution of transboundary water conflicts. *Water Resources Bulletin,* Vol. 30, No. 3, pp. 375–383.

McCormick, Z., 1994a. Interstate water allocation compacts in the western United States—some suggestions. *Water Resources Bulletin,* Vol. 30, No. 3, pp. 385–395.

McCormick, Z., 1994b. Institutional barriers to water markets in the West. *Water Resources Bulletin,* Vol. 30, No. 6, pp. 953–960.

McKay, R. R., 1996. The Conservation Reserve Program and recent efforts to conserve soil and water resources in Pennsylvania. *The Pennsylvania Geographer,* Vol. XXXIV, No. 2, Fall/Winter, pp. 98–112.

McNally, M., 1994. Water marketing: The case of Indian reserved rights. *Water Resources Bulletin,* Vol. 30, No. 6, pp. 963–970.

Meko, D., C. W. Stockton, and W. R. Boggess, 1995. The tree-ring record of severe sustained drought in the Southwest. *Water Resources Bulletin,* Vol. 31, No. 5, pp. 789–801.

Meyers, C. J., A. D. Tarlock, J. C. Corbridge, and D. H. Getches, 1988. *Water Resource Management,* Third Edition, The Foundation Press, Mineola, New York.

Metropolitan Denver Water Study Committee (MDWSC), 1975. *Metropolitan Water Requirements and Resources 1975–2000,* Vol. 1, Text, Denver, Colorado.

Michelsen, A. R., 1994. Administrative, institutional, and structural characteristics of an active water market. *Water Resources Bulletin,* Vol. 30, No. 6, pp. 971–982.

Miller, G. T., Jr., 1993. *Environmental Science,* Fourth Edition, Wadsworth Publishing Company, Belmont, California.

Miller, G. T., Jr., 1995. *Environmental Science,* Fifth Edition, Wadsworth Publishing Company, Belmont, California.

Miller, G. T., Jr., 1997. *Environmental Science,* Sixth Edition, Wadsworth Publishing Company, Belmont, California.

Mitchell, B. 1989. *Geography and Resource Analysis,* Longman Scientific & Technical, New York.

Musgrave, G. W., 1955. How much of the rain enters the soil? In *Water: The Yearbook of Agriculture, 1955,* USGPO, Washington, D.C.

Nash, R., 1973. Rivers and Americans: A century of conflicting priorities. In C. R. Goldman, J. McEvoy, III, and P. J. Richardson (eds.), *Environmental Quality and Water Development,* W. H. Freeman and Company, San Francisco, California.

National Resources Board (NRB), 1934. *A Report on National Planning and Public Works in Relation to Natural Resources and Including Land Use and Water Resources with Findings and Recommendations,* USPGO, Washington, D.C.

National Resource Committee (NRC), 1935 (revised 1936). *Drainage Basin Problems and Programs,* USGPO, Washington, D.C.

National Water Commission (NWC), 1973. *Water Policies for the Future: Final Report,* USGPO, Washington, D.C.

Nelson, J. O., 1997. *WaterWiser, 1997 Residential Water Use Summary,* American Water Works Association (AWWA), online information at *http://www.waterwiser.org/wateruse/main.html*

New York State Office of Parks, Recreation and Historic Preservation, 1989. *People, Resources, Recreation,* Appendix E, Albany, New York.

New York Times, 1997. Breaching of dams would let endangered fish swim to sea. Article retrieved online from the world wide web, April 21, 1997.

Oglesby, R. J. and D. J. Erickson, III, 1989. Soil moisture and the persistence of North American drought. *Journal of Climate,* Vol. 2, pp. 1362–1380.

Omernik, J. M. and R. G. Bailey, 1997. Distinguishing between watersheds and ecoregions. *Journal of the American Water Resources Association,* Vol. 33, No. 5, pp. 935–949.

Osborne, C.T., F. Llacuna, and M. Linsenbigler, 1992. The Conservation Reserve Program, Statistical Bulletin No. 843, U.S. Department of Agriculture, Washington D.C.

Palmer, W. C., 1965. *Meteorological Drought,* Research Paper 45, U.S. Weather Bureau, Washington, D.C.

Penman, H. L., 1948. Natural evaporation from open water, bare soil, and grass. *Proceedings of the Royal Society of London, Ser. A,* Vol. 139, pp. 120–146.

Petersen, M. S., 1984. *Water Resource Planning and Development,* Prentice-Hall, Englewood Cliffs, New Jersey.

Petulla, J. M., 1988. *American Environmental History,* Merrill Publishing Company, Columbus, Ohio.

Platt, R. H., T. O'Riordan, and G. F. White, 1997. Classics in human geography revisited. *Progress in Human Geography,* Vol. 21, No. 2, pp. 243–250.

Postel, S., 1992. *Last Oasis: Facing Water Scarcity,* W. W. Norton, New York.

Powell, J. W., 1879. *Report on the Lands of the Arid Region of the United States with a More Detailed Account of the Lands of Utah,* USGPO, Washington, D.C.

President's Water Resources Policy Commission (PWRPC), 1950. *A Water Policy for the American People* (Vol. 1), *Ten Rivers in America's Future* (Vol. 2), and *Water Resources Law* (Vol. 3). USGPO, Washington, D.C.

Public Land Law Review Committee (PLLRC), 1970. *One Third of the Nation's Land,* A Report to the Congress by the Public Land Law Review Committee. USGPO, Washington, D.C.

Reisner, M., 1986. *Cadillac Desert,* Viking, New York.

Reuss, M., 1992. Coping with uncertainty: Social scientists, engineers, and federal water resources planning. *Natural Resources Journal,* Vol. 32, No. 1, pp. 101–135.

Ruckdeschel, F. R., 1981. *BASIC Scientific Subroutines,* Vol. II, Byte/McGraw-Hill, Peterborough, New Hampshire.

Savini, J. and J. C. Kammerer, 1961. Urban growth and the water regimen. U.S. Geological Survey Water Supply Paper 1591-A. USGPO, Washington, D.C.

Shallat, T., 1992. Water and bureaucracy: Origins of the federal responsibility for water resources, 1787–1838. *Natural Resources Journal,* Vol. 32, No. 1, pp. 5–25.

Scientific American, 1995a. It's melting, Its' melting. Vol. 273, No.1, July, p. 28.

Scientific American, 1995b. High tidings. Vol. 273, No. 2, August, pp. 21–22.

Schlesinger, W. H., J. F. Reynolds, G. L. Cunningham, L. F. Huenneke, W. M. Jarrell, R. A. Virginia, and W. G. Whitford, 1990. Biological feedbacks in global desertification. *Science,* Vol. 247, pp. 1043–1048.

Sewell, W. R. D., J. Davis, A. D. Scott, and D. W. Ross, 1965. A guide to benefit-cost analysis. In I. Burton and R. W. Kates (eds.), *Readings in Resource Management,* University of Chicago Press, Chicago, Illinois.

Sherk, G. W., 1994. Resolving interstate water conflicts in the eastern United States: The re-emergence of the federal-interstate compact. *Water Resources Bulletin,* Vol. 30, No. 3, pp. 397–408.

Shiklomanov, I. A., 1993. World fresh water resources. In P. E. Gleick (ed.), *Water in Crisis,* Oxford University Press, New York.

Shupe, S. J. and J. F. Folk-Williams, 1988. *The Upper Rio Grande: A Guide to Decision Making,* Western Network, Santa Fe, New Mexico.

Simon H. A., 1959. Theories of decision making in economics and behavioral science. *American Economic Review,* Vol. 49, pp. 253–283.

Smith L. G., 1982. Mechanisms for public participation at a normative planning level in Canada. *Canadian Public Policy,* Vol. 8, pp. 561–572.

Solley, W. B., R. R. Pierce, and H. A. Perlman, 1993. *Estimated Use of Water in the United States in 1990,* U.S. Geological Survey Circular 1008. USGPO, Washington, D.C.

Speidel, D. H. and A. F. Agnew, 1988. The world water budget. In Speidel, D. H., L. C. Ruedisili, and A. F. Agnew (eds.), *Perspectives on Water,* Oxford University Press, New York.

Statistical Abstract of the United States, 1997. USGPO, Washington, D.C.

Stein, B. A. and S. R. Flack, 1997a. *1997 Species Report Card: The State of U.S. Plants and Animals,* The Nature Conservancy, Arlington, Virginia.

Stein, B. A. and S. R. Flack, 1997b. Conservation priorities: The state of U.S. plants and animals. *Environment,* Vol. 39, No. 4, pp. 6–11 and 34–39.

Teclaff, L. A. and E. Teclaff, 1973. A history of water development and environmental quality. In C. R. Goldman, J. McEvoy, III, and P. J. Richardson (eds.), *Environmental Quality and Water Development,* W. H. Freeman and Company, San Francisco, California.

John Muir Institute, 1980. *Western Water Institutions in a Changing Environment,* Volumes 1 and 2. Napa, California.

Thornthwaite, C. W., 1948. An approach towards a rational classification of climate. *Geographical Review,* Vol. 38, pp. 55–94.

Thompson, S. A., 1998. *Hydrology for Water Management,* Balkema Publishers, Rotterdam, Netherlands.

Thompson, S. A., 1992. Simulation of climate change impacts on water balances in the Central United States. *Physical Geography,* Vol. 13, No. 1, pp. 31–52.

Thoreau, H. D. 1854, *Walden and Civil Disobedience,* Harper & Row, New York, 1965.

Tourbier, J. T. and R. Westmacott, 1981. *Water Resource Protection Technology: A Handbook of Measures to Protect Water Resources in Land Development,* Urban Land Institute, Washington, D.C.

Tourbier, J. T. and R. Westmacott, 1992. *Lakes and Ponds,* Urban Land Institute, Washington, D.C.

Trelease, F. J., 1979. *Water Law.* West Publishing Co., St. Paul, Minnesota.

Trenberth, K. E., 1997. Commentary. *Environment,* Vol. 39, No. 4, pp. 5 and 40–41.
U.S. Army Corps of Engineers, 1977. *Estimate of National Hydrolelectric Power Potential at Existing Dams,* Institute for Water Resources, Fort Belvoir, Virginia.
U.S. Bureau of the Census, 1990. *Census of Population and Housing,* series CPH-2, USGPO, Washington, D.C.
U.S. Bureau of the Census, 1994. *1994 Farm and Ranch Irrigation Survey,* USGPO, Washington, D.C.
U.S. Bureau of Indian Affairs, 1996. *Department of the Interior Lands Under Jurisdiction of the Bureau of Indian Affairs,* online data downloaded from *http://www.doi.gov/bia/realty/state.html*
U.S. Bureau of Reclamation, 1997. *A Brief History,* online information at *http://www.usbr.gov/history/borhist.htm.*
U.S. Congress, 1961. *Report of the Senate Select Committee on National Water Resources,* Senate Report No. 29, with 32 Committee Prints. 86th Congress, 1st Session. USGPO, Washington, D.C.
U.S. Congress, 1966. Task Force on Flood Policy, *A Unified Program for Managing Flood Losses,* House Document 465. USGPO, Washington, D.C.
U.S. Department of Agriculture, Soil Conservation Service, 1970. Irrigation Water Requirements, Technical Release No. 21. USGPO, Washington, D.C.
U.S. Department of Commerce, National Oceanic and Atmospheric Administration, National Climate Data Center, 1990. *National Climate Information Disc,* Volume 1, Asheville, North Carolina.
U.S. Environmental Protection Agency (USEPA), 1987 and 1993. *Guidelines for Delineation of Wellhead Protection Areas,* USGPO, Washington, D.C. Available from U.S. Department of Commerce, National Technical Information Service, Springfield, Virginia.
U.S. Environmental Protection Agency (USEPA), 1992. *Secondary Treatment and Benefits Study,* Office of Wastewater Enforcement and Compliance, Washington, D.C.
U.S. Environmental Protection Agency (USEPA), 1994a. *Drinking Water Regulations and Health Advisories,* Office of Water, Washington, D.C.
U.S. Environmental Protection Agency (USEPA), 1994b. *The Quality of Our Nation's Water: 1994,* EPA-841-S-94-004. Washington, D.C.
U.S. Environmental Protection Agency (USEPA), 1995. How to conserve water and use it effectively, *Cleaner Water Through Conservation,* EPA 841-B-95-002. Washington, D.C.
U.S. Environmental Protection Agency (USEPA), 1996a. *Safe Drinking Water Act Amendments of 1996, http://www.epa.gov/OGWDW/SDWAsumm.html* and *http://www.epa.gov/OGWDW/SDWAthem.html*
U.S. Environmental Protection Agency (USEPA), 1996b, *National Water Quality Inventory: 1994 Report to Congress,* online information at *http://earth1.epa.gov/305b/*
U.S. Environmental Protection Agency (USEPA), 1996c. *Pointer No. 1, Nonpoint Source Pollution: The Nation's Largest Water Quality Problem,* EPA-841-F-96-004A. Washington, D.C.
U.S. Environmental Protection Agency (USEPA), 1996d. *Pointer No. 4, The Nonpoint Source Management Program,* EPA-841-F-96-004D. Washington, D.C.
U.S. Environmental Protection Agency (USEPA), 1996e. *Why watersheds?,* Office of Water, EPA-800-F-96-001. Washington, D.C.
U.S. Environmental Protection Agency (USEPA), 1997. *Drinking Water Infrastructure Needs Survey,* First Report to Congress, Office of Water Quality, EPA-812-R-97-001, Washington, D.C.
U.S. Fish and Wildlife Service, 1992. Annual Report of Lands Under Control of the Fish and Wildlife Service, USGPO, Washington, D.C.
U.S. Forest Service, 1993. *Report of the Forest Service—Fiscal Year 1992,* USGPO, Washington, D.C.
U.S. Geological Survey, 1996. *National Water Quality Assessment Program* home page, Worldwide Web address, *http://wwwrvares.er.usgs.gov/nawqa_home.html*
U.S. Geological Survey, 1998. Annual flood data downloaded from the USGS's homepage at *http://water.usgs.gov/*
U.S. National Parks Service, 1992. *The National Parks: Index 1991,* USGPO, Washington, D.C.
Viessman, W., Jr. and C. Welty, 1985. *Water Management: Technology and Institutions,* Harper & Row, New York.

Water Resource Council (WRC), 1968. *The Nation's Water Resources,* Part 1, *First National Water Assessment,* USGPO, Washington, D.C.

Water Resources Council (WRC), 1973. Principles and standards for planning water and related land resources. *Federal Register,* Vol. 38, No. 174, Part III, September 10.

Water Resources Council (WRC), 1976. *A Unified Program for Floodplain Management,* USGPO, Washington, D.C.

Water Resource Council (WRC), 1978. *The Nation's Water Resources 1975–2000,* USGPO, Washington, D.C.

Water Resources Council (WRC), 1983. Repeal of water and related land resources planning principles, standards, and procedures; and adoption of economic and environmental principles and guidelines for water and related land and resources implementation studies, *Federal Register,* Vol. 48, No. 48.

Walz, M. 1993. Map adapted from *National Geographic* magazine for use by the Geography Education Program, National Geographic Society, Washington, D.C.

Ward, R. C., 1996. Water quality monitoring: Where's the beef? *Water Resources Bulletin,* Vol. 32, No. 4, pp. 673–680.

Warrick, R. A., 1975. *Drought Hazard in the United States: A Research Assessment,* IBS Monograph NSF-RA-E-75-004. University of Colorado, Boulder.

Warrick, R. A. and M. J. Bowden, 1981. The changing impacts of droughts in the Great Plains. In M. P. Lawson, and M. E. Baker (eds.), *The Great Plains: Perspectives and Prospects,* University of Nebraska Press, Lincoln.

Weston, R. T., 1984. Delaware River Basin: Courts vs. compacts, paper presented for the *American Society of Civil Engineers' Symposium on Social and Environmental Objectives in Water Resources Management: The Courts as Water Managers,* Atlanta, Georgia.

White, G. F., 1945. *Human Adjustment to Floods,* Research Paper No. 29. University of Chicago, Department of Geography, Chicago, Illinois.

White, G. F., 1969. *Strategies in American Water Management,* Ann Arbor Paperback. The University of Michigan Press, Ann Arbor, Michigan.

Wilhite, D. A., 1997. State actions to mitigate drought: Lessons learned. *Journal of the American Water Resources Association,* Vol. 33, No. 5, pp. 961–968.

Wilson, H. and J. Hansen, 1994. Global and hemispheric temperature anomalies from instrumental surface air temperature records. In T. A. Boden, D. P. Kaiser, R. J. Sepanski, and F. W. Stoss (eds.), *Trends '93: A Compendium of Data on Global Change,* Carbon Dioxide Analysis Center, Oak Ridge National Laboratory, Oak Ridge, Tennessee.

Wood, D. B., 1991. Silt is obstructing Oakland shipping traffic. *The Christian Science Monitor,* December 11, p. 9.

World Resources Institute (WRI), 1992. *World Resources: 1992–1993,* Washington, D.C.

World Resources Institute (WRI), 1994. *World Resources: 1994–1995,* Washington, D.C.

Young, R. A., 1995. Coping with a severe sustained drought on the Colorado River: Introduction and overview. *Water Resources Bulletin,* Vol. 31, No. 5, pp. 779–788.

Zich, A. and B. Sacha, 1997. China's Three Gorges before the flood. *National Geographic,* Vol., 192, No. 3, September, pp. 3–33.

Zinser, C. I., 1995. *Outdoor Recreation: United States National Parks, Forests, and Public Lands,* John Wiley & Sons, New York.

Zorpette, G., 1996. Hanford's nuclear wasteland. *Scientific American,* Vol. 274, No. 5, May, pp. 88–97.

CASES CITED

State of New Mexico v. Aamodt, 618 F. Supp. 993, 1009 (D.N.M. 1985)

Arizona v. California, 373 U.S. 546 (1963).

Arizona v. California, 460 U.S. 605 (1983).

Baker v. Ore-Ida Foods, Inc., Supreme Court of Idaho, 95 Idaho 575, 513, P.2d 627 (1973).

City and County of Denver v. Northern Colorado Water Conservancy District, Supreme Court of Colorado, 130 Colo. 375, 276 P.2d 992 (1954).

Clark v. Peckham, 10 R.I. 35 (1871).

Coffin v. Left Hand Ditch Company, 6 Colo. 443 (1882).

Colorado v. New Mexico, 459 U.S. 178 (1982).

Colorado v. New Mexico, 467 U.S. 310 (1984).

Connecticut v. Massachusetts, 228 U.S. 660 (1931).

Gibbons v. Ogden, 22 U.S. (9 Wheat.) 1 (1824).

In re application of Howard Sleeper, Case No. RA-84-53(C), Rio Arriba County, First Judicial District, April 16, 1985. Rev'd, 107 N.M. 494, 760 P.2d 787 (Ct. App. 1988).

Irwin v. Phillips, Supreme Court of California, 5 Cal. 140 (1855).

National Audobon Society v. Superior Court, 33 Cal. 3d 419, 658 P.2d 509 (1983).

New Jersey v. New York, 283 U.S. 336 (1931).

New York v. Baumberger, 7 Robt. 219 (1867).

Sporhase v. Nebraska ex rel. Douglas, 458 U.S. 941 (1982).

State of New Mexico ex rel. Reynolds v. Lewis, 1989. Nos. 20294, 22600 (Chaves County 1956) (consolidated) (decision of the Court filed January 26, 1989).

State of New Mexico ex rel. Reynolds v. Aamodt, 537 F.2d 1102 (10th Cir. 1976) also at 618 F. Supp. 993 (D.N.M. 1985).

Stratton v. Mt Hermon Boys' School, Supreme Judicial Court of Massachusetts, 216 Mass 83, 103, N.E. 87 (1913).

United States v. New Mexico, 438 U.S. 696 (1978).

Winters v. United States, 207 U.S. 564 (1908).

Wyoming v. United States, 42 U.S. 406 (1989).

INDEX

308 reports, 43
305(b) reports, 298
404 permits, 63
Acidity, 274
Acid mine drainage, 288
Acid precipitation, 289
Aquifers, 21
Arizona v. California, 1963, 102, 112, 116
Arizona v. California, 1983, 102
Autoregressive (AR) model, 183

Baker v. Ore-Ida Foods, Inc., 96
Barrows, Harlan H., 46, 50, 308
Basin–wide management, 41, 70, 301, 326
Beneficial use in prior appropriation, 86
Beneficial uses for water quality assessment, 298
Bennett, Hugh Hammond, 48
Best Management Practices (BMPs), 296
Biochemical oxygen demand, 278
Blaney-Criddle method, 12, 234
Boulder Canyon Project Act, 44, 112, 116, 197
Bounded rationality, 206
Bureau of Land Management and recreation, 266
Bureau of Reclamation, 40, 43, 56, 59, 60, 64, 68, 262, 267

California Debris Commission, 37
California gold mining, 34, 81
California State Water Project, 197
Canal building era, 32
Carter, Jimmy, 66
Case law, 72
Cash flow diagram, 155
Central Arizona Project, 116, 197
City and County of Denver v. Northern Colorado Water Conservancy, 85
Clark v. Peckham, 37
Clean Water Act, 63, 292, 295
Climate change, 24
Coffin v. Left Hand Ditch Company, 191
Colorado–Big Thompson Project, 149
Colorado River, 56, 112, 196
Colorado River Aqueduct, 195, 197
Colorado River Compact, 108, 115, 197
Colorado River Storage Project Act, 56
Colorado v. New Mexico, 105
Columbia River, 259
Commerce power, 33, 97
Common law, 72, 81
Comprehensive Environmental Response, Compensation, and Liability Act, 67, 291
Conjunctive use, 94
Connecticut v. Massachusetts, 105
Conservation, 39
Constraints to decision making, 208
Consumptive use, 12, 234
Contingent valuation, 165, 169
Corps of Engineers, 32, 43, 60, 65, 262, 267, 268

Cost–benefit analysis, 153
  direct benefits, 163
  Green Book, 52,167
  indirect benefits, 163
  intangibles, 163
  optimality criteria, 162
  problems with the method, 163
  Senate Document 97, 59, 167
  wellhead protection, 225
Crop evapotranspiration, 234

Dams and reservoirs, 56, 57, 178
  environmental impacts, 190
  evaporation, 179
  sedimentation, 179, 180
  yield reliability, 180
Decision making, 200
Defendant, 72
Demand management, 174
Desalination, 178, 198
Desert Land Act, 35
Desertification, 3
Discharge, 4
Discount factors, 155
Discount rate, 154
Disease–causing organisms, 222, 225, 276
Dissolved oxygen, 278, 279
Domestic water use, 212
Drainage basin, 7
Drinking water standards, 64, 220
Drought, 326
  analysis, 328
  definition, 326
  human adjustments, 336
Dust Bowl, 48, 332–335
Dynamic equilibrium system, 23
Echo Park Dam controversy, 56
Economically efficient irrigation, 239
Economically efficient pollution removal, 170
Economics, 143
Ecosystems and water quality, 303
Effective precipitation, 238
Effluent trading, 303
Electrical power categories, 249
El Niño and flooding, 69
Endangered species, 259, 303, 304
Endangered Species Act, 62
Environmental economics, 168
Environmental Impact Statement, 62
Environmental values of instream flows, 270
Epilimnion, 280
Equitable apportionment, 105
Eutrophication, 283
Evaporation, 8, 10
Evaporation pans, 9
Evapotranspiration, 8, 11, 234
Everglades, 58
Exceedence probability, 186, 316
Expected value, 165
Externality, 18, 168
Extinction of species, 303

Federal Inter-Agency River Basin Committee, 51
Federal powers for water resources, 96
Federal reserved water right, 98
Federal subsidies, 152
Federal Water Pollution Control Act, 54, 62
Federal Water Pollution Control Act Amendments, 63
Federal Water Power Act, 40
Fish and Wildlife Coordination Act, 54
Flood control, 34, 320
Flood Control Act of 1928, 43
Flood Control Act of 1936, 49
Flood Control Act of 1944, 51
Flood control paradox, 60, 320
Flood frequency analysis, 316
Foreign water, 89
Frequency analysis for reservoir storage, 186

Gallatin Report, 32
General Reservation Act, 39
Geographic Information Systems (GIS), 136
Gibbons v. Ogden, 33
Glen Canyon Dam, 56, 191, 257
Global average temperature anomalies, 27
Global circulation models, 25
Global warming, 24
Green Book, 52,167
Groundwater, 20
Groundwater law, 92
Groundwater mining, 95
Groundwater withdrawals, 122, 133, 136

Heat pollution, 289
Hetch-Hetchy Valley and Dam, 40
Hoover Dam, 44, 116, 248
Human adjustments to droughts, 336
Human adjustments to floods, 321

Human–ecological model of hazards, 308
Human modification of runoff, 18
Hydroelectric dams, 255
Hydroelectric power, 246
  water use, 253
Hydroelectric power potential, 251
Hydrograph, 313
Hydrologic cycle, 5–7
Hypolimnion, 280

Ickes, Harold, 45
Indian water rights, 98, 149
Indoor water use, 215
Infiltration and soil moisture, 16
Inland Waterways Commission, 41
Inorganic chemical pollutants, 288
Instream flow protection, 271
Instream flow values, 270
Instream water uses, 119, 245
Intangibles, 163
Interbasin transfers, 191
Interstate compacts, 98, 106–112
Interstate waters, 105
Iron triangle, 51
Irrigation, 227
  environmental impacts, 242
  historical development, 35, 40, 227
  water conservation, 241
  water withdrawals, 130, 132, 133, 230
  in the United States, 229
Irrigation application methods, 231
Irrigation efficiency, 238
Irrigation water requirement, 238
Irwin v. Phillips, 82
Isohyet, 12

Karez irrigation system, 228

Lake Erie and eutrophication, 286
Leaching requirement, 238
Lead pollution, 288
Load factor, 250
Los Angeles Aqueduct, 194, 195
Los Angeles water supply development, 193
Lysimeters, 11

Markov model, 183
Marsh, George Perkins, 37
Middle Rio Grande Conservancy District, 202
Mississippi Valley Committee, 47
Missouri River, 53
Muir, John, 40
Mulholland, William, 194
Municipal wastewater treatment, 293
Multiobjective planning, 55, 325
Multipurpose dam, 116, 188, 189
Multipurpose planning, 41

National Audubon Society v. Superior Court, 90
National Conservation Commission, 41
National Environmental Policy Act, 61
National Flood Insurance Program, 61, 319
National Forest Service and recreation, 262, 263
National Monitoring Program (for BMPs), 296
National Parks Service and recreation, 265
National Resources Board, 47
National Resources Committee, 47
National Resources Conservation Service, 48
National Resources Planning Board, 47
National Water Commission, 64
National Water Quality Assessment Program, 297
National Waterways Commission, 41
Natural hazards, 307
Navigable waters, 33, 97
NAWAPA, 192
New Deal, 45
New York v. Baumberger, 37
Niagara Falls, 258
Nile River and irrigation, 227
Nonpoint source pollution, 54, 295
Nontributary ground water, 92, 95
NPDES permit, 63, 293
Nutrient management, 284
Nutrients, 283

Objectives and procedures for water planning, 166
Offstream use, 119, 211
Ogallala Aquifer, 55, 192
Organic Act, 39
Organic chemical pollutants, 285
Outdoor water use, 218
Overland flow, 17
Overturn (lakes), 280
Owens River, 194
Oxygen-demanding wastes, 277

Palmer Drought Severity Index, 330
Park City Principles, 69

Perception, 206, 209
Persistence (hydrological), 182
Pesticides, 287
Pick-Sloan Plan, 53
Pinchot, Gifford, 39
Plaintiff, 72
Planning, 200
Point source pollution, 54, 292
Pollutant categories, 276
Pollutant concentration, 276
Population growth, 176
Populist movement, 39
Potential evapotranspiration, 12
Powell, John Wesley, 36
Practicably irrigable acreage, 102
Precipitation, 12, 13
 measurement, 14
Present value, 154
Presidential Advisory Committee on Water Resources, 55
President's Water Resources Policy Commission, 54
Price elasticity of demand, 146
Price theory, 145
Primary drinking water standards, 64, 220
Primary wastewater treatment, 293
Prior appropriation doctrine, 80
 abandonment, 86
 beneficial use, 86
 due diligence, 84
 flow right, 83
 relation back, 84
 return flow, 86
 storage right, 83
 water right transfer, 89
Progressive period, 39
Public domain, 81
Public participation in decision making, 206
Public trust, 90
Public water system withdrawals, 125, 130, 132, 133, 220
Public water systems, 211
Public Works Administration, 45
Pueblo Indians, 103

Quality of the Nation's waters, 298
 Estuaries, 301
 Lakes and reservoirs, 300
 Rivers and streams, 299

Radioactive pollutants, 290
Radon-222, 222, 290
Raster GIS, 139
Rational method, 311
Rational model, 203
Reagan, Ronald, 66
Reclamation Act, 40
Reclamation Service, 40
Recreation, 260
Recreation on federal public lands, 262
Recreation visitor day, 262
Refuse Act, 37, 62
Regulated riparianism, 91
Regulatory taking, 63, 68
Reservoir storage classification, 189
Resource allocation—market versus government, 150
Resource Conservation and Recovery Act, 64, 67
Return period, 316
Riparian doctrine, 75
Roosevelt, Franklin D., 45
Roosevelt, Theodore, 39
Runoff, 17, 19
Runs analysis of drought, 328

Safe Drinking Water Act, 64, 220
Saint Francis Dam failure, 196
Salinity, 117
Salinization, 243
Secondary wastewater treatment, 293, 294
Sediment, 36, 179, 180, 281
Senate Document 97, 59,167
Senate Select Committee on Water Resources, 59
Sequent peak method, 181
Shad restoration, 259
Single–purpose water management, 42
Sleeper decision, 90
Soil conservation, 282
Soil Conservation Service (SCS), 48, 59, 60, 283
Special districts, 201
Sporhase v. Nebraska ex rel. Douglas, 97
State of New Mexico ex rel. Reynolds v. Lewis, 102
Statutory law, 72
Stratton v. Mt Hermon Boys' School, 78
Streamflow simulation, 182
Superfund Amendment and Reauthorization Act (SARA), 67
Supply and demand, 145
Surface water law, 74
Surface water withdrawals, 122, 132, 134
Sustainable water use, 70, 209

Synthetic streamflow simulation, 182

Taylor Grazing Act, 48
Tennessee Valley Authority, 48, 61, 262
Three Gorges Project (China), 252
Timber Culture Act, 2
Transboundary resources, 105
Tributary groundwater, 92, 95

United States v. New Mexico, 104
Unit hydrograph, 313–316
Unit systems for water measurement, 3
Upper Colorado River Basin Compact, 56, 116
Urban water conservation, 215–219
Urban water supply, 211
U.S. Environmental Protection Agency, 63, 219, 220, 225, 296, 298
U.S. Fish and Wildlife Service and recreation, 267
Usufructuary right, 73

Vector GIS, 137

Water availability per capita, 120, 123, 124
Water balance, 22
Water conservation, 198, 215, 241
Water law, 71
  federal law, 96
  state law, 74
Water markets, 89, 148
Water measurement, 3
Water meters, 175
Water molecule structure, 274
Water pollution, 276
Water quality management, 292
Water quality monitoring, 296
Water recycling, 127, 177, 178
Water resource regions, 127, 128
Water Resources Council, 60, 66, 167, 168
Water Resources Planning Act, 60
Water Resources Research Act, 60
Water rights transfer, 89
Watershed, 7
Watershed approach to water quality, 301
Water supply costs, 178
Water supply management, 177
Water supply treatment, 212, 213
Water use per capita, 213, 214
Water withdrawals, 122
  by region, 127
  by state, 131
  by water-use sector, 125
Weeks Act, 42
Welfare economics, 151
Wellhead protection, 225
Wetlands, 34, 58, 322, 323
White, Gilbert F., 46, 50, 54, 60, 308
White House Conference of Governors, 41
Wild and Scenic Rivers Act, 61
Winters v. United States, 98, 99
Wise use, 40
Wyoming v. United States, 102

Yield, 120, 180